An

Alfred Russel Wallace

Companion

Alfred Russel Wallace circa 1877. From a photograph by W. Usherwood, Dorking, Surrey; later reset as the frontispiece to Cope (1891).

An Alfred Russel Wallace *Companion*

Edited by
Charles H. Smith, James T. Costa,
and David Collard

The University of Chicago Press
Chicago and London

The University of Chicago Press, Chicago 60637
The University of Chicago Press, Ltd., London

Published 2019
Printed in the United States of America

28 27 26 25 24 23 22 21 20 19   1 2 3 4 5

ISBN-13: 978-0-226-62210-1 (cloth)
ISBN-13: 978-0-226-62224-8 (e-book)

DOI: https://doi.org/10.7208/chicago/9780226622248.001.0001

Library of Congress Cataloging-in-Publication Data

Names: Smith, Charles H. (Charles Hyde), 1950- editor, author. | Costa, James T., 1963- editor, author. | Collard, David A., editor, author.
Title: An Alfred Russel Wallace companion / edited by Charles H. Smith, James T. Costa, and David Collard.
Description: Chicago : The University of Chicago Press, 2019. | Includes bibliographical references and index.
Identifiers: LCCN 2018045049 | ISBN 9780226622101 (cloth : alk. paper) | ISBN 9780226622248 (e-book)
Subjects: LCSH: Wallace, Alfred Russel, 1823-1913. | Naturalists—Great Britain—Biography. | Evolution (Biology)
Classification: LCC QH31.W2 A74 2019 | DDC 508.092—dc23
LC record available at https://lccn.loc.gov/2018045049

♾ This paper meets the requirements of ANSI/NISO Z39.48-1992 (Permanence of Paper).

Contents

Introduction

CHARLES H. SMITH, JAMES T. COSTA, AND DAVID COLLARD

Although the name Alfred Russel Wallace is slowly becoming more familiar again, there remains much to do, in terms both of making his very interesting biography better known, and of understanding the basis of his many intellectual explorations. In the present work we concentrate on the latter subject. While a volume of this size cannot hope to cover in detail the full range of Wallace's thought, we hope that its level of treatment will stimulate the interested reader to further investigations.

For readers who are largely unfamiliar with Wallace, we begin with a brief biographical summary. For those interested in more detail, a number of very good full-scale biographies are available.[1]

A Capsule Biography

Alfred Russel Wallace was born 8 January 1823 at Usk, Monmouthshire, Wales, to English parents, the eighth of nine children.[2] The Wallaces moved to Hertford, England, when Alfred was about five years of age; apparently some inexpensive lodging had become available as the result of the death of a relative. The family had been on very shaky financial grounds for many years, as Wallace Sr. was not much of a provider, despite having been trained for the law. Wallace's upbringing to the age of fourteen was not very remarkable, but in late 1836 the family experienced a final financial meltdown, and he was forced to leave school and move to London to join an older brother (John) who was already employed by a builder. Less than a year later, Wallace was apprenticed as a surveyor to another older brother (William), working in southwestern England and South Wales, and apart from one brief interlude remained with him until late 1843. It was during this period that he began to take a spare-time interest in natural history subjects, especially botany. Soon he was also giving lectures on related subjects at local mechanics' institutes, and quite possibly even doing some volunteer curatorial and library work. But this period of his life was ended by a slowdown in surveying work, which forced his brother to let him go.

Undaunted, Wallace found a position teaching a variety of basic technical skills at a private school in Leicester. He was there for only about fifteen months, but it was an important period in his development. The area had good libraries, and he took full advantage of them, digesting several works that would feature importantly in his subsequent activities (see chapter 1). During this period he also met the naturalist Henry Walter Bates, who though two years Wallace's junior was already an accomplished collector (mainly, of beetles) and a published writer. Wallace himself began to collect local beetles. Meanwhile, in early 1845, he attended a lecture by a skilled phreno-mesmerist, Spencer T. Hall. Wallace was intrigued, especially after he found himself able to put his students into various forms of trance; at the time mesmeric trance was usually regarded as a hoax, and this experience led him to consider more carefully the possible limitations of materialistic science.

Also in early 1845, Wallace's surveyor brother William suddenly died, and Alfred felt obliged to return to Wales to take over his business. Things did not go very well, and by late 1847 Wallace was ready to try something new. One of the books he had read while at Leicester, *Vestiges of the Natural History of Creation*, was a new treatise on the subject of the transmutation—evolution—of life. Enthralled, he began to consider how he might investigate the subject on his own. An obvious strategy became apparent: he would travel to some tropical locale where the imposing diversity of species and landscapes might inspire some answers as to how this alleged transmutation of species took place. He would pay his way by selling some of the specimens he collected. Bates, also locked into a dull existence (as a factory clerk), expressed an interest to join him, and after a fairly short period of preparation the two young men sailed for Brazil in the spring of 1848.

On reaching Pará (now Belém) in April, the two quickly set to work. It was not long before they split up and went their separate ways, however,[3] with Bates remaining mostly in the lower Amazon valley, and Wallace moving upstream and even into the upper reaches of the Rio Negro, the Amazon's largest tributary. All told, both were extremely successful in their collecting activities (for Bates, mostly insects; for Wallace, largely insects and birds), but tragedy would strike Wallace twice during his last two years in the area. In 1851 Wallace's younger brother Herbert, who had also decided to try the collector's life and had come out two years earlier, died of yellow fever at Pará. Wallace, meanwhile, was also experiencing various ailments, and decided to return to England with the bulk of his previous two years of collections. He managed to find a vessel that would take him, but in early August 1852, midway out in the Atlantic, it caught fire and sank, taking almost all of Wallace's collections with it. Wallace and the crew found themselves in a precarious position in lifeboats, but finally, after ten days at sea, they were seen and picked up by a passing vessel. All told it would take him eighty days at sea to reach home.

Although Wallace's four years in the Amazon Valley had convinced him he was on the right track as regards a causal relationship between geography and evolution, his thoughts on the mechanism of transmutation had actually not advanced much, nor did he now have collections he could study toward that end. He rested, wrote some scientific papers (and two books), and contemplated his future. Now a skilled field collector, he was able to secure a grant from the Royal Geographical Society to pay the transportation costs to his next destination, the Malay Archipelago.[4]

Using the same basic strategy of financing his research efforts through sales of duplicate or pricey coveted specimens, Wallace was able to remain in the Far East for a full eight years. Despite frequent illnesses and travel delays, the expedition went well. Apart from an astonishing harvest of specimens, he recorded characteristics of individual plants and animals, the habits and languages of native peoples, and his impressions on the local geology and physical geography. And in 1858, halfway through the adventure, there came that communication to Darwin revealing his solution to the mechanism of evolution. A year later he sent off for publication his full description of what has become known as Wallace's Line, a sharp discontinuity in the characteristics of geographical ranges of species in the region that became one of the important steps in the development of the modern approach to the science of biogeography (see chapter 10). By the time he returned home in 1862, he was no longer an obscure figure toiling in obscure places for obscure reasons.

The astonishing success of Wallace's time in the Malay Archipelago can be attributed to a number of factors, including luck: this time all or almost all of the specimens he collected actually made it back to England.

But it wasn't all luck. By paying careful attention to opportunities as they emerged, he deftly managed to visit almost all the main islands in the region. Further, one suspects that he had a real knack for dealing with native peoples: in neither the Amazon nor the Malay Archipelago did he apparently ever experience a true threat to his person, a matter for some wonder.

Although Wallace counted his twelve years in the tropics as his single most shaping experience, it is yet true that more than fifty further years of colorful events would fill his remaining days, right up to his death in 1913. For several years after his return he concentrated on studying his collections, meanwhile attending numerous scientific meetings and doing his best to defend Darwinian tenets. In the mid-1860s he became involved with spiritualism,[5] and by the late 1860s was beginning to publicly express concerns over some of Darwin's positions (e.g., on sexual selection, and the origins of human consciousness). With the 1870s came explorations in a variety of subjects, including physical geography and biogeography, the theory of adaptive coloration, and, ultimately, economic and other social questions.

A steady stream of technical and popular works in these decades made his name well known across Britain and the world. His *Malay Archipelago*, published in 1869, was immensely popular, and was succeeded by *Contributions to the Theory of Natural Selection* (1870), *The Geographical Distribution of Animals* (1876), *Tropical Nature and Other Essays* (1878), and *Island Life* (1880)—all of which achieved generally rave reviews. Even his *On Miracles and Modern Spiritualism* (1875c) proved a hit, though appealing to a more specialized readership.

By the end of this period Wallace was beginning to tire of the London environment, and began a series of moves that would lead him and his family[6] farther and farther into the English countryside. His financial situation, meanwhile, became something of a problem, as he had invested poorly in the 1860s and lost much of the profits from his collecting years. But he managed to keep his head above water, taking on various writing, editing, and lecturing assignments as they emerged. By the mid-1880s he had practically become an institution, having succeeded Darwin (who died in 1882) as the most famous naturalist in Britain.

Wallace kept as busy as ever in his last two decades. Although he was no longer spending time in the field and reporting his own finds, his abilities as a synthesizer came even more to the fore, as did his defiance of politically correct attitudes, both in science and society. He stepped down from physically active participation in some social movements, but remained penetratingly active as a writer on those same subjects. In the natural sciences, there were new editions of his *Island Life* and *Darwinism* (first edition 1889), two new books on astronomy and one on biology, and an array of commentary essays and letters to the editor. In his ninth and final decade of life he published new (and new editions of) works that in total surpassed 4,500 pages in print, an astonishing figure for a man of that age.

Wallace spent his last eleven years in Wimborne, near Bournemouth, in a house he built within sight of the English Channel. It was a generally happy time for him—but then again, most of his times had been generally happy ones. He died in his sleep on 7 November 1913, after a short illness. As one of his friends, Edward Poulton, summed things up in a remembrance printed some days later, "How can I best speak of the long, happy, hard-working, many-sided life that has just come to a close?"[7]

Even so, memory of Wallace's efforts and contributions faded rather rapidly; a few months later the Great War was underway, and attention was focused elsewhere. Beginning with the 1958 centennial celebrations of the introduction of the natural selection model, however, he has been slowly "working his way back up."

Reading Wallace

The "digestion" of Wallace's thought involves some challenges for the consumer—but the effort can be very rewarding, if sometimes presenting "glass half full/glass half empty" dilemmas. First off, his combination of evolutionary, spiritualistic, socialist, and libertarian perspectives is utterly unique in the history of thought, and sometimes strains the reader's patience for its rank unfamiliarity. At the same time, this is what makes it fun to explore, for one not given to predisposition! Even Wallace's supporters will admit that at times his ideas strike one as sounding rather naïve, though often behind this naïveté lies a clarity of thought that signals an uncompromising attention to elementals: a sincere drive to uncover fundamental truths. On more than one occasion, Wallace's "naïve" thoughts have proved to be valid anticipations for which he has not always received credit.

The reader also encounters a tenacity of argumentation in Wallace's writings that is rarely matched elsewhere in the literature. In his science writings, this is especially evident in his ability to argue logically upon the given facts, and secondarily in his thoroughness in seeking out those facts to begin with (i.e., in his "marshalling of evidence"). On the social studies side, we often witness a level of passion seldom seen in essays and public letters designed for reading by professionals and laypersons alike. The philosopher Charles Sanders Peirce, in a review of Wallace's autobiography, *My Life*, remarked that Wallace believed "in all he believes down to the very soles of his boots."[8] This might be interpreted as an accusation of stubbornness or inflexibility on Wallace's part, but as we shall see, Wallace was guilty of nothing of the kind. True, he often stuck to his guns, but he was a fair and generous adversary, and capable of changing his mind as new evidence came into play.

Our concentration on his intellectual side in no way argues against our making a few observations on some of Wallace's personal attributes, especially those that surfaced as part of his public face.[9] Take for example his exemplary modesty. No one could ever accuse Wallace of conceit or self-obsession. His deferral to Charles Darwin on the discovery of the natural selection concept is perhaps the most famous instance of self-effacing restraint in the entire history of science, and it was not his only such act. Late in life he remarked,

> I have long since come to see that no one deserves either praise or blame for the *ideas* that come to him, but only for the *actions* resulting therefrom. Ideas and beliefs are certainly not voluntary acts. They come to us—we hardly know *how* or *whence*, and once they have got possession of us we cannot reject or change them at will. It is for the common good that the promulgation of ideas should be free—uninfluenced by either praise or blame, reward or punishment. But the *actions* which result from our ideas may properly be so treated, because it is only by patient thought and work, that new ideas, if good and true, become adopted and utilised; while, if untrue or if not adequately presented to the world, they are rejected or forgotten.[10]

With the help of these words, one can begin to understand how Wallace could be, on the one hand, shy and self-effacing in dealing with personal matters, and on the other, a dynamo in the pages of the public literature. But most telling in this direction was his supreme confidence in his reasoning abilities. In 1905 he wrote, "I have never hesitated to differ from Lyell, Darwin, and even Spencer, and, so far as I can judge, in all the cases in which I have so differed, the weight of scientific opinion is gradually turning in my direction. In reasoning power upon the general phenomena of nature or of society, I feel able to hold my own with them; my inferiority consists in my limited knowledge, and perhaps also in my smaller power of concentration for long periods of time."[11]

This is a sound recipe for intellectual success—a certain fearlessness, combined with an attention to the basics of logical inference—but in Wallace's case some further qualities and abilities helped shape the final product. Most elementally, perhaps, he possessed a very strong native curiosity. There have been many figures over the years who have sought for truth, but few who have looked for it in so many directions. Perhaps this came merely as a result of his "smaller power of concentration for long periods," as he notes in the quote above, but perhaps not: most of the subjects he took up became, as he aged, permanent wells on his intellectual palette.

It was much to Wallace's advantage that he was an excellent, if sometimes somewhat undisciplined, writer. Organizing his thoughts and analyses into just the right stream of words seemed to come easily to him; in the same book review of *My Life* cited above, Charles Sanders Peirce also opined that "Wallace never wrote a dull line in his life, and couldn't if he tried." As a result, he was much in demand as an essayist and book reviewer,[12] while his hundreds of letters to the editor were welcomed by newspapers and other publications across the English-speaking world,[13] and occasionally beyond it. Wallace was never afraid to express a thought, if he felt it was in the interest of "the promulgation of ideas."

One further element of Wallace's persona that merits special mention was his eternal optimism. Despite the fact that he spent a good deal of time complaining about perceived social ills—how society was, unfortunately, "rotten at the core"—he never gave up on us. He almost never was merely critical of what he viewed to be an evil, instead accompanying his complaints with some suggestion for remedial action. His attitude was neatly summed up in the final words of a short essay, "Is Britain on the Down Grade?," which he prepared in 1899 for a magazine called *The Young Man*: "Truly, we will *not* despair of the Republic of Humanity."[14]

This mix of abilities produced an individual uniquely suited to communicating, and not only in his own time and place. We the editors believe his words still reach out to engage, and in this review we attempt to survey those main areas of thought in which they remain relevant.

*

The plan of chapters we have adopted here is somewhat arbitrary, but in outline it proceeds from the more general to the more specific. Chapters 1 and 2 present a model of the foundations of Wallace's thought, concentrating on the evolutionary cosmology he developed over the first half of his life. The next three chapters focus on his biological studies, especially as related to his thoughts on systematics, natural selection, and adaptation. Chapter 6 discusses the main elements of his evolutionary anthropology, in effect continuing from chapter 2. Chapters 7 and 8 deal with his social evolution schemes, as manifest in his attention to "well-being" and governmental fairness. The last four chapters return to a consideration of his natural science explorations, featuring the more specific subjects of physical geography, biogeography, conservation, and astronomy. Throughout we seek to demonstrate how so many of Wallace's intellectual contributions continue to resonate with concerns of our modern world. The essays mostly provide a review of these thoughts and their context and what people have said of them, but here and there some new material has been introduced, pursuant to the "there remains much to do" remark made earlier. The reader should understand, of course, that Wallace was a product of his times, and that a few of his ideas, clever as they may be adjudged, no longer have much direct relevance in our day and age. But most of them quite clearly do.[15] And perhaps even more significantly, some of the ones he set forth in his own time that were ignored then and ever since, may yet find application, if we take the time to give them a bit more consideration.

One notation convention: in the endnotes to each chapter, references to archival letters collected in the Wallace Correspondence Project and the Darwin Correspondence Project give their assigned numbers preceded by the tags "WCP" and "DCP," respectively.[16]

Charles H. Smith
James T. Costa
David Collard

Notes

1. Among English titles, we especially recommend Raby (2001), Shermer (2002), Fichman (2004), and Slotten (2004). Other recent biographies include works published in Spanish, Italian, German, Indonesian, French, and Dutch. One also should not forget Wallace's very informative autobiography, *My Life*, published in 1905. Despite these efforts, we fully believe that the *definitive* Wallace biography remains to be written.

2. There is some heated debate as to whether Wallace should be considered a Welshman; we think this claim is tenuous, for reasons stated at http://people.wku.edu/charles.smith/wallace/FAQ.htm#Welsh. Of course he might simply be referred to as "British," or, descriptively, as "an Englishman born in Wales."

3. The exact reasons for the Wallace-Bates split remain uncertain. Probably they simply felt they could gather more material that way, but some personal issue might also have been involved.

4. Roughly, the Indonesian archipelago, plus the Philippines and New Guinea islands.

5. One should not assume a simple relationship between spiritualism and those concerns, however, as we will see.

6. Wallace married Annie Mitten, daughter of botanist William Mitten, in 1866. They had three children, the first of whom died in childhood in 1871.

7. Poulton (1913), p. 347.

8. Peirce (1906), p. 160.

9. For those interested in a more psychological kind of approach, we recommend the biography by Shermer (2002), which explores some related theories.

10. Wallace (1909a), p. 10.

11. Wallace (1905b), vol. 2, p. 42.

12. Wallace was also a very effective, though often restrained, public lecturer. See Smith and Derr (2013), which includes many reviews of the lectures he delivered during a tour of North America in 1886–1887.

13. See Smith and Patterson (2014), a collection of more than two hundred such items.

14. Wallace (1899), p. 224; emphasis in the original.

15. The degree to which Wallace's thoughts still enter into present-day discussions is well documented by a feature titled "Wallace-Related Research Threads" at the Alfred Russel Wallace Page, http://people.wku.edu/charles.smith/wallace/threads.htm.

16. The Wallace Correspondence Project (WCP) oversees Wallace Letters Online, a publicly accessible electronic archive of Wallace-related documents, including metadata, digital scans, and transcripts of all known letters sent to and written by Wallace, as well as a comprehensive selection of other important Wallace-related manuscripts. The Darwin Correspondence Project (DCP) is a University of Cambridge–based digital archive of the approximately 15,000 letters of Charles Darwin. Directed by J. Secord et al., this archive forms the basis of both the printed Correspondence of Charles Darwin volumes (Cambridge University Press) and the annotated letters of the Darwin Correspondence Project website.

‹ 1 ›

The Early Evolution of Wallace as a Thinker

CHARLES H. SMITH

Apart from his obvious importance to the development of evolutionary biology and biogeography, one of the main reasons that Wallace has been the subject of more than twenty books over the past several years is that he was, simply, one interesting character. Actually, there has never been another quite like him: respected naturalist and scientist, leading social critic, inspiring explorer and collector, intrepid defender of spiritualism, contributor to a dozen fields of knowledge, and ultimately earner of "The Grand Old Man of Science" tag out of respect for his nearly fatherly position within the Victorian intellectual community. With all these (and more) credits to his biography, the writer assigned to providing a succinct but insightful introduction to his early evolution as a thinker may be pardoned for accepting his charge with some trepidation. But on close examination Wallace's intellectual trajectory seems to follow a logical progression, most of the elements of which were established long before his gaining fame as codiscoverer of natural selection at the age of thirty-five in 1858. Not all the relevant influences and their manner of synthesis are completely understood as yet, to be sure, but we are making progress.

In surveying Wallace's intellectual career, one is struck first by the sheer range of subjects he took up, extending from evolutionary theory to economics, biogeography to socialism, astronomy to descriptive statistics, physical geography to spiritualism, and beyond. He was both a major voice for Darwinism and natural selection, and a well-known defender of social evolution and spiritual agendas at a time when the latter were not always universally popular topics. One may reasonably wonder at the outset how such a person could come to be: perhaps all this creativity was the product of a totally detached and isolated mind, one that did not follow the usual paths merely because of a lack of exposure to them? But this was not the case.

Actually, Wallace appears to have been enamored from his earliest adult years with thoughts of "the advantages of varied knowledge." Portions of his early essay with this title, perhaps drafted for presentation at a local mechanics' institute around 1843, were reprinted in his autobiography, *My Life*, in 1905.[1] It contains the following remarks:

> There is an intrinsic value to ourselves in the[se] varied branches of knowledge, so much indescribable pleasure in their possession, so much do they add to the enjoyment of every moment of our existence, that it is impossible to estimate their value, and we would hardly accept boundless wealth, at the cost, if it were possible, of their irrecoverable loss. And if it is thus we feel as to our general store of mental acquirements, still more do we appreciate the value of any particular branch of study we may ardently pursue. What pleasure would remain for the enthusiastic artist were he forbidden to gaze upon the face of nature, and transfer her loveliest scenes to his canvas? or for the poet were the means denied him to rescue from oblivion the passing visions of his imagination? or to the chemist were he snatched from his laboratory ere some novel experiment were concluded, or some ardently pursued theory confirmed? or to any of us were we compelled to forego some intellectual pursuit that was bound up with our every thought? And here we see the advantage possessed by him whose studies have been in various directions, and who at different times has had many different pursuits, for whatever may happen, he will always find something in his surroundings to interest and instruct him.

He continues,

> Can we believe that we are fulfilling the purpose of our existence while so many of the wonders and beauties of the creation remain unnoticed around us? While so much of the mystery which man has been able to penetrate, however imperfectly, is still all dark to us? While so many of

> the laws which govern the universe and which influence our lives are, by us, unknown and uncared for? And this not because we want the power, but the will, to acquaint ourselves with them. Can we think it right that, with the key to so much that we ought to know, and that we should be the better for knowing, in our possession, we seek not to open the door, but allow this great store of mental wealth to lie unused, producing no return to us, while our highest powers and capacities rust for want of use?

One might imagine from these words that Wallace had been, early on, a close admirer of the thought of Benedict de Spinoza, but there is no indication he ever read anything by the great Rationalist philosopher. A few years later, however, he probably did encounter similar ideals, as expressed by the German geographer and naturalist Alexander von Humboldt (see below). With such an impressive figure on whom to model himself (at that point Humboldt was probably the most famous naturalist in the world)—to reinforce his innate enthusiasm—Wallace's exuberance in exploring each and every new subject that came to his attention is easily understood. His was in no sense a genius born of isolation; on the contrary, he took every opportunity to sift through the literature surrounding any idea that seemed relevant to his subject of the moment.

Two further ingredients contributing to the basic "Wallace formula" can be identified. First, as mentioned in the introduction, he was supremely confident of his ability to reason from the basic facts of a matter. In *My Life* he wrote, "This rather long digression may be considered to be out of place, but it is given in order to illustrate the steps by which I gradually acquired confidence in my own judgment, so that in dealing with any body of facts bearing upon a question in dispute, if I clearly understood the nature of the facts and gave the necessary attention to them, I would always draw my own inferences from them, even though I had men of far greater and more varied knowledge against me."[2] This in good part explains the ease with which he immersed himself in so many kinds of discussions. And, of course, his continuing successes did little to stem the tide.

It should further be noted that Wallace's father Thomas V. (1771–1843), though not much of a family provider, likely encouraged his young son's varied interests—especially his literary ones, as Wallace Sr. was himself at various times a dabbler in literary projects, and for some years a town librarian. In *My Life* Wallace discusses some of the many books he digested as a boy and adolescent, ranging from *The Boy's Own Book*, to Shakespearean plays, to the novels of James Fenimore Cooper. From that time on, he remained a ravenous reader across many subjects. This was his habit right up to his final illness.

Before attempting to survey Wallace's intellectual evolution from his youthful years through to the milestone essay of 1858, it seems prudent to devote a few pages to the main cast of characters involved. I proceed roughly chronologically, in the approximate order in which Wallace became exposed to their influence.

Wallace's Influences prior to 1858

Wallace's immediate family. Wallace's father was just one family member who was an influence. In 1837, through his older brother John (1818–1895), Wallace first became acquainted with the problems experienced by working men, and had his initial taste of the philosophy of socialism (John's friends in London included a group that had fallen in with the Owenist movement; see below). Brother William (1809–1845), for whom Wallace worked most of the next several years, taught him the technical skills needed to be a surveyor, and provided him with a responsible adult model during his late teen years; he also effectively became Wallace's first teacher on subjects of geological, geographical, and astronomical nature. Wallace's only surviving sister Fanny (1812–1893),[3] meanwhile, was an inspiration for independent action: after spending some time in France in the late 1830s and opening a school back in England, she left home to teach students in Georgia and Alabama from 1844 to 1847. Fanny was also the person most centrally responsible for getting Wallace to investigate spiritualism in 1865, and helped fuel his early interest in photography when she married one of London's first professional practitioners of the art, Thomas Sims (1826–1910).

Robert Owen (1771–1858). The socialist-utopian ideals of Robert Owen had generated a full-scale movement by the time of Wallace's brief residence in London in 1837, and Wallace was quickly caught up in it. Owen's philanthropic ideas on social organization foreshadowed secular humanism, Owen himself being a strong religious skeptic. Like many other dissenters of his time, in his late years he became a spiritualist. Wallace remained receptive to Owenism for the rest of his life, and many of his social sympathies ultimately may be traced to this source: "His great fundamental principle, on which all his teaching and all his practice were founded was that the character of every individual is formed *for* and not *by* himself, first by heredity, which gives him his natural disposition with all its powers and tendencies, its good and bad qualities; and, secondly, by environment, including education and surroundings from earliest infancy, which always modifies the original character for better or for worse."[4]

Thomas Paine (1737–1809). According to *My Life*, Wallace read Paine's *Age of Reason* during his London stay in 1837. Its criticism of organized religion and rejection of miracles inspired many a budding free-thinker, as Wallace was at this time. And, it might be noted, one of the major influences on Paine's assessment of religion was Spinoza, so perhaps it could be said Wallace had been exposed to his philosophy, if indirectly, after all.

Robert Dale Owen (1801–1877). Robert Owen's son Robert Dale helped further his father's projects in a number of ways, but also had a successful political and writing career of his own. Wallace was particularly impressed early on by Dale Owen's *Consistency*, released in late 1839 as a pamphlet (but was predated by similar writings of his that Wallace may also have seen), which criticized the doctrine of eternal punishment as "degrading and hideous."[5] Dale Owen was also a spiritualist; he produced two important midcentury surveys of the subject that Wallace would later praise.

George Combe (1788–1858). Combe, a lawyer by profession, became one of the early nineteenth century's leading promoters of phrenology. It is not certain when Wallace was first exposed to his writings, especially *The Constitution of Man*, but this might have come as early as 1837, again during his months in London.[6] Wallace was undoubtedly impressed with Combe's operating philosophy, which espoused a naturalistic model of cause and effect for human endeavor. Further, its secular and humanitarian implications—especially the notion that natural science had "outstripped" moral evolution—undoubtedly appealed to Wallace as he drifted away from conventional beliefs.

The enclosure movement. For several hundred years, extending into the nineteenth century, a legal process of combining small landholdings into larger ones had been forced upon the rural working class in England. This often involved hardships, and young Wallace witnessed them during his years as an apprentice surveyor in the late 1830s and early 1840s. He never forgot these experiences.

Charles Darwin (1809–1882). Darwin's pre-1858 influence on Wallace was actually rather small, restricted to the latter's appreciation of the older man's *Journal of Researches*.[7] Judging from comments in *My Life*,[8] Wallace first read Darwin's book in the early 1840s.

Henry Walter Bates (1825–1892). Wallace met Bates in Leicester in 1844. It was Bates who, by his example, turned Wallace into an insect collector, a critical step in his ascendance to naturalist superstardom. Bates was in fact his main sounding board for the next several years, and the younger man's elucidation of the concept of protective mimicry was quickly extended by Wallace on his return to England from the East in 1862.

Charles Lyell (1797–1875). Lyell, the leading geologist of his time, presents an interesting case in terms of his influence on Wallace. As the leading proponent of geological uniformitarianism—the doctrine that geological/geographical structures evolve gradually, and through processes observable at the present time—Lyell provided Wallace with a model of earth evolution that appealed to the latter's conservative, "no first causes" approach to things, even early on. At the same time, however, Lyell viewed the dynamics of the living world through creationist lenses—so, ultimately, his position became a challenge Wallace had to overcome. Lyell's writings might also have been the source of Wallace's first exposure to full-blown evolutionism, in the form of his description of the views of the French zoologist Lamarck.

Alexander von Humboldt (1769–1859). Wallace first read the great German geographer's *Personal Narrative of Travels* (in South America) while at Leicester, around 1845. This work (along with Darwin's *Journal*) became a potent influence on his desire to travel to the tropics. English editions of Humboldt's *Cosmos* and *Aspects of Nature*,[9] moreover, may have provided Wallace with an earth-systems view guiding his later (late 1840s and 1850s) thoughts on physical geography, biogeography, and evolutionary biology (see below).

Thomas Malthus (1766–1834). According to *My Life*, Wallace first read Malthus's *Essay on the Principle of Population* during his time in Leicester. Malthus's notion of the "positive checks" on population growth would, of course, become a central element in the natural selection model, both for Darwin and for Wallace.

Robert Chambers (1802–1871). Chambers and his brother William were a force in the literary world, the publishers of an array of magazines, reference books, and other works. Robert himself was an educated amateur who wrote or edited a large number of titles (mostly on subjects of historical, biographical, geological/geographical, and popular nature). One of these stands out: *Vestiges of the Natural History of Creation*. Published anonymously in 1844, it advanced a theory of progressive transmutation of species, and Wallace was quickly taken with the discussion, if not with its inability to identify particular agents of change.

Herbert Spencer (1820–1903). By the time Wallace returned from the

Amazon in 1852, Spencer's career as a philosopher and sociologist was already on the ascent. His *Social Statics*, published in 1851, made an immediate impression, not least with Wallace, who first read it around 1853. Two themes that Spencer explored especially resonated with Wallace, later becoming central elements of his views on social evolution. These were (1) that everyone should enjoy no more or less than what was truly due them on the basis of their abilities and efforts, and (2) that access to the land should be considered a fundamental right of the individual, not just another commodity that is bought and sold.

Other influences. A good number of other early influences on Wallace's thought have been identified. I briefly note just a few of these: Sir Humphry Davy and Justus von Liebig (Wallace read their works on agricultural chemistry, influencing his later ideas on land reform and his early thoughts on the effect of environment), Sir William Lawrence (who wrote early works, for example *The Natural History of Man*, that anticipated later evolutionary models), and Hugh Strickland and William Swainson (who excited Wallace's interest in classification).

With these introductions made, we can proceed to some particulars.

The Evolving Wallace: A Framework

Although it is clear enough that Wallace absorbed influences from a wide range of sources, observers have been less successful in coming to grips with the redirections he fashioned from them. To understand these, one must look at some of the basic guiding principles he adopted. Among the most important and long-lasting of these was his acceptance of a "no merit to belief" ethical foundation, seemingly derived originally from his reading of Robert Dale Owen. This began with the idea, inherent in the teachings of the older Owen, that we are largely a product of our environment, and that it is institutions, not individuals, that are most to blame when things go wrong. Beliefs were therefore transient and could be modified; we are responsible for our actions, but often not their causes.[10] This position is most beautifully expressed in one of Wallace's signature writings, a letter from the field he sent to his brother-in-law Thomas Sims in 1861:

> You intimate that the happiness to be enjoyed in a future state will depend upon, and be a reward for, our belief in certain doctrines which you believe to constitute the essence of true religion. You must think, therefore, that belief is *voluntary* and also that it is *meritorious*. But I think that

a little consideration will show you that belief is quite independent of our will, and our common expressions show it. We say, "I wish I could believe him innocent, but the evidence is too clear"; or, "Whatever people may say, I can never believe he can do such a mean action." Now, suppose in any similar case the evidence on both sides leads you to a certain belief or disbelief, and then a reward is offered you for changing your opinion. Can you really change your opinion and belief, for the hope of reward or the fear of punishment? Will you not say, "As the matter stands I can't change my belief. You must give me proofs that I am wrong or show that the evidence I have heard is false, and then I may change my belief"? It may be that you do get more and do change your belief. But this change is not voluntary on your part. It depends upon the force of evidence upon your individual mind, and the evidence remaining the same and your mental faculties remaining unimpaired—you cannot believe otherwise any more than you can fly.

Belief, then is not voluntary. How, then, can it be meritorious? When a jury try a case, all hear the same evidence, but nine say "Guilty" and three "Not guilty," according to the honest belief of each. Are either of these more worthy of reward on that account than the others? Certainly you will say No! But suppose beforehand they all know or suspect that those who say "Not guilty" will be punished and the rest rewarded: what is likely to be the result? Why, perhaps six will say "Guilty" honestly believing it, and glad they can with a clear conscience escape punishment; three will say "Not guilty" boldly, and rather bear the punishment than be false or dishonest; the other three, fearful of being convinced against their will, will carefully stop their ears while the witnesses for the defence are being examined, and delude themselves with the idea they give an honest verdict because they have heard only one side of the evidence. If any out of the dozen deserve punishment, you surely agree with me it is these. Belief or disbelief is therefore not meritorious, and when founded on an unfair balance of evidence is blameable.

Now to apply the principles in my own case. In my early youth I heard, as ninety-nine-hundredths of the world do, only the evidence on one side, and became impressed with a veneration for religion which has left some traces even to this day. I have since heard and read much on both sides, and pondered much upon the matter in all its bearings. . . . I think I have fairly heard and fairly weighed the evidence on both sides, and I remain an *utter disbeliever* in almost all that you consider the most sacred truths. I will pass over as utterly contemptible the oft-repeated accusa-

> tion that sceptics shut out evidence because they will not be governed by the morality of Christianity. You I know will not believe that in my case, and *I* know its falsehood as a general rule. I only ask, Do you think I can change the self-formed convictions of twenty-five years, and could you think such a change would have anything in it to merit *reward* from *justice*? I am thankful I can see much to admire in all religions. To the mass of mankind religion of some kind is a necessity. But whether there be a God and whatever be His nature; whether we have an immortal soul or not, or whatever may be our state after death, I can have no fear of having to suffer for the study of nature and the search for truth, or believe that those will be better off in a future state who have lived in the belief of doctrines inculcated from childhood, and which are to them rather a matter of blind faith than intelligent conviction.[11]

Thus, progress-serving values could be established only on the basis of a continuing, unbiased—Spinozian, one might say—examination of facts. Those who indulged in superficial or prejudging evaluations were likely to suffer accordingly—that is, to have their actions rejected by the greater reality, which usually will have nothing of preconceived ends or means.

I believe that the entire track of Wallace's intellectual career is directly relatable to this basic principle and its correlates.[12] For individuals, "progress" was only possible through the combination of a willingness to re-evaluate positions with a receptivity to constructive change. The same formula could be applied to whole societies, which should seek to structure themselves accordingly. By extension (or analogy), success as a biological entity could be maintained as a function of a well-rounded adaptation to multiple influences. Among thinking beings, the necessary "re-evaluations of position" represented exertions of will. In a strictly biological context, however, the same effect could be achieved by rote, through trial and error, as guided by ambient ecological/environmental circumstances: it seemed that those organisms or populations capable of productive response to the widest range (and/or most severe) of relevant constraints were the ones that ended up prevailing in the struggle for existence.

From all evidence the influence that permanently converted Wallace to transmutationism was his 1844 or 1845 reading of *Vestiges of the Natural History of Creation*. Understand, however, that at that point he had little idea of what immediate causal agencies might exist to produce such a process. His early rejection of Lamarckism was a good starting point, as was his realization that there were many "facts" that could be interpreted as being the *result* of an evolutionary process; further, his correspondence with Bates from that period shows he understood that the anonymous author (Chambers) had not made much progress on the subject beyond theory spinning. Still, the natural history activities Wallace had conducted in his spare time while working as a surveyor and teacher in the early to mid-1840s had suggested a framework that might be exploitable, one connecting the facts of distribution and diversity with biological evolution. Thereafter, he would make constructing a connection between evolution and the distribution of organisms a conscious goal of his collecting activities, ultimately including his expeditions to South America and the Malay Archipelago. Evidence for his continuing attention comes from words presented in his "Sarawak law" essay of 1855:

> The great increase of our knowledge within the last twenty years, both of the present and past history of the organic world, has accumulated a body of facts which should afford a sufficient foundation for a comprehensive law embracing and explaining them all, and giving a direction to new researches. It is about ten years since the idea of such a law suggested itself to the writer of this paper, and he has since taken every opportunity of testing it by all the newly ascertained facts with which he has become acquainted, or has been able to observe himself. These have all served to convince him of the correctness of his hypothesis.[13]

Despite the reasonably auspicious start to his thought process, as of 1848 and his removal to South America Wallace seems to have figured out relatively little about how further to proceed toward his goal. Transmutation was being discussed in the general literature, but, as no one was offering up any specific models, he was left largely to his own devices.

Others in his position might have considered organic change on the basis of theistic models, but Wallace had apparently rejected the orthodox Christian understanding of nature and society long before reading *Vestiges*. Certainly, he would have had nothing of a God figure who had personally created, and was still intervening in and/or supervising, a basically unprogressing (or, for that matter, progressing) natural existence.[14] But considering the little evidence we have, this is as far as we dare speculate on this matter. Wallace later looked back at himself as being an agnostic—not an atheist—during this period. "God" was something that, as he put it, he "cared and thought nothing about."[15] As a result of this open-ended position, we cannot conclude that at that point he had also rejected the possibility that some kind of supraphysical or godly entity (as suggested in *Vestiges*) might constitute the *final* cause of natural existence.[16] Elsewhere I have noted,

> By 1871, of course, he undoubtedly accepted that behind all "universal forces and laws" there lay "the will or power" of a "Great Mind" or "Supreme Intelligence." It seemed, however, that the "law" of such a will or entity, probably being "connected with the absolute origin of life and organization," was likely "too deep for us to discover." Such "law" was not, it need be emphasized, to be interpreted as one operating through event-specific Godly interventions. Rather, the "Great Mind" represented, in some sense, a universal source of will as an expression of which the laws of the universe operated, in fully uniformitarian fashion, to evolve "special" (i.e., individual) effects: ". . . only in reference to the origin of universal forces and laws have I spoken of the will or power of 'one Supreme Intelligence.'" Again, within this framework, a final cause was in theory thought to be operating, but probably was too remote from human appreciation ever to be fully understood. Nevertheless, the "will" of the "Supreme Intelligence" was manifest as an ordered, changing existence encompassing all of reality (including itself, whatever it might be), and operating according to describable laws of interaction (e.g., natural selection and gravitational attraction). These laws combined in such a fashion—quite possibly itself formally describable, and at the least recognizable—making ordered change the necessary consequence of their operation.[17]

Given no evidence to the contrary, Wallace may well have been entertaining similar thoughts, if not yet so worked out, as early as the 1840s. These would have fit in relatively well with the ideas of a thinker who must have had considerable influence on him during the 1840s and 1850s: Alexander von Humboldt.

We have already noted that Wallace mentioned Humboldt's travel writings as an inspiration for his own tropical explorations, but it is now time to suggest that Humboldt had an even larger influence on his philosophical leanings. English translations of Humboldt's books *Cosmos* and *Aspects of Nature* were released in England in the late 1840s, and there is a strong likelihood that Wallace saw the first before he left for the Amazon in 1848, and the second either during his trip or shortly after he returned to England in 1852. In contrast with Humboldt's largely journalistic *Personal Narrative*, these works feature a good deal of what has come to be called Humboldtian science, including elements of his "terrestrial physics" concept (loosely, physical geography) and a variety of points that Wallace had expressed interest in even before seeing them in Humboldt. One of these was the "advantages of varied knowledge" theme that runs through several of Wallace's earliest writings; another was Humboldt's surmise that "more recondite forces" are invariably acting to make things come out as they are observed.[18] Humboldt also emphasized the value of collecting empirical facts in the field, and their representation through conceptual tools (especially maps and graphs, in both of which Wallace took a lifelong interest). Accuracy and precision of measurement of natural phenomena were also a part of the program, as was a call for the invention of instrumentation that could make such possible.

Wallace—and just about every other ground-breaking naturalist of his era, including Darwin and Lyell—took heed of most of these arguments (and others besides), incorporating them into his own research agendas. Beyond practical considerations, moreover, Humboldt provided Wallace with a general model of environment that helped frame his search for the mode of organismal change. This was the "general equilibrium of forces" ideal, summarized in the following (recently) translated passage from Humboldt's 1807 work "Essai sur la Géographie des Plantes":

> This science [la physique générale], which without doubt is one of the most beautiful fields of human knowledge, can only progress . . . by the bringing together of all the phenomena and creations which the earth has to offer. In this great sequence of cause and effect, nothing can be considered in isolation. The general equilibrium, which reigns amongst disturbances and apparent turmoil, is the result of an infinity of mechanical forces and chemical attractions balancing each other out. Even if each series of facts must be considered separately to identify a particular law, the study of nature . . . requires the bringing together of all the forms of knowledge which deal with the modifications of matter.[19]

Wallace may well have read this in the original French (he could read the language decently, and both Bates and Wallace's sister Fanny were fluent), but in any case similar thoughts are expressed in both *Cosmos* and *Aspects of Nature*. And remember the lofty ideals stated earlier in Wallace's essay on the advantages of varied knowledge? Consider the following quotation from Humboldt's *Aspects of Nature*:

> [W]ho does not feel himself differently affected in the dark shade of the beech, on hills crowned with scattered fir-trees, or on the turfy pasture, where the wind rustles in the trembling foliage of the birch? These trees of our native land have often suggested or recalled to our minds images and thoughts, either of a melancholy, of a grave and elevating, or of a cheerful character. The influence of the physical on the moral world,—that reciprocal and mysterious action and reaction of the material and the immaterial,—gives to the study of nature, when regarded from higher points of view, a peculiar charm, still too little recognised.[20]

There was in Humboldt something beyond simple materialism, a romantic element in part derived from one of his mentors, Goethe. He took interest in all the peoples of the world, their political organization, psychology, and sociology, and how all these were affected, and enhanced, by environment.[21] He has been regarded not only as a founder of modern geographical studies, but also as one of the first true ecologists. It is apparent from even a cursory examination that Wallace and Humboldt had a lot in common (at least, intellectually!), and bits and pieces of Humboldtian thinking turn up in Wallace on numerous later occasions (for one obvious instance, in the latter's frequent use of the somewhat awkward term "aspects of nature" in his writings from the 1850s and 1860s). An important theme in both of their writings is the poor treatment of native peoples by Westerners, and an abhorrence of slavery. And Wallace probably would have been encouraged by Humboldt's approach to the subject of biological evolution: the older man was not a full-blown transmutationist, but he seemed, at least, to be potentially open to the idea.

Perhaps the single most useful thing Wallace would have gotten from Humboldt was his account of biogeographical phenomena, especially the recognition that a multitude of interacting forces underlie what we would now term the "dynamic equilibrium" of animal and plant distributions.[22] Wallace looked to Lyell for much of his approach to geological science, but to Humboldt for geographical and ecological models. In his pre-1858 writings he cites Lyell just once, but Humboldt more than twenty times. He was also aware of the work of some of Humboldt's protégés, including the botanist Franz Julius Ferdinand Meyen (1804–1840) and the great chemist Justus von Liebig (1803–1873), who developed the "law of the minimum" that became the basis for the limiting factors concept that dominated ecological studies for nearly a hundred years.[23] Wallace's interest in this body of work is evidenced by his words in an 1852 paper on the distribution of Amazonian monkeys: "What physical features determine the boundaries of species and of genera? Do the isothermal lines ever accurately bound the range of species, or are they altogether independent of them?"[24]

I surmise that Wallace's changing worldview from 1845 to 1870 responded mainly to his shifting opinions on how the fundamental processes of nature (1) responded to remote, final, causes, and (2) did so in a manner respecting a basic harmony of integration.[25] Through the late 1840s and most of the 1850s he apparently felt that biological/social evolution might be explained through laws of interaction directly analogous to those governing the physical world, but eventually he would conclude that (at least) three, rather than two, general domains of interaction pursuant to such laws existed: the third domain consisted of a realm of "spirit." So while at first Wallace felt that biological and social evolution were forced by physical (i.e., geophysical/geographical) relationships analogous to, but of greater complexity than, say, gravitational attraction, later he would conclude that a kind of mutual causality prevailed.

To be more specific, it appears Wallace was envisioning what we would now term a "push-pull" kind of causality, a term Magoroh Maruyama employed in 1963 to describe a coupling of positive and negative feedback relations that yields irreversible forms of change.[26] Wallace himself, of course, was not directly engaged in any attempt to develop abstract theory of this kind, yet his efforts may be viewed as a precursor to such thinking.[27] At the least it seems that he came to believe that after prebiological forces had "prepared" a world condition permitting biological organization to come into being, a co-evolution and stabilization had taken place that eventually permitted the increasing involvement of a domain of purely psychic organization. The growth of this domain within human consciousness was dependent on our increasing transcendence of the "everyone for himself" sentiments rooted in basic survival instincts. The transcendence process produced an accelerating "pull" effect on the operation of the full physical-biological-spiritual hierarchy: consciously willed acts by humans had the potential of enhancing those causal chains dependent on rotely operating physical and biological processes alone (but, of course, they could also—hopefully temporarily—ruin things through selfish actions leading to destabilizations).

Before returning to Wallace's early struggles to comprehend, we should draw attention to a fundamental difference between the ways Darwin and Wallace would treat the concept of adaptation. Darwinian natural selection represents a *process*: one in which characters are selected in such a fashion that they modify or replace previous characters. To cut a long story short, the Darwinian approach is to accept historical continuity of organismal form as signifying continuity of process; thus both natural selection specifically and evolution in general are considered demonstrated if speciation, including its biogeographical ramifications, can be synonymized with chronologies of adaptive change.

In contrast, Wallace *never* recognized a *necessary* connection between process and *particular* structure. As I have noted elsewhere,[28] Wallace's natural selection is a "law" of natural interaction: there is no generalizable "process" of adaptation implicit in it. There is instead only the ongoing state of "being adapted." There is much in Wallace's writings that gives evidence of this difference in the two men's approach, but for now we can point only to a few hints. One is provided by a letter Wallace sent to Darwin on 2 July 1866.[29] He complains that Darwin had been applying the term "natural selection" in two separate senses: as the "survival of the fittest," *and* as the changes produced *through* the survival of the fittest.[30] Some years later, Wallace wrote the following in an essay entitled "The Origin of Species and Genera":

> By means of whatever laws we suppose living things first to have originated, why should not the primeval germs have appeared many times over, and in forms determined or modified by the infinitely varied chemical and physical conditions to be found in the crust of the earth? The identity of ultimate structure and wonderful similarities of development of all organisms may be due to the unity of the laws by which organic life was first produced; the diversity of the great types of animal and vegetable forms may be due to the operation of those laws at different places, acting on different combinations of elements, which are subject to unlike physical conditions.
>
> The point here insisted upon is, that the origin of all organisms, living and extinct, by "descent with modification," is not necessarily the same thing, and is not included in, "the origin of species by means of natural selection." The latter we not only know has occurred, but we can follow the process step by step by means of known facts and known laws; the former, we are almost equally certain, has occurred, but we cannot trace its steps, and there may have been facts and laws involved of which we have no certain knowledge. The terms "laws of growth," "laws of development," "laws of inheritance," "laws of variation," "laws of correlation,"

> "direct action of the environment," "laws of habit and instinct," with some others, are used to express the action of causes of which we are almost wholly ignorant, as we are of the nature of life itself. Now Mr. Darwin has himself admitted that there are these unknown causes at work, and that "natural selection is the most important but not the exclusive means of modification" . . . [but] even if it should be ever proved that higher laws than "natural selection" have brought about the more fundamental divergences of the animal and vegetable kingdoms, this will not be held to detract in any way from the greatness and the value of Mr. Darwin's work, any more than it will be held to detract from the greatness of Newton, if it should some day be demonstrated that the law of gravitation as expressed by him is not absolutely true, but that (as some physicists now suppose) it should be found to be subject to a higher law for remote stellar distances.[31]

As one further indication of Wallace's feelings in this direction, note his frequent use over the years of the term "accumulate" to describe his view of how favorable variations were added to a population. "Accumulation" implies some combination of ordered and haphazard, idiosyncratic addition—perhaps probabilistically, under the imposed influence of regular forces, as in the production of an alluvial fan under the influence of gravity.[32]

Inescapably one concludes that Wallace felt characters were undergoing selection, *from whatever external causes*, and it was this *fact*, not any process, that most closely corresponded to Darwin's term "natural selection." The final proof of this is Wallace's continuing referral to natural selection as a "law." The most direct instance of this is his single-page 1870 "demonstration" of natural selection in *Contributions to the Theory of Natural Selection*, where he connects "proved facts" to their "necessary consequences," then describes them as "afterwards taken as proved facts."[33]

Wallace's views on *evolution* (as a cosmological concept) are quite distinct, as I have commented elsewhere:

> "Evolution," by contrast, is regarded by Wallace as a more general process fueled by the simultaneous operation of all such laws of interaction, any of which might be secondarily influencing natural selection (or each other) over long periods of time in any number of ways. Thus, each such law could be observed to produce certain classes of immediate effects, but no one actor on the evolutionary stage could be considered "caused" in its entirety by any of these laws individually. Moreover, Wallace was not above assigning phenomena to the program of evolution whose causes

> were either unknown or poorly known, as long as there were facts available to prove their existence and an appropriate chain of logic could be constructed to indicate their relevance. The two best examples of this were his treatments of the origin of variations, and spiritualism. In the first instance, he (and other Darwinians) argued that it was not necessary to the natural selection model to know the ultimate origin of variations to understand how the fact of variation contributed to understanding its basic operation. In the case of spiritualism, spiritualistic "manifestations" were considered a proof of the existence of a level of super-physical organization, and here too Wallace had an argument for how it fit into the overall program.[34]

Wallace's mature thoughts on the concept "evolution" are best summarized in an essay of his from 1900 entitled, appropriately, "Evolution":

> Evolution, as a general principle, implies that all things in the universe, as we see them, have arisen from other things which preceded them by a process of modification, under the action of those all-pervading but mysterious agencies known to us as "natural forces," or, more generally, "the laws of nature." More particularly the term evolution implies that the process is an "unrolling," or "unfolding," derived probably from the way in which leaves and flowers are usually rolled up or crumpled up in the bud and grow into their perfect form by unrolling or unfolding. Insects in the pupa and vertebrates in the embryo exhibit a somewhat similar condition of folding, and the word is therefore very applicable to an extensive range of phenomena; but it must not be taken as universally applicable, since in the material world there are other modes of orderly change under natural laws to which the terms development or evolution are equally applicable. The "continuity" of physical phenomena, as illustrated by the late Sir William Grove in 1866, has the same general meaning, but evolution implies more than mere continuity or succession—something like growth or definite change from form to form under the action of unchangeable laws.
>
> The point to be especially noted here is, that evolution, even if it is essentially a true and complete theory of the universe, can only explain the existing conditions of nature by showing that it has been derived from some pre-existing condition through the action of known forces and laws. It may also show the high probability of a similar derivation from a still earlier condition; but the further back we go the more uncertain must be

our conclusions, while we can never make any real approach to the absolute beginnings of things.[35]

Natural selection was, of course, one of these "natural forces," these "laws of nature." Darwin followed, as might be expected considering his geology background, a more historical/process-oriented understanding of the place of natural selection in evolution, viewing it in terms of origination phenomena and, ultimately, the shaping of adaptive characters. Wallace, by contrast, had implicitly created for himself a framework within which the forces of geological and ecogeographical evolution were the prior considerations, their actions ultimately manifest in (some kind of) biological change. In the late 1850s and 1860s he found it easier simply to defend the fundamental proposition that biological evolution occurred at all by accepting the Darwinian lead: that evolution corresponded to the observed record of character divergence.[36] On narrower questions of process, mimetic resemblance for example, this was an easy concession, as there were many examples of adaptation that seemed directly relatable to particular classes of natural selection-mediated causation. On the "bigger" questions, that is, those related to evolution in general, however, he and Darwin began quite far apart, and remained so.

The Evolving Wallace: The Thought Process, 1845 to 1858

When we left the youthful Wallace, he had just been bowled over by the discussion in *Vestiges*, and was beginning to consider how to proceed toward a process model accounting for evolutionary change. The book's author, Robert Chambers, argued that evolution would eventually develop "godly" beings, a view that carried with it notions of progressive social advance. Wallace would have found this a convenient starting point: one could imagine a progressive individual and social evolution (Owen and Chambers) moving ahead in the form of a uniformitarian process (Lyell) operating in response to generally acting physical laws (Humboldt). The investigative approach taken would focus on the spatial/historical evidence of divergence of species lines. But immediately there was a problem.

This was, simply, that it was one thing to present evidence suggesting evolution had actually taken place, and quite another to demonstrate how this happened at the level of individual creatures and/or populations. After reading *Vestiges*, Wallace may have felt he had some handle on the first matter, but as to how, specifically, plants and animals "became adapted," he had no clue. With a limited number of options available to him, he quickly rejected creationism and Lamarckism, instead concentrating on Chambers's position that evolving systems must be self-regulating, characterized by an intelligible relation of cause to effect. But here, probably also influenced by Humboldtian ideas, he set off in a largely unproductive direction. Deterred by creationist arguments that adaptations had occurred to fulfill a prior, predetermined, utilitarian function, he rejected the notion of necessary utility and instead took the position that bodily structures emerged, perhaps randomly, and were later subjected to some kind of superpopulation environmental influence that forced change.[37] He accordingly fell into the working hypothesis that many extant characters actually served "no material or physical purpose," remarks included in an essay on the orangutan written as late as 1856:

> Do you mean to assert, then, some of my readers will indignantly ask, that this animal, or any animal, is provided with organs which are of no use to it? Yes, we reply, we do mean to assert that many animals are provided with organs and appendages which serve no material or physical purpose. The extraordinary excrescences of many insects, the fantastic and many-coloured plumes which adorn certain birds, the excessively developed horns in some of the antelopes, the colours and infinitely modified forms of many flower-petals, are all cases for an explanation of which we must look to some general principle far more recondite than a simple relation to the necessities of the individual. We conceive it to be a most erroneous, a most contracted view of the organic world, to believe that every part of an animal or of a plant exists solely for some material and physical use to the individual—to believe that all the beauty, all the infinite combinations and changes of form and structure should have the sole purpose and end of enabling each animal to support its existence—to believe, in fact, that we know the one sole end and purpose of every modification that exists in organic beings, and to refuse to recognize the possibility of there being any other. Naturalists are too apt to *imagine*, when they cannot *discover*, a use for everything in nature. . . . [W]e believe that the constant practice of imputing, right or wrong, some use to the individual, of every part of its structure, and even of inculcating the doctrine that every modification exists solely for some such use, is an error fatal to our complete appreciation of all the variety, the beauty, and the harmony of the organic world.[38]

For those used to hearing Wallace typified as a "hyperselectionist" who rigorously applied the necessary utility argument, these words will come as something of a shock. But there they are. A year earlier, in what is known as the "Sarawak law" essay, he had also observed, in his discussion of "rudimentary organs," that these "in most cases have no special function in the animal oeconomy."[39] So it appears that in the years before 1858 Wallace believed that physical characters were not *necessarily* directly causal in the evolutionary process, but only in some sense "correlated" with it.[40] Instead, and as I have put it elsewhere, "such organs [i.e., 'rudimentary' organs] seemed entirely inexplicable in the absence of forces yielding slow, gradational changes in organisms (forces so pervasive, one supposes, as to override the significance—whatever that might be—of those structures having no immediate utility)."[41]

The most "pervasive" environmental force that might have been responsible was climate, and Wallace kept an eye out for possible related causalities. But in the few years before 1858 he came across various pieces of evidence that seemed to suggest climate was not the main controlling element. In the late 1857 paper "On the Natural History of the Aru Islands" he writes, "We can hardly help concluding, therefore, that some other law has regulated the distribution of existing species than the physical conditions of the countries in which they are found, or we should not see countries the most opposite in character with similar productions, while others almost exactly alike as respects climate and general aspect, yet differ totally in their forms of organic life."[42] Instead, he concludes, "[i]t is evident that, for the complete elucidation of the present state of the fauna of each island and each country, we require a knowledge of its geological history, its elevations and subsidences, and all the changes it has undergone since it last rose above the ocean."[43]

Meanwhile, Wallace was also considering the adaptive world of the native societies with which he was interacting. It was apparent, for instance, that the various groups responded quite differently to the same challenges (e.g., in the degree of efficiency of design of their building structures): nothing preordained there. In the orangutan essay of 1856 he takes this to the point of considering the role of the human appreciation of beauty in nature: "Naturalists are too apt to *imagine*, when they cannot *discover*, a use for everything in nature: they are not even content to let 'beauty' be a sufficient use, but hunt after some purpose to which even that can be applied by the animal itself, as if one of the noblest and most refining parts of man's nature, the love of beauty for its own sake, would not be perceptible also in the works of a Supreme Creator."[44]

By early 1858 Wallace was collecting in the Moluccas, and it was there that things finally sorted themselves out.

The Ternate Essay, February 1858

Compared to the mix of thoughts that were occupying Wallace's mind by 1857, Darwin's arrival at the concept of natural selection years earlier had been a relatively simple process. Swayed by the congruence of a number of forms of evidence—most notably, from the fossil record, adaptive trends, and the apparent analogy of domestication processes—Darwin could imagine a relatively straightforward process of divergence from common types, into new adaptive strategies that, effectively, represented new species. Wallace, to a considerable degree, was in agreement with this thinking, but a number of things were still bothering him. His letter to Darwin dated 10 October 1856 shows, for example, that he did not believe that domestication provided a reliable analogy to natural process.[45] Meanwhile, in the 1856 paper on the orangutan, he had reiterated his earlier suspicions regarding the relation of adaptation to necessary utility. Further, he was still not sure, after witnessing countless examples of "above-necessity" human abilities in his dealings with aboriginals, how the utility matter played out in the case of human beings.[46] Neither had his Humboldt-influenced effort to identify overriding environmental forces that might act to force a generalizable process made any headway. It still appeared that "more recondite" forces were operating, but what were these, and what was the mechanism?

An important step toward reconciling this swirl of difficulties came after Wallace's completion of his collecting activities in the Aru Islands during the first six months of 1857. He quickly put out several papers on their natural history, the most important of which was "On the Natural History of the Aru Islands."[47] In this work Wallace uses his "Sarawak law" model of 1855[48] to work up a challenge to Lyellian biogeography (see chapters 4 and 10). In a letter to Charles Darwin dated 27 September 1857 he drew attention to this work, probably guessing it would appear in print right around the time (late 1857 or early 1858) that Darwin actually received the letter. Darwin wrote back on 22 December 1857 and Wallace received his words in Ternate through the 9 March 1858 mail delivery there,[49] but by that point Wallace most likely had already set into motion a history-setting event.

In late February 1858 he was suffering from a bout of malaria when he had a revelation. Recalling the ideas of Thomas Malthus on the "checks" to population growth, it occurred to him that a mechanism of organic change could be envisioned wherein "better adapted" individuals might be expected to live longer on the average, thus having a greater chance to pass along their particular characters to offspring. This required there be variation among individuals (and he well knew there was), and that rates of reproduction much exceeded actual carrying capacities of the environment (this also seemed obvious). The sum result could be understood as a "law of nature" to the extent that it would apply to all circumstances of organismal response to its surrounding conditions.[50]

According to all six known accounts of the event written by Wallace, he drafted an essay on the subject as soon as the immediate symptoms of his illness had subsided, and within a few days sent this off to Darwin, ostensibly via the 9 March 1858 mail.[51] Actually, he probably really wanted feedback from Lyell, as the accompanying note to Darwin asked that he pass his materials along to him if Darwin felt them worthy. One can easily understand this in view of the just-recently published Aru Islands analysis; now Wallace had a process model that might explain the actual biogeographical patterns he discussed in it.[52]

So there it was: an intelligible theory of how organisms might change over time. But just how completely stated in the essay was Wallace's understanding of the overall process?

What Wallace Knew—or Thought He Knew—as of March 1858

Right from the beginning the common appreciation of Wallace's Ternate essay on natural selection has been that it represented a nearly perfect abstract of Darwin's ideas on the subject. This interpretation has been fueled by Darwin's own comment that "even his terms now stand as heads of my chapters."[53] I feel, however, that we have taken far too much for granted in this regard.

It is true that, like Darwin, Wallace had identified the immediately signal components of the model: the Malthusian constraint on unlimited growth, the inherent character variation within populations, the potential of superabundant organismal fecundity, and the net result of the differential passing on of traits through the generations. Beyond this, he had finally been successful in identifying a Humboldtian kind of "terrestrial physics" relationship between populations and their surroundings: in the Ternate essay he famously likens selection to the way a governor operates on a steam engine, in so doing implying the existence of an ongoing species-environment equilibrium.[54]

To accomplish this advance, he had been forced to give up a position he had been holding for more than ten years, that adaptive characters were not necessarily utilitarian. It now became apparent that characters *had* to have adaptive utility, or at the very least be closely linked to some other character that *was* utilitarian. This was not a terrible concession to the degree that emerging adaptive characters could be viewed as immediately—and in some cases productively—engaging, *somehow*, with *whatever* conditions the surrounding environment might present. This was a far cry from thinking that such structures had been preordained to fill a *particular* purpose, that is, to "fit into" the surroundings in some preplanned way. Yet this comfortable admission still left Wallace with some problems to deal with.

It reasonably can be argued that Wallace saw, right from those early days of 1858, that natural selection consisted, in essence, of a weeding-out operation. Over the years he would increasingly refer to it as an "elimination of the unfit"[55] as opposed to a "survival of the fittest." Darwin apparently distinguished little between the two concepts, and this difference in perspective lies at the bottom of their opposing views as to the usefulness of domestication as an analog for natural forms of selection.[56] It is in connection with this schism that a Pandora's box of difficulties opens, as follows.

As noted earlier, a core idea of Humboldt's that appears regularly in Wallace's thought is the notion that ever more recondite forces typify the operation of nature. Natural selection à la Wallace is, as just suggested, a conservative process in which every time a population system strikes an imbalance with its environment, selection occurs to bring the two back into (Humboldtian) harmony. Change—that is, irreversible *evolutionary* change—takes place when the new harmony differs structurally (and informationally?) from the old harmony. Wallace's aversion to first causes–based thinking led him to believe, both before and after 1858, that the striking up of new harmonies originated primarily with influences emanating from outside the population: that is, with the environment (both physical and biological). It should therefore come as no surprise that a nod to environmental (in the broadest sense of the word) causation would remain a permanently central element of Wallace's understanding of biological evolution.

In the case of domestication, the changes produced were *out* of harmony with nature. One of the manifestations of this "non-harmony" was human-induced attempts to produce *particular* characters in the plants and animals so treated, a violation of the "whatever" understanding discussed above. Darwin used the artificial selection model to better understand how change per se might be effected, and in this effort he was quite successful. But in so doing, he largely lost track of any possible "more recondite forces" that might be operating. The net result was a Darwinism in which natural selection was treated as an ongoing *process*; meanwhile, Wallace continued to view it as an ecological state-space, that is, true to his Humboldtian roots, as a natural balance between opposing forces.[57]

In practice, the two approaches coincided well enough to produce agreements between the two men on the likely course of adaptation in many immediate contexts. A particularly fine example of this was the application of selection principles to the meaning of color in the natural world: for example, to the relationships producing protective coloration, and mimetic resemblance (see chapters 4 and 5). Wallace's old comrade Bates came out with a major work in 1862 that introduced the concept of protective mimicry,[58] and his splendid enlistment of the still-new natural selection model to explain it all pleased Darwin and Wallace immensely, and equally.

There was still that "more recondite forces" matter to deal with, however—at least in Wallace's mind. The rest of his intellectual life was spent trying to sort out the possible ramifications, and there were many.

For one, it took a long time for Wallace to convince himself that even the immediate conditions of adaptive response could be understood through the operation of natural selection alone. While in the East in the 1850s, he noticed some characteristics of butterfly morphology that he was forced to admit might be due to "local causes";[59] now and then over the next fifteen years he continued to exhibit this uncertainty, even preparing an entire lecture on the subject that he delivered as president of Section D, Biology, at the 1876 annual meetings of the British Association for the Advancement of Science.[60] A few years later, however, he learned of some new ideas on mimicry advanced by the German naturalist Fritz Müller (1821–1897) that seemed to explain most of the phenomena he had earlier attributed to "local causes,"[61] and from that point on there were few if any references to possible influences of this kind in his writings.

Evidence of a continuing interest in "more recondite forces" may be found in many other Wallace natural science writings. For example, he attached much importance to the influence of continental position on biogeographic patterns (see chapter 10); so too there was his approach to continental glaciation (chapter 9), and the constraining effect of astronomical position on life (chapter 12). Further, he continued to recognize various instances of incipient processes in nature.[62]

And, worst of all, there was still that nagging problem of how to explain the higher attributes of thought and emotion in humankind. Wallace simply did not believe the Darwinian approach to the matter, which recognized no real difference between the emergence and evolution of higher consciousness and that of any more directly observable adaptive character. Many years of travel in the company of what Wallace referred to (as did most others, during his time) as "savages" had convinced him (some would now say wrongly) that there was no reason for them to have evolved capacities within their milieu that had no application there (e.g., mathematical or musical genius). Thus it seemed unlikely that they had been "selected for" through natural selection. This brings up an important point. Analysis of Wallace's intellectual trace after 1862 has long been dominated by the assumption that his natural selection paper of 1858 had been meant to apply to the higher attributes of humans in exactly the same way it did to the adaptations of plants and other animals. But there is no evidence of this, certainly not in the paper itself. Neither did he himself ever claim later that he had meant the argument to extend to an understanding of the origins of higher consciousness.[63] We will speak more of this in chapter 2.

Finally, a continuing, annoying, reality: neither Wallace nor Darwin had any idea as to how the natural variation within populations came about or, worse yet, what exact mechanism permitted its faithful transport across generations. And, in fact, neither man was ever able to make any lasting progress on either question.

I believe that the directions Wallace's studies took after his return to England in 1862 are entirely predictable, based on the considerations presented above.[64] Most centrally, Wallace still had to work out to his satisfaction how self-aware beings—*human* beings, that is—could emerge through a process that until then had depended strictly on a utilitarian trajectory. Chapter 2 continues this discussion, in its connection with Wallace's adoption of spiritualistic beliefs.

Acknowledgments

Much of this chapter summarizes my earlier writings on this subject. Thanks go to my co-contributors and two referees for their constructive remarks on earlier manuscript versions of this essay.

Notes

1. Wallace (1905b), vol. 1, pp. 201–204.

2. Ibid., vol. 2, pp. 41–42. That others had faith in Wallace's ability to reason and marshal evidence is evident in many secondary-source writings of the time, including the following words from a review of his *Geographical Distribution of Animals* (Wallace 1876a): "In a review of this nature it is of course impossible to do more than allude to the immense mass of facts from which Mr. Wallace draws these and other deductions. They have been collected with an industry and discrimination, and are marshalled with a clearness and conciseness, which probably his great colleague Mr. Darwin alone could rival" (Alston 1876, p. 64).

3. Four others died before Wallace reached his teens.

4. Wallace (1905b), vol. 1, p. 89.

5. Ibid., vol. 1, p. 88.

6. In *My Life* (vol. 1, p. 234), Wallace reports reading Combe's book while "with my brother," but doesn't note *which* brother.

7. Kottler (1985, p. 370) notes, however, that Darwin discusses Malthusian thoughts in his second edition of the work, which Wallace had with him while in the East. This discussion might have been important in making Wallace think of Malthus at the critical moment. Still, this would be something of an indirect influence.

8. Wallace (1905b), vol. 1, pp. 255–256.

9. Humboldt (1846, 1849).

10. In 1908 Wallace (1909a, p. 10) put it this way: "I have long since come to see that no one deserves either praise or blame for the *ideas* that come to him, but only for the *actions* resulting therefrom. Ideas and beliefs are certainly not voluntary acts. They come to us—we hardly know *how* or *whence*, and once they have got possession of us we cannot reject or change them at will. It is for the common good that the promulgation of ideas should be free—uninfluenced by either praise or blame, reward or punishment."

11. Marchant (1916a), pp. 65–67.

12. I have written about this more extensively elsewhere (see Smith 2003–2006, 2008b).

13. Wallace (1855b), p. 185. See Smith (2014b, 2015a, 2016a) for further discussion.

14. Wallace (1871b), p. 372A: "I reject the hypothesis of 'first causes' for any and every special effect in the universe, except in the . . . sense that the action of man or of any intelligent being is a first cause." This would remain Wallace's mantra for the rest of his life.

15. Wallace (1905b), vol. 1, p. 228.

16. A conclusion that has also been drawn by Malinchak (1987). Note in this regard the final sentence in the Sims letter quoted earlier.

17. Smith (2003–2006), chapter 2.

18. "Recondite," referring to something of unusual complexity, difficulty, or obscurity, was a term Wallace applied often over the full span of his career (e.g., in no fewer

than twelve of his books). For an example of how he used it, see the middle of the first long quotation in the section "The Evolving Wallace."

19. This translated sequence appears in Nicolson (1990), p. 172.

20. Humboldt (1849), p. 16.

21. Humboldt (1849, p. 202) discusses the matter of "uniting natural knowledge with poetry and artistic feeling," a theme present in many of his works, including *Cosmos*.

22. In the version of his essay "A Theory of Birds' Nests" printed in *Contributions* (Wallace 1871b, p. 257), Wallace notes, "Nature is such a tangled web of complex relations, that a series of correspondences running through hundreds of species, genera, and families, in every part of the system, can hardly fail to indicate a true causal connexion." In the same collection he writes, "I believe . . . the universe is so constituted as to be self-regulating; that as long as it contains Life, the forms under which that life is manifested have an inherent power of adjustment to each other and to surrounding nature; and that this adjustment . . . does depend on general laws" (p. 268).

23. Wallace was a subscriber to Meyen's work *Outlines of the Geography of Plants* when its first English translation was published in 1846; he mentions his debt to Liebig in a Land Nationalisation Society address delivered in 1885 (Wallace 1885b). It is well known that Lyell and Darwin themselves had been highly influenced by Humboldt.

24. Wallace (1852b), pp. 109–110. "Isothermal lines" is an obvious reference to Humboldt, who had invented the concept many years earlier.

25. By illustration, in Wallace's famous paper on the origin of human races (Wallace 1864b), he uses the word "harmony" and its direct variants nineteen times; for example, "Now, health, strength, and long life are the results of a harmony between the individual and the universe that surrounds it" (p. clx).

26. Maruyama (1963).

27. A conclusion also reached by Gregory Bateson (1972, p. 435; 1979, p. 43).

28. Smith (2003–2006, 2012a, 2013a, 2016b).

29. Marchant (1916a), pp. 140–143.

30. Wallace's remarks to this effect were later discussed by Morgan (1888a, 1888b); Wright (1870, p. 293) observed, "Strictly speaking, Natural Selection is not a cause at all, but is a mode of operation of a certain quite limited class of causes." This general line of criticism has since been used by many critics to portray Darwinian natural selection as tautological reasoning, or even as teleological (Reiss 2009).

31. Wallace (1880a), pp. 95–96. Einstein's theories, of course, did this very thing. Here we see more of Humboldt's philosophical influence; Kleiner (1981, 1985) has noted the resemblance of Wallace's "universal laws" approach to science to Newton's studies on gravitation.

32. It should not be surprising that Wallace would come to favor the planetesimal accretion theory in explaining the evolution of planets. See chapter 12.

33. Wallace (1870b), p. 302. A more elaborate development of this idea is found in "Creation by Law," the reprinted essay in *Contributions* to which his "demonstration" is appended.

34. Smith (2003-2006), chapter 2.

35. Wallace (1901a), pp. 3–4.

36. This is still the tendency: "evolution" is commonly portrayed in terms of tree diagrams conveying divergences of species lines over a time frame.

37. Ironically, this point of view is not so far from today's understanding that character variation is mostly the product of mutation and/or biochemical processes, the resulting characters then being selected for through the action of natural selection.

38. Wallace (1856b), pp. 30–31.

39. Wallace (1855b), p. 195.

40. As H. Lewis McKinney first observed in 1972 (p. xii) of his introduction to a new edition of Wallace's *A Narrative of Travels on the Amazon and Rio Negro*.

41. Smith (2003-2006), chapter 3. In that same chapter, I suggest that Wallace "imagined the confining/promoting influences to be of grander scale, overriding individual effects through continuity of influence, just as Newtonian forces had supported the original consolidation of the solar system. In short, he thought the direction of the continuing individual 'accumulations' of characters was being influenced by very general—and prior—properties of environmental organization." This ultimately proved to be a partial truth only, as the evolution of a new species was just as importantly caused by more remotely based geographical changes that promoted or restricted contacts among populations, leading to competition and differential selection. On "rudimentary" organs see Smith (2003-2006, 2004a, 2008b).

42. Wallace (1857d), p. 481.

43. Ibid., p. 483.

44. Wallace (1856b), p. 30. In these remarks the philosophy of Humboldt is again palpable.

45. This letter has been lost, but Darwin's reply of 1 May 1857 makes it clear what Wallace was saying (Darwin 1887, vol. 1, p. 453).

46. In his autobiography Wallace (1905b, vol. 1, pp. 361–362) notes how prior to his Ternate epiphany he was thinking of Malthus and the constraints on human increase, but this was in the context of "animal evolution," as he puts it, and not necessarily the evolution of the higher faculties.

47. Wallace (1857d). The special issue, though given as a supplement to volume 20, was actually published on 1 January 1858.

48. Wallace (1855b).

49. Ternate, a small island in the Moluccas near Halmahera, was Wallace's base of operations from January 1858 to May 1859. Here he had a place where he could organize and pack specimens, rest and recover from illness, and send and receive mail.

50. Wallace referred to natural selection as "the law of natural selection" just about as often as he called it "the theory of natural selection."

51. This time line has been disputed, especially by van Wyhe and Rookmaaker (2012), but see Smith (2013b, 2014b, 2015a, 2016a).

52. Wallace (1858). Wallace might just as easily have sent the essay directly to Lyell,

but he had never met or corresponded with him previously, and Lyell was such a major figure that Wallace probably felt it prudent to get to him through Darwin, with whom he had at least already corresponded, and who seemingly would also have been much interested in the subject of the paper.

53. Darwin letter to Charles Lyell (Darwin 1887, vol. 1, p. 473).

54. To be sure, this is a dynamic equilibrium, not a static one. But the term "dynamic" only implies a *changing* condition, not necessarily an *evolving* one. Humboldt surely recognized that earth systems changed over time, thus (in today's terms) describing a dynamic condition, but had not recognized a mechanism for irreversible development *within* individual populations. In a 1902 letter to James Mark Baldwin (Wallace 1926, vol. 2, p. 246) Wallace noted, "I am myself so impressed by the extreme rigidity of natural selection in keeping up each species to a high standard of adaptability to its environment."

55. Smith (2012a).

56. This difference was well recognized back in the nineteenth century. See Morgan (1888a); also note the review of *The Origin of Species* mentioned in chapter 3 of Smith (2003–2006).

57. The ecological basis of Wallace's understanding of biological change is just as strong as the historical divergence-of-populations one. But through the first, an exploitable thermodynamics direction is implicit, whereas in the second only change per se is manifest.

58. Bates (1862).

59. These "local causes" are described in Wallace (1865a).

60. Wallace (1876c).

61. Wallace (1882a).

62. In his "Sarawak law" paper (Wallace 1855b), Wallace discusses the necessity of "some great natural law" that explains, among other things, the presence of what he terms "rudimentary organs." These are now known as "vestigial organs" (such as the "anal spurs" of boid snakes), the remnants of earlier morphology modified by selective forces, but Wallace thought they were anticipatory structures, an indication of future directions. He would come to recognize many such "incipient structures," for example the evolution of migration systems, morphological degeneration trends, the separation of varieties through the infertility of intercrossings, and the occasional appearance of a sport (e.g., specimens of fowls with horns; Smith 2016b). Most famously, he would describe how the "rapid progress of civilization under favourable conditions, would not be possible, were not the organ of the mind of man prepared in advance, fully developed as regards size, structure, and proportions" (Wallace 1871b, p. 358), and even came to believe that people displaying special mental abilities (e.g., those recognized as spirit mediums and witches) were in so doing exhibiting an incipient pattern. In an 1894 book review he wrote, "All selection seems . . . to have tended to the extermination of the possessors of humane and altruistic sentiments, not to their continuous preservation and increase. Yet nothing is more certain than that they *do*

now prevail to an extent never before known, and if they have not been developed by selection they must have been inherent in the race, developed perhaps at some earlier period, and have lain dormant till a more peaceful and more intellectual epoch called for their manifestation" (Wallace 1894c, p. 550).

63. For further discussion see Smith (2003–2006, 2004a, 2008b).

64. Not everyone agrees with this assessment; see chapter 6.

‹ 2 ›

Wallace and the "Preter-normal"

CHARLES H. SMITH

> We . . . note with satisfaction that, in these days when the divorce between political responsibility and intellectual responsibility is rapidly growing more complete and universal, when men, aware of the existence of great speculative subjects and of their importance, are afraid of the conclusions to which free inquiry might bring them and dread nothing so much as making up their mind, we find in Wallace, a robust thinker, who not merely takes the trouble to acquire views on great questions, but who also does not suppress his conclusions for the sake of keeping spurious peace on earth and superficial goodwill among men.
> —*Indian Review*, "The autobiography of Alfred Russell Wallace: A review"

Probably the most contentious of Wallace's many involvements was his fascination with spiritualism. Wallace was indeed a full-blown spiritualist from about the end of 1866, or a little more than half of his life. In his own time and ever since, he has been subjected to considerable ridicule for his acceptance of the belief, and this criticism has done much damage to his overall reputation. Yet his active endorsement of spiritualism seemingly saw little expression in the work he is most famous for as a man of science. Indeed, it is not obvious that his belief contradicted many, if any, of his thoughts on other matters; instead, a kind of symbiosis seemed to be involved. The central questions entertained here are, then, why did he turn to spiritualism, and how did his eventual acceptance of the belief serve his intellectual development?

It should first be noted that spiritualism itself, at least the "modern" type emerging in the late 1840s from the alleged phenomena surrounding the Fox sisters in upstate New York,[1] is not truly a religion in the classic sense. In 1892, for an encyclopedia, Wallace himself defined spiritualism as "the name applied to a great and varied series of abnormal or preter-normal phenomena purporting to be for the most part caused by spiritual beings, together with the belief thence arising of the intercommunion of the living and the so-called dead." Directly he goes on to quote a definition from another source:

> Spiritualism is a science based solely on facts; it is neither speculative nor fanciful. On facts and facts alone, open to the whole world through an extensive and probably unlimited system of mediumship, it builds up a substantial psychology on the ground of strictest logical induction. Its cardinal truth, imperishably established on the experiments and experiences of millions of sane men and women, of all countries and creeds, is that of a world of spirits, and the continuity of the existence of the individual spirit through the momentary eclipse of death; as it disappears on earth reappearing in that spiritual world, and becoming an inhabitant amid the ever-augmenting population of the spiritual universe.[2]

It is not my goal here either to dismiss this idea or to try to defend it, but we should at least begin our evaluation with an allowance: that Wallace took these notions seriously, and believed they bore on the state of the world, and its evolution. Accordingly, we must try to understand his point of view, and how it simultaneously developed with his ideas on other matters. We begin by exploring Wallace's early influences in this direction, and then move on to the immediate factors that led him to investigate the subject in detail in 1865.

Early Lessons

Wallace, as he reports in his autobiography, had a "thoroughly religious" early upbringing, but one that was "by no means rigid." He and his family only irregularly attended church on Sundays, and on that day "the only books allowed to be read . . . were the 'Pilgrim's Progress' or 'Paradise Lost,' or some religious tracts or moral tales, or the more interesting parts of the

Bible."[3] Thus he was little exposed to the "mystery, the greatness, the ideal and emotional aspects of religion,"[4] and, as discussed in chapter 1, drifted into a philosophical skepticism as he came under the spell of Owenism in 1837.

In the late 1830s or early 1840s, Wallace read two works by George Combe and developed an interest in phrenology, which, beyond its dubious positions on the relation of cranial physiognomy to personality traits (but see chapter 6), espoused a kind of humanism that recognized holistic approaches to intellectual and moral knowledge, and to human physiology and health. Wallace's interest intensified some years later when he attended a demonstration in Leicester by a traveling phreno-mesmerist, Spencer T. Hall (1812–1885), who connected the phenomena of trance to particular areas of the head. Wallace quickly discovered that he himself could elicit such responses from some of his Leicester students, going so far as to report a few of the results to a London newspaper in May 1845.[5] Later, while on expedition in the Amazon valley, he would perform similar experiments on native subjects. At that time the scientific community was in general very skeptical about the reality of these phenomena, but once Wallace had elicited them himself, he felt he had "learnt my first great lesson in the inquiry into these obscure fields of knowledge, never to accept the disbelief of great men or their accusations of imposture or of imbecility, as of any weight when opposed to the repeated observation of facts by other men, admittedly sane and honest."[6] Actually, there were two lessons: (1) that in general, new discoveries, no matter how unusual, are deserving of rational attention, not ignorant rejection, and (2) more specifically, that there might be more to what we now term "the paranormal" than meets the eye.

After returning to England from the Amazon in 1852, Wallace seems to have taken relatively little further direct interest in mesmerism (though he remained a strong advocate of phrenology for the rest of his life, and defended both it and hypnotism, a closely related phenomenon, in separate chapters of his *Wonderful Century* in 1898). But his years (1854–1862) in the Malay Archipelago did produce at least two influences on his thinking that would later prove important to his adoption of spiritualism.

The first involved a continuation of observations from his Amazon days, during which he witnessed rituals performed by shamans and other medicine men. These sometimes involved apparent communications with otherworldly entities during induced trance states. While in the Malay Archipelago he encountered more phenomena of this kind, frequently hearing from the residents explanations for mundane events based on elements of spiritism.[7] A second influence came, as he reports in his autobiography, in the mail: through the magazines and newspapers supplied to him from back home he began to hear of the rapidly expanding spiritualism movement and its various claims. For the time being he just took notice, but the seeds of curiosity had been sown.

Wallace's Worldview Beginning in 1862

At the end of chapter 1 I left the reader with the idea that Wallace's model of evolution by natural selection was probably a good deal less complete in 1858 than generally has been acknowledged. This bears directly on his eventual adoption of spiritualism.

One of the most surprising things about Wallace's writings in 1859 and the early 1860s is their complete lack of any further development of the natural selection concept. Indeed, beyond an occasional mere allusion to Darwin's work during this period, it was not until late 1863—a full four years after the publication of *On the Origin of Species,* and five and a half years after the Ternate essay had been mailed to Darwin—that Wallace finally produced an analysis employing natural selection-based theory.[8] Considering Wallace's usual habit of plunging into things, some explanation for this is required. Perhaps he was just too busy, but there is no real evidence for that conclusion, especially as the 1858 to 1862 period was for him relatively sparse in publications. Or maybe he was content to let Darwin have the limelight; this is possible, but it doesn't explain the turmoil he would create after 1862, with his various expressions of divergence of viewpoint. And one should remember how he reacted to Lyell's ideas on biogeography—that is, stating his contrary views as fast as they came to him.

I believe this behavior indicates a temporary post-1858 uncertainty on Wallace's part. It has long been assumed by just about everyone who has written on this subject that the ideas in Wallace's Ternate essay were meant to apply to human beings in the same way they did to animals and plants. As noted in chapter 1, there is no actual evidence of this.[9] Wallace does not mention humans in the essay, nor did he on a later occasion suggest that he intended to do so at the time. There is no reason to believe he doubted that the historical evolution of the human body had been pursuant to the same forces shaping other animals, but there was still the nagging question of why there should be "supra-utilitarian" human abilities. The concept of natural selection had likely done nothing to change his mind on the

Humboldtian principles that eventually led him to it; the "more recondite forces" understanding remained among the most important of these. Natural selection was more the *outcome* of these forces in operation than it was their cause.[10] Again, it is for this reason that he was ready to invoke environmental causalities to explain organic change. Beyond these, natural selection was merely elimination of the unfit, and could not be expected to do more than maintain an equilibrium between population potential and its surroundings. Thus, as of 1858, even with his newfound way to understand the role of utility in adaptation, he likely was still asking himself the question, "How do we explain the utility of abilities that seem to have no necessary reason for coming into being?"

I have discussed Wallace's continuing concern over this issue elsewhere;[11] for now we can refer to just one related event, reported in *My Life*, that occurred soon after his return from the East. He recounts that he and Bates went to visit Herbert Spencer, already one of his idols, to seek illumination on the matter of the origins of life. Spencer was not much help, but Wallace remained enthusiastic enough about Spencer's approach to use it as the basis for one of his most famous essays, "The Origin of Human Races and the Antiquity of Man Deduced from the Theory of 'Natural Selection,'" read before the Anthropological Society of London on 1 March 1864. The essay, which attempts to reconcile the views of those who believed in separate origins for the races of humankind with those who didn't (see chapter 6), was in general a substantial success, and pleased Darwin immensely. But it was the end of Wallace's efforts in this direction.

Much later, in *My Life*, Wallace described how he temporarily had become infatuated with Spencer's "individualistic" form of materialism at this time, only to subsequently reject it. The rejection was not abrupt (or even total, for that matter); it developed over the period of his concerted study of spiritualism in 1865–1866, his adoption of the belief late in the latter year, and two further years of contemplation. But in the summer of 1868 he was decided enough to state at the annual meeting of the British Association for the Advancement of Science, "With regard to the moral bearing of the question as to whether the moral and intellectual faculties could be developed by natural selection, that was a subject on which Mr. Darwin had not given an opinion. He (Mr. Wallace) did not believe that Mr. Darwin's theory would entirely explain those mental phenomena."[12] The final blow came eight months later with the now-famous remarks closing his review of two new editions of Lyell's geology texts, in which he states,

> [B]ut enough, we think, has now been said, to indicate the possibility of a new stand-point for those who cannot accept the theory of evolution as expressing the whole truth in regard to the origin of man. While admitting to the full extent the agency of the same great laws of organic development in the origin of the human race as in the origin of all organized beings, there yet seems to be evidence of a Power which has guided the action of those laws in definite directions and for special ends. And so far from this view being out of harmony with the teachings of science, it has a striking analogy with what is now taking place in the world, and is thus strictly uniformitarian in character. Man himself guides and modifies nature for special ends. The laws of evolution alone would perhaps never have produced a grain so well adapted to his uses as wheat; such fruits as the seedless banana, and the bread-fruit; such animals as the Guernsey milch-cow, or the London dray-horse. Yet these so closely resemble the unaided productions of nature, that we may well imagine a being who had mastered the laws of development of organic forms through past ages, refusing to believe that any new power had been concerned in their production, and scornfully rejecting the theory that in these few cases a distinct intelligence had directed the action of the laws of variation, multiplication, and survival, for his own purposes. We know, however, that this has been done; and we must therefore admit the possibility, that in the development of the human race, a Higher Intelligence has guided the same laws for nobler ends.[13]

So what happened in 1864–1866 to cause him to modify his position?

The Events Leading to Wallace's Adoption of Spiritualism circa December 1866

Among the many problems associated with the "change (reversal) of mind" interpretation of Wallace's shift in thinking in the mid-1860s is that at heart it is based on negative evidence: the assumption that the 1858 Ternate essay was meant to apply to humans (actually, human consciousness) as it did to animals and plants. Beyond this, the pieces of so-called evidence that have been advanced to back the position are easily disposed of on other grounds.[14] My interpretation of the chain of events is that Wallace did, in fact, "modify" his position, but that in no sense did he "reverse" it.

It has not been noticed sufficiently that the 1864 paper on human races is mostly about how adaptive abilities, *once in existence*, will modify previous ones. It says nothing about the origins of higher consciousness, instead concentrating on what effects higher consciousness might have on subsequent selection events. Thus it still did not address how, and to what ends, that higher consciousness came about.

Two extensive biological papers occupied most of Wallace's time during the rest of the spring and early summer of 1864,[15] but as fall approached, the focus of his attention changed sharply. Over the next several months, he wrote three important essays that provide strong evidence he was still seeking some way of integrating extenuating explanations for humankind's higher abilities into his general evolutionary model. The first of these was "On the Progress of Civilization in Northern Celebes."

This work, presented at the annual British Association for the Advancement of Science meetings on 19 September 1864, describes what Wallace terms "true savage life" in Celebes (Sulawesi), and how the Dutch had modified this (for the better, he implies) by introducing coffee plantation culture there. He suggests that an attitude of mild despotism might initially benefit completely uncivilized peoples, arguing that

> there is in many respects an identity of relation between master and pupil, or parent and child, on the one hand, and an uncivilised race and its civilised rulers on the other. We know, or think we know, that the education and industry, and the common usages of civilised man, are superior to those of savage life; and, as he becomes acquainted with them, the savage himself admits this. . . . But as the wilful child or the idle schoolboy, who was never taught obedience and never made to do anything which of his own free will he was not inclined to do, would in most cases obtain neither education nor manners; so it is much more unlikely that the savage, with all the confirmed habits of manhood, and the traditional prejudices of race, should ever do more than copy a few of the least beneficial customs of civilisation, without some stronger stimulus than mere example.[16]

This essay has a lot of thoughts being expressed between the lines. Centrally, we see Wallace tossing around ideas on what kinds of forces might be applied to raise societal consciousness levels; was he trying to tease some new meaning out of his long-held feelings on "informed belief"? In late January 1865 he re-presented the paper, this time to the Ethnological Society of London, which published a full version of it one year later in its *Transactions* series. Themes present in "The Progress of Civilization" are developed further in his next significant writing, "Public Responsibility and the Ballot," a long letter to the editor of *Reader* printed in its issue of 6 May 1865.[17] The following excerpt is long, but crucial to understanding why Wallace would soon buy into the spiritualism concept.

> Mr. Mill truly says, that a voter is rarely influenced by "the fraction of a fraction of an interest, which he as an individual may have, in what is beneficial to the public," but that his motive, if uninfluenced by direct bribery or threats, is simply "to do right," to vote for the man whose opinions he thinks most true, and whose talents seem to him best adapted to benefit the country. The fair inference from this seems to be, that if you keep away from a man the influences of bribery and intimidation, there is no motive left but to do what he thinks will serve the public interest—in other words, "the desire to do right." . . . Instead of drawing this inference, however, it is concluded that, as the "honest vote" is influenced by "social duty," the motive for voting honestly cannot be so strong "when done in secret, and when the voter can neither be admired for disinterested, nor blamed for selfish conduct." But Mr. Mill has not told us what motive there can possibly be to make the man, voting in secret, vote against his own conviction of what is right. Are the plaudits of a circle of admiring friends necessary to induce a man to vote for the candidate he honestly thinks the best; and is the fear of their blame the only influence that will keep him from "mean and selfish conduct," when no possible motive for such conduct exists, and when we know that, in thousands of cases, such blame does not keep him from what is much worse than "mean and selfish conduct," taking a direct bribe?[18]
>
> Perhaps, however, Mr. Mill means (though he nowhere says so) that "class interest" would be stronger than public interest—that the voter's share of interest in legislation that would benefit his class or profession, would overbalance his share of interest in the welfare of the whole community. But if this be so, we may assert, first, that the social influence of those around him will, in nine cases out of ten, go to increase and strengthen the ascendency of "class interests," and that it is much more likely that a man should be thus induced to vote for class interests as against public interests, than the reverse. In the second place, we maintain that any temporary influence whatever, which would induce a man to vote differently from what he would have done by his own unbiassed judgment, is bad—that a man has a perfect right to uphold the interests of his class, and that it is, on the whole, better for the community that he should do so. For, if the voter is sufficiently instructed, honest, and far-seeing, he

will be convinced that nothing that is disadvantageous to the community as a whole can be really and permanently beneficial to his class or party; while, if he is less advanced in social and political knowledge, he will solve the problem the other way, and be fully satisfied that in advancing the interests of his class he is also benefiting the community at large. In neither case, is it at all likely, or indeed desirable, that the temporary and personal influence of others' opinions at the time of an election, should cause him to vote contrary to the convictions he has deliberately arrived at, under the continued action of those same influences, and which convictions are the full expression of his political knowledge and honesty at the time?

It seems to me, therefore, that if you can arrange matters so that every voter may be enabled to give his vote uninfluenced by immediate fear of injury or hope of gain (by intimidation or bribery), the only motives left to influence him are his convictions as to the effects of certain measures, or a certain policy, on himself as an individual, on his class, or on the whole community. The combined effect of these convictions on his mind will inevitably go to form his idea of "what is right" politically, that idea which, we quite agree with Mr. Mill, will in most cases influence his vote, rather than any one of the more or less remote personal interests which have been the foundation of that idea. From this point of view, I should be inclined to maintain that the right of voting is a "personal right" rather than a "public duty," and that a man is in no sense "responsible" for the proper exercise of it to the public, any more than he is responsible for the convictions that lead him to vote as he does.[19] It seems almost absurd to say that each man is responsible to every or to any other man for the free exercise of his infinitesimal share in the government of the country, because, in that case, each man in turn would act upon others exactly as he is acted upon by them, and thus the final result must be the same as if each had voted entirely uninfluenced by others. What, therefore, is the use of such mutual influence and responsibility? You cannot by such means increase the average intelligence or morality of the country; and it must be remembered, that the character and opinions, which really determine each man's vote, have already been modified or even formed by the long-continued action of those very social influences which it is said are essential to the right performance of each separate act of voting. It appears to me that such influences, if they really produce any fresh effect, are a moral intimidation of the worst kind, and are an additional argument in favour of, rather than against, the ballot.

In sum, the only way to effect a productive change in the implications of a vote is to increase the number of voters who are "sufficiently instructed, honest, and far-seeing, [as to] be convinced that nothing that is disadvantageous to the community as a whole can be really and permanently beneficial to his class or party." Once more there is much to be found between the lines; specifically, the "no merit to uninformed belief" position is being used to address the question of what it will take to "raise the average intelligence or morality" of people. There being no merit to uninformed belief, people have to have *conviction* that, in words Wallace used later, "the thoughts we think and the deeds we do here will certainly affect our condition and the very form and organic expression of our personality hereafter."[20]

Wallace was not yet done with this general line of thought; only a few weeks later, in the 17 June 1865 issue, he contributed another essay to *Reader*, this one titled "How to Civilize Savages."[21] Some excerpts again give the flavor of the argument.

> Are the dogmas of our Church adapted to people in every degree of barbarism, and in all stages of mental development? . . . Can the savage be mentally, morally, and physically improved, without the inculcation of the tenets of a dogmatic theology? . . . If the history of mankind teaches us one thing more clearly than another, it is this—that true civilization and a true religion are alike the slow growth of ages, and both are inextricably connected with the struggles and development of the human mind. . . . A form of religion which is to maintain itself and to be useful to a people, must be especially adapted to their mental constitution, and must respond in an intelligible manner to the better sentiments and the higher capacities of their nature. . . . We are told that the converted savages are wiser, better, and happier than they were before—that they have improved in morality and advanced in civilization. . . . No doubt, a great deal of this is true; but certain laymen and philosophers believe that a considerable portion of this effect is due to the example and precept of civilized and educated men . . . [and] it may fairly be doubted whether some of these advantages might not be given to savages without the accompanying inculcation of particular religious tenets. . . . We believe that the purest morality, the most perfect justice, the highest civilization, and the qualities that tend to render men good, and wise, and happy, may be inculcated quite independently of fixed forms or dogmas, and perhaps even better for the want of them. . . . [However], the practices of European settlers are too often so diametrically opposed to the precepts of Christianity, and so deficient in humanity,

> justice, and charity, that the poor savage must be sorely puzzled to understand why this new faith, which is to do him so much good, should have had so little effect on his teacher's own countrymen. . . . It seems desirable, therefore, that our Missionary Societies should endeavour to exhibit to their proposed converts some more favorable specimens of the effect of their teaching.

While outwardly a discussion of the behavior of missionaries, Wallace's general train of thought has now advanced to a consideration of the kinds of "model institutions" (as he refers to them in the essay) that might serve what I have elsewhere termed "believable example":[22] that is, that would provide a foundation for informed belief. The arguments presented in the previous essays suggest that agents featuring simple inculcation would not do the trick, nor would a mere heeding of the opinions of the masses. So what "model institution" could be depended on to "increase the average intelligence or morality of the country"?

As part of his debt to Herbert Spencer, Wallace believed that people were due no more nor less than the implicit consequence of their actions, this being nothing more than simple justice. While ignorant beliefs often resulted in actions that were counterproductive, one could apply oneself (in a near-Spinozian manner) to a program of self-instruction designed to broaden one's mind and ultimately produce fewer inappropriate actions. Thus, "intelligent conviction," as he referred to it in the Sims letter mentioned in chapter 1, could be progress-serving. The problem was to find some societal influence and/or body of teachings that at once promoted intelligent, dispassionate examination of the facts *and* a sense of commitment to the notion that one's current actions determined, or at least strongly affected, the quality of later experiences (thus productively exercising what Wallace termed "faculties which enable us to transcend time and space"[23]). Perhaps, indeed, this was the way to evolve citizens who were "convinced that nothing that is disadvantageous to the community as a whole can be really and permanently beneficial to his class or party."

It was at just this time, June 1865,[24] that Wallace heeded his sister Fanny's suggestion that he look into the world of spiritualism. Fanny was already a follower, and Wallace probably thought enough of his sister's opinion to at least consider the possibility that something worthwhile was there. He began to attend séances, but at first witnessed few "manifestations." However, it has been largely ignored that he also entered into a thorough literary review of the subject, from which he would have learned that spiritualists claimed there was a "natural" afterlife in which the implications of one's biological life experience were simply further lived out, a bill closely paralleling the Spencerian tenet he held most dear, that concerning the continuity of due action. Along these lines, spiritualist spokespersons such as William Stainton Moses and Hudson Tuttle alleged that important causal influences radiated from the "Spirit Realm," mostly as subtle forms of communication (e.g., dreams[25]) between the "spirits" populating it and the still living. These "communications" were supposed to contribute to the learning experiences of those still living and breathing. The general literature complains that a relationship of this kind reduces to a violation of free will, and leads to an understanding that humans are no more than "God's domestic animals."[26] This argument, however, neglects something important: specifically, that the pangs of conscience (and other subtle feelings) invoked are not deterministic, possibly or possibly not leading to a productive response—an "evolutionary" change—on the part of the individual. Thus this was not inculcation of the type Wallace objected to in conventional religion; instead it seemed more to resemble an ongoing flow of usable advice.[27] Wallace also would have liked the fact that spiritualists encouraged the interested to investigate the matter for themselves, and draw their own conclusions.

Other similarities between spiritualism and a more conventional brand of naturalism must also have interested Wallace. The process of spiritualism, as he later described it,[28] paralleled natural selection. He may have entertained the thought, for example, that the advice relayed from the "Spirit Realm" resembled the variation acted upon by biological processes, in the end sometimes generating changes, other times not, and even when followed sometimes leading in longer-term productive directions, and sometimes not. At another level, the whole structure of the alleged "Spirit Realm" might be viewed, in good Humboldtian style, as a new element of "terrestrial physics" operating as a body of "more recondite forces."

Clearly, Wallace was entertaining at least some such thoughts during his early (1865 and 1866) investigations into spiritualism. Beginning in June 1865 he appears suddenly to have ceased just about all of his professional involvements for a full year.[29] Not only did he publish nothing, but his contributions to professional meetings ended for the time being. Nevertheless, he was no instant convert. Even after several months of investigation, he apparently was still in an evaluative mode: in 1870 spiritualist spokesperson and event-organizer Benjamin Coleman recalled that Wallace had attended an event that took place on 6 November 1865, and how "Mr. Wallace was present as a strong disbeliever."[30] The event itself would prove a decisive influence on him, however, as it featured the first of a series of new lectures by the renowned spiritualist trance speaker Emma Hardinge (later Emma Hardinge Britten, 1823–1899).[31]

Hardinge's lectures were spread out over a period ending in June 1866. Her presentation that first night consisted as usual of extemporaneous responses to questions raised on the spot by members of the audience. Portions of these were published within weeks.[32] Consider the following three samples:

> In pointing to the analogy that exists between the great physical and spiritual laws of Earth, together with the modes in which they act, I have sought to shew you that all that man has called the supernatural, and classes as miracle, is but the out-working of an harmonious plan, which the mighty Spirit reveals through eternal laws; and the Spiritualism at which you marvel, and the Christianity before which you bow, are but parts of the same divine law and alternating life of order, which ever sees the day spring out of the darkest night.

> By Chemistry, man learns through scientific processes, to dissolve and re-compose in changed form, every existing atom. Time, instruments, and material processes alone are asked for the chemistry of science to accomplish these results. To the Spirit (whose knowledge comprehends all laws revealed to man) such chemistry is possible, and truly is achieved, *without* the lapse of time, or the aid of human science yet known *as such* to Man.

> Translated through the solemn utterance of dim antiquity all this is "Miracle"—in simple modern science, it is "Chemistry," requiring only knowledge to effect these changes; in modern spiritualistic phrase 'tis mediumship, or chemistry employing subtler forces to effect in yet more rapid time and simpler modes than man's, the self-same changes which man can make by science. To-day you listen to the tap, tap, of the electric telegraph of the soul; you translate into sentences that strange and grotesque form of telegraphy; you behold inscribed on the blank page the name of some beloved one written with no mortal hand; you feel the baptism of the falling water, you know not from whence; and the fragrance of flowers not gathered by mortal power appeals to your startled senses. You call this Spiritualism; and what is this but the chemistry of the spirit?[33]

It can only be concluded that Wallace took these utterances, and most likely those spoken by Hardinge at the later lectures, to heart. Within weeks he was beginning to prepare *The Scientific Aspect of the Supernatural,* which first saw light in serialized form in August and September 1866.[34] A close reading of this long essay reveals an elaborate argument as to why spiritualism should receive formal study: that is, the advocacy present is not of a kind offered up by a firm believer, but instead the reasoned recommendations of an investigator. Wallace's feelings on the issue had apparently not shifted by as late as October or early November, when he circulated a pamphlet version of the study among his friends and colleagues to try to drum up support. The immediate response was, as they say, underwhelming.

At just about the same time (or a little later), early November 1866, Wallace's sister introduced him to a new medium named Nichol[l] (later better known as Mrs. Guppy). Beginning in late November, Nichol[l] produced some sensational phenomena in Wallace's own residence;[35] after these there was no looking back. But even then it was more than two years before he found an opportunity, in the Lyell review in 1869, to fully declare his belief that "we must therefore admit the possibility, that in the development of the human race, a Higher Intelligence has guided the same laws for nobler ends."

Needless to say, Wallace's revised position shocked a lot of people. Many felt, and have felt to this day, that he abandoned natural selection in favor of supernatural causality.

The Effects of Spiritualism on Wallace's Later Work

A dispassionate look at most of Wallace's writings after late 1866, however, forces one to conclude that his adoption of spiritualism had, for the most part, little if any direct effect on his positions either as a natural scientist or as a social critic. Most sources who have written on this subject seem to believe otherwise,[36] but they have been seduced by the facile understanding that his acceptance of the belief directly caused him to abandon part of his natural selection model. In my opinion, nothing of the sort occurred; instead, spiritualism was *added* to his overall model of evolutionary causation to provide an explanation for those elements of nature that he never felt could be explained by it. The Spirit Realm was regarded, simply, as a newly discovered component of the natural world.

Thus I would argue that the entry of spiritualism into Wallace's life was not so much the *cause* of any new developments in his general intellectual worldview, as it was the *result* of already existing ones. This is too large a subject to discuss here in any detail, but I can at least offer a summary of the most important connections.

- The notion of hierarchical causality in nature (and thus, possibly, a domain of nonphysicality beyond the material) had long been an element of Wallace's thinking, originating in good part with his acceptance of the ideas of Humboldt and Combe.
- The idea that a "just" or "due" form of causality should characterize the personal, social, and natural order—now to include the afterlife—had been with him since his first contacts with the writings of Spencer.
- Wallace's long-held objections to inculcation, extending back to his school days,[37] were met by spiritualism, which espoused little in the way of theocracy. Over the whole of his adult life Wallace had little sympathy for religious doctrine per se, at least anything having to do with institutional delivery systems and their trappings, especially idolatry and ceremony.
- As will be highlighted in chapter 8, a central theme in Wallace's social theory explorations was the objective of improving "well-being" of the population. This was certainly one of Owen's influences on him, but others, including George Combe, Thomas Paine, Robert Dale Owen, and even Herbert Spencer, provided additional models.[38] An obvious example of his efforts in this direction may be viewed in the 1898 address he gave at an international spiritualists' convention; in this he tries to unite spiritualist and socialist objectives.[39]

Wallace is most famous for his scientific work in the fields of evolutionary biology and biogeography. In most of his studies in these directions, no accusations of interference involving spiritualistic concepts have ever been leveled at him; this extends to his work on geological subjects and physical geography. Of course, on the matter of human evolution, and by extension human social evolution, it did influence his thought, but as I have just argued it did not cause him to *reverse* a former position, but instead to augment one. This leaves one area where its effects seem uncertain: astronomy.

It has sometimes been argued that Wallace's anthropocentric view of the universe is a manifestation of his spiritualist beliefs, as will be discussed in chapter 12. But here the evidence seems slender to me. For one thing, many spiritualists (including movement inspirers such as Emanuel Swedenborg and Andrew Jackson Davis) held views contrary to his on the unlikelihood of life on other planets. Thus, it doesn't seem that the tenets of spiritualism force such a link. Further, Wallace had been aware of "plurality of worlds" arguments since at least the mid-1840s.[40]

As far as Wallace's efforts in the realms of social science and social criticism go, one seems to see more of a parallel in purpose between them and his spiritualism than sequential influence. Again, most of the positions he held that underlay his socially conscious writings had been part of his basic philosophy from early in life. Thus, the better argument is that his spiritualistic beliefs and social perspectives were simply complementary to one another, with the latter helping to lead to his adoption of the former. There will be more on this subject in chapters 6 and 7.

Perspective

I feel obliged to end this chapter by posing a weighted, and as it turns out rather complex, question: if, as the majority of the population (at least the intellectual population) believes, spiritualism is little more than naïve nonsense, what does this say about Wallace's belief in it, and how in the end does that belief reflect on the validity and/or usefulness of his ideas and opinions on these and other subjects?

This is not the place to attempt an impassioned plea (à la Wallace 1866) for the study of purported nonphysical realities. Yes, a fair number of astonishing séance manifestations have been produced by individuals who never were exposed as frauds, but even in these instances one is hard put to give the benefit of the doubt.

Spiritualism presents some additional issues, however, perhaps more closely paralleling alleged instances of clairvoyance, out-of-body travel, near-death experiences, and past-lives recall. In these latter, there are no physical manifestations, only claims of mental events. In support of these claims, passable forms of evidence do exist, though so far these are not striking enough to position the phenomena themselves within the confines of confident knowledge.[41] Adherents to spiritualism understandably wished to establish the legitimacy of their belief on the basis of "manifestations" that could be observed and "tested," but perhaps this was asking too much.

This returns us to Wallace, and some of the things he believed about the nature of various phenomena.

Obviously, Wallace was open to the possibility of existence of what we would now usually term paranormal phenomena. Make no mistake, however: those phenomena he accepted as actually having occurred were to be considered a part of the "natural" world, and not as "supernatural" (i.e., literally "above nature") events.[42] Thus, in one of his best-known essays, he took the position that well-witnessed instances of apparently miraculous occurrences should not be ridiculed, but instead be investigated on the possibility they may represent natural phenomena observing laws of nature not yet identified.[43] Truly, in Wallace's mind the very word "miracle" was a contradiction in terms: in his opinion everything that happens (at least in "physical" space) is "real," and with sophisticated enough understandings can be placed within the scope of *natural* (i.e., "law-based") organization.

Along related lines, Wallace was inclined to question whether many myths and legends constitute simple literary fictions. In a book review of Edward Tylor's *Primitive Culture* printed in *Nature* in 1872, he wrote,

> A recognition of the now well-established phenomena of mesmerism would have enabled Mr. Tylor to give a far more rational explanation of were-wolves and analogous beliefs than that which he offers us. Were-wolves were probably men who had exceptional power of acting upon certain sensitive individuals, and could make them, when so acted upon, believe they saw what the mesmeriser pleased; and who used this power for bad purposes. This will explain most of the alleged facts without resorting to the short and easy method of rejecting them as the results of mere morbid imagination and gross credulity.[44]

In the same notice Wallace criticizes Tylor's views on other instances of animism (such as beliefs in witchcraft, or the devil):

> We are constantly told that each such belief or idea "finds its place," with the implication that it is thus sufficiently accounted for. But this capacity of being classified necessarily arises from the immense variety of such beliefs and from the fact that they are founded on natural phenomena common to all races, while the faculties by which these phenomena are interpreted are essentially the same in every case. Any great mass of facts or phenomena whatever can be classified, but the classification does not necessarily add anything to our knowledge of the causes which produced the facts or phenomena.[45]

In 1904 Wallace made another interesting attempt to argue for the real-world foundations of myth and legend, in particular that some of the content of the Arabian Nights stories could be linked to actual places and living things present in the islands of the Malay Archipelago where he spent so many years. The very first words of this lengthy essay read, "A considerable experience among savage and barbarous peoples, and some acquaintance with the records of past ages and the beliefs of unlettered peasants in all parts of the world, have convinced me that, in the great majority of cases, beliefs or legends referring to natural phenomena are founded on facts, and are for the most part actual descriptions of what has been observed, though often misinterpreted, and sometimes overlaid with supernatural accessories."[46]

Nevertheless, and despite Wallace's suspicions on such counts, he was unwilling to accept theories of the esoteric for which no concrete evidence was being presented. Considering his acceptance of the existence of a "Spirit Realm," for example, it seems somewhat surprising that he had little enthusiasm for the parallel body of knowledge being espoused by the followers of Helena Blavatsky: theosophy. On several occasions, especially, he railed against the likelihood of there being a cyclic reincarnation process. He offered up reasons of a general nature,[47] and even an argument against, based on natural selection theory![48] Wallace was also no fan of the "telepathic impressions" theory of early psychical researchers such as Frank Podmore and Frederic Myers; as with his feelings on instinct, he saw no reason to adopt hypotheses of "unseen forces" where pertinent evidence seemed lacking.[49]

One final element of Wallace's approach to the "preternatural" deserves some comment here: his appreciation of the nature of space. It is difficult to fully comprehend Wallace's adoption of spiritualism without this information, yet the subject has completely escaped notice.

In his mid- and late maturity Wallace believed there existed, most fundamentally, three domains of natural existence: the physical/abiotic, the biological/living, and the spiritual/nonphysical. The first two comprised what we term "physical," or "extended," (or even "material") space, and the last unextended space. But what exactly might he have understood as "unextended space"? In 1894 the spiritualist journal *Light* hosted a discussion on what the "fourth dimension" might represent; Wallace advanced the following interesting argument:

> Sir,—The discussion on this subject seems to me to be wholly founded upon fallacy and verbal quibbles. I hold, not only that the alleged fourth dimension of space cannot be *proved* to exist, but that it *cannot* exist. The whole fallacy is based upon the assumption that we *do* know space of one,

> two, and three dimensions. This I deny. The alleged space of one dimension—lines—is not space at all, but merely directions in space. So the alleged space of two dimensions—surfaces—is not space, but only the limits between two portions of space, or the surfaces of bodies in space. There is thus only one Space—that which contains everything, both actual, possible, and conceivable. This Space has no definite number of dimensions, since it is necessarily infinite, and infinite in an infinite number of directions. Because mathematicians make use of what they term "three dimensions" in order to measure certain portions of space, or to define certain positions, lines, or surfaces in it, that does not in any way affect the nature of Space itself, still less can it limit space, which it must do if any other kind of space is possible which is yet not contained in infinite Space. The whole conception of space of different *dimensions* is thus a pure verbal fantasy, founded on the terms and symbols of mathematicians, who have no more power to limit or modify the conception of Space itself than has the most ignorant schoolboy. The absolute unity and all-embracing character of Space may be indicated by that fine definition of it as being "a sphere whose centre is everywhere and circumference nowhere." To anyone who thus thinks of it—and it can be rationally thought of in no other way—all the mathematicians' quibbles, of space in which parallel lines *will* meet, in which two straight lines *can* enclose a definite portion of spaces, and in which knots *can* be tied upon an endless cord, will be but as empty words without rational cohesion or intelligible meaning.[50]

Wallace found the subject so fascinating that he prepared a medium-sized essay that extended his discussion, but apparently this was never published.[51] Whether one accepts his arguments above or not, it becomes obvious from them how Wallace could contemplate the existence of a realm of nature that exhibited no material characteristics: that is, spiritualism. One wonders for how long he had held such an opinion: that is, had he come to this conclusion *before* or *after* he adopted spiritualism?[52]

Interestingly, Wallace's thoughts on this suggest possible directions for theoretical exploration. The sum of Wallace's writings on natural science, including those on spiritualism, clearly indicate his sense that the whole of reality consists of laws-based phenomena; this is one of his central debts to Humboldtian philosophy. His later writings (such as *The World of Life*) confirm a belief in the existence of final causes; he felt that some general and inexorable process was leading in a progressive direction, through the medium of evolution. The "Spirit Realm" was part of this overall structure, including being connected, perhaps at the interface of conscious awareness, to the material realms of nature. Otherwise put, "certain portions of space" (as he terms them) are spatially extended, observing some of the full set of laws of nature, and generating dimensional structures, whereas other portions are not, and observe some different, or possibly additional, laws. Yet the whole structure represents a uniformly moving, and uniformly directioned, phenomenon.[53]

*

Before attempting a brief summary, I need, finally, to tackle head-on the question of why Wallace, brilliant observer that he was, seemingly was fooled time and time again at the séances he attended. I can suggest a few fairly obvious reasons, and a couple that are not so obvious.

Under the "obvious" category, we must acknowledge that, careful as Wallace probably thought he was being in most instances, his powers of observation were simply no match for the deceptive skills of the mediums he was dealing with. Add to this his general naïvéte as to the basically good intentions of people, and we have a recipe for disaster in this direction. Undoubtedly, he was easily taken in, though perhaps no more so than many other of his colleagues.

It must also be remembered that he felt, a priori, that remarkable preternatural phenomena existed, based on his experience as a skilled mesmerist back in the 1840s. Surely these new manifestations were just more of the same . . . ? Further, it is undoubtedly true that he did not trust the critical powers of dissenters (especially men of science), at the same time believing that even though some instances of fraud were likely taking place, there was enough genuine evidence that this didn't really matter.

Less obviously, this whole body of spiritualistic phenomena and process seemed to provide just the patch that would account for those aspects of evolution that natural selection was unable to explain. The fit was irresistible, and once it was in place nothing was going to budge it.

Still, it is difficult to come to grips with Wallace's obstinate resistance to each and every accusation made of the mediums of his time, including those situations where anyone could see that an individual had been caught red-handed. Perhaps he truly felt that they had all been mistreated, frauds or not, or reasoned that a solid front was needed to keep up morale among spiritualists in general. Maybe, but how could it help morale to know there was fakery going on? And how could he state with such *certainty*, as in a letter he once sent to John Tyndall, that "I *know* that the facts are real natural phenomena, just as certainly as I know any other curious facts in nature."?[54] Surely something is going on here that does not quite meet the eye![55]

To summarize:

- After an eighteen-month study of the subject, Wallace wholeheartedly, and permanently, adopted spiritualism around December 1866.
- The details of Wallace's adoption are quite transparent, more than evident from his writings (published and otherwise) from 1864 through 1869.
- The first main influence on Wallace's adoption was his already-existing philosophical and moral beliefs, which were nearly perfectly consistent with spiritualistic philosophy and doctrine. The second main influence was his conclusion that spiritualism (i.e., all the "natural" phenomena thought to be connected with it) provided a vehicle for understanding elements of the overall process of evolution that natural selection could not account for. Short of the marshaling of a good deal more evidence than now exists, his conversion to the belief should not be understood as related to a primary desire to contact dead relatives, depression over the failure of his first marital engagement, or other such trivial connections.
- An overly simplistic emphasis on the sensational aspects of Wallace's attachment to spiritualism and the paranormal in general has possibly diverted attention from derivative questions of greater interest. Recalling the warnings of Humboldt, existing theories of nature are doubtlessly incomplete, and perhaps incomplete in unforeseen ways. We have always gained from thinking outside the box, and Wallace's example provides an inspiring model on that count.

Acknowledgments

Some of this chapter derives from my earlier writings on this subject.[56] Thanks to my co-contributors, two referees, and an additional outside source for their constructive remarks on earlier manuscript versions of this essay.

Notes

1. See Brandon (1983), Hardinge (1870), and Chapin (2004).

2. Wallace (1892c), pp. 645–646. Nelson (1988, p. 108) describes recent British spiritualist belief as following the model of J. Arthur Findley, that it "can be conceived as a comprehensive system of values and beliefs that, starting from conceptions of the nature of reality and of the nature of human beings, explains the meaning of existence and provides individuals with a set of values that define 'the Good' life and by implication a code of ethics through which this may be implemented in practice."

3. Wallace (1905b), vol. 1, pp. 226, 227.

4. Ibid., vol. 1, p. 227.

5. Wallace (1845). In his autobiography Wallace reports 1844 as the year of Hall's demonstration, but local newspaper reports of the time indicate that Hall's only appearance in the area during Wallace's stay there was in February 1845.

6. Wallace (1893d), p. 440.

7. See, for example, Wallace (1869c, pp. 190–191; 1868b).

8. This was "Remarks on the Rev. S. Haughton's Paper on the Bee's Cell, and on the Origin of Species" (Wallace 1863c). From 1858 through 1863 Wallace wrote several papers that included mentions of Darwinian thinking and several others that explored biogeographical and faunal subjects, but elements of the natural selection model per se were not taken up.

9. See discussion in Smith (2003–2006, 2008b).

10. This kind of system again relates to "push-pull" causal forces, (evolutionary) complexification based on the integration of negative and positive feedback processes, as discussed in chapter 1.

11. Smith (2003–2006, 2008b).

12. Wallace (1869a).

13. Wallace (1869d), pp. 393–394.

14. See Smith (2003–2006, 2004a, 2008b).

15. Wallace (1864c, 1865a).

16. Wallace (1865b), p. 67.

17. Wallace (1865c). Wallace was responding to opinions set out in the previous issue by John Stuart Mill, the famous philosopher.

18. Note the similarity of this argument to points made by Wallace in the Thomas Sims letter discussed in chapter 1.

19. Again, the resemblance to the arguments offered in the Sims letter is palpable.

20. Wallace (1892c), p. 648.

21. Wallace (1865d).

22. Smith (2008b).

23. Wallace (1870c), p. 359.

24. See discussion in Smith (2003–2006, 2008b).

25. Wallace (1891c) provides a spiritualist's interpretation of the role of dreams and premonitions.

26. For example, Kottler (1974). In a 1910 interview Wallace was quoted as saying, "I do not mean that the control is absolute or that it is of the nature of interference. The control is evidently bound by laws as absolute and irrefragable as those which govern man and his universe. It is certainly dependent on us in a very large measure for its success. I believe we are influenced, not interfered with, and that the management of our bodies is at least as difficult, for those charged with it, as, let us say, the cultivation of this planet for us" (Wallace 1910b).

27. This perspective is well exemplified by a short comment Wallace printed in 1900 (Wallace 1900e): "If you will continually keep *this duty* [i.e., protesting against various 'national crimes'] before you, asking yourselves *how* you can best further this great Cause, your spirit-guides will, I feel sure, impress you how you should act so that the New Century may witness the birth, and perhaps even the maturity of a truly moral and spiritual civilisation."

28. Wallace (1866a, p. 50) writes, "The organic world has been carried on to a high state of development, and has been ever kept in harmony with the forces of external nature, by the grand law of 'survival of the fittest' acting upon ever varying organisations. In the spiritual world, the law of the 'progression of the fittest' takes its place, and carries on in unbroken continuity that development of the human mind which has been commenced here." Later, in 1900, he wrote, "Spiritualism is not an end in itself, but a means of advancing humanity both morally and materially" (Wallace 1900e).

29. Smith (2008b).

30. Wallace (1870f).

31. Much is still uncertain about Hardinge's early life, but apparently she was born in London, where she allegedly developed mediumistic qualities at an early age, and took up acting. She then emigrated to the United States, where she gained fame as a leading voice of the spiritualism movement. Many of her lectures were published, and she was equally well known in both countries. She spent her later years back in England.

32. Portions of the talk were published shortly thereafter in a feature story in the *Spiritual Magazine* ("Miss Emma Hardinge" 1865).

33. Ibid., pp. 531–532.

34. Wallace (1866a). It is notable that this work contains many more quotations from Hardinge than from any other source.

35. These included the production of "apports," in this case materialized flowers.

36. See, for example, Kottler (1974), Benton (2008, 2013), and Malinchak (1987).

37. Wallace (1905b), vol. 1. In his autobiography Wallace writes with obvious disgust of his school lessons, most of which were little more than memorization exercises.

38. Throughout the post-1862 period Wallace continued to assimilate influences in this direction—a generally humanitarian one—for example, from Edward Bellamy (on socialism) and Tolstoy (on ethics).

39. Wallace (1898b).

40. Wallace had thoroughly digested the plurality-of-worlds discussion in *Vestiges*, later bringing up the matter in one of his articles on the orangutan (Wallace 1856b,

pp. 30–31). In a 1903 letter regarding the reasons for his astronomical views, published after his death (Wallace 1913d), Wallace wrote, "You mistake in thinking that the suggestion in my art. on the 'Universe and Man' was written for the purpose of *proving* anything or for answering the objections of agnostics. In studying modern astronomy for another purpose, I came across it as it were, and was so struck by it as a remarkable *fact*, that I looked at it carefully, and brought together various other facts bearing upon the same view. Of course, *I* was specially interested in it because it *does* accord with views I held previously, that the *earth* exists for the development of man; and to those inclined to hold that view the facts I now adduce render it more probable."

41. For studies involving past-lives regressions and reports of near-death experiences, see Stevenson (1987) and Ring (1984), respectively. As far back as 1902 the skeptic Frank Podmore wrote, "Inadequate, as we have endeavoured to show, as an explanation even of the physical phenomena, deliberate fraud is seen to be preposterous as a final solution of what are conveniently called the mental manifestations" (Podmore 1902, vol. 2, p. 329). Other forms of "mental manifestations," for example dreams and premonitions, are often portrayed by spiritualists as representing subtle, subliminal ties between the alleged Spirit Realm and the mental processes of individual living persons.

42. Later Wallace (1905b, vol. 2, p. 280) commented on the confusion involved in use of the term "supernatural": "It was called 'The Scientific Aspect of the Supernatural,' a somewhat misleading title, as in the introductory chapter I argued for all the phenomena, however extraordinary, being really 'natural' and involving no alteration whatever in the ordinary laws of nature."

43. Wallace (1870e).

44. Wallace (1872a), p. 70.

45. Note in this context an unpublished typescript in the Alfred Russel Wallace Collection at the Natural History Museum, London (NHM WP2/6/3/11), prepared in 1904 by the biologist Wilmatte P. Cockerell after a visit to the Wallaces, and providing a loose transcript of their conversation; in it she writes, "Dr. Wallace, being himself what we should call a spiritualist, sees in most ghost stories a foundation of fact; and pointing as they must (according to his ideas) to the world's dawning consciousness of a spiritual universe, he finds them of greatest interest." See http://people.wku.edu/charles.smith/wallace/Visit_to_Alfred.htm.

46. Wallace (1904a), p. 379. The final sentence in the work (pp. 570–571) recapitulates his discussion: "Considering the length and complexity of this story, filled from beginning to end with magic, and mystery, and the powers of magicians and demons; considering, further, that the scene of the story ranges overland from Baghdad, through Central Asia to China, then to Malaya, and thence to the Aru Islands, a distance altogether not far short of ten thousand miles, over lands and seas at that time most imperfectly known; considering also, the nature of the collection of stories of which it forms a part, which nowhere profess to be more than imaginative tales to pass away idle hours, it is really most surprising and instructive to find throughout, from the Castle of the Seven Princesses to the Land of Camphor, and from the Land of the Beasts, through the country of the Jinns, to the mysterious and magical islands of Wák-

Wák, everywhere a basis of recognisable fact—of geographical and biological truth." Wallace's sympathies in this direction are evident in other places as well: for example, in his discussion of the origins of stories of female Amazon warriors (Wallace 1889d, pp. 343–344), his belief that Maori tales of giant birds (moas) were based on actual associations with living animals (Wallace 1893j, p. 446), and his conclusions on the origins of the anthropoid ape images of the Pacific Northwest (described in Wallace 1891b).

47. Wallace (1878a, 1890b, 1890c).

48. Wallace (1904e).

49. Wallace (1890e). He did however on various occasions argue that dreams, premonitions, and prayer all represented functional elements of the spiritualization process.

50. Wallace (1894e).

51. The handwritten manuscript ("Supposed 'Dimensions' of Space as Possible Realities," NHM WP7/116[1]) has been transcribed and can be accessed at http://people.wku.edu/charles.smith/wallace/Supposed_Dimensions.htm. Nor is this notion entirely new: in a review of "A Defence of Modern Spiritualism" (Wallace 1874c), the anonymous author notes that spiritualism "asserts that the spiritual and the material being merely different sides of the same nature, spiritualism and materialism are no more than different provinces of the same scientific domain—the one appointed to complement and illustrate the other; so that, in process of time, a body of philosophy complete in all its parts and including all that is the subject of knowledge shall be the result" ("Modern Spiritualism" 1875, p. 204).

52. Unfortunately, limited evidence exists on this matter. Did he perhaps first grapple with the issue as a teen or young man, during his surveying days? In *My Life* (vol. 1, p. 110), he notes his early interest in the rules of trigonometry: "for some years I puzzled over these by myself, trying such simple experiments as I could, and gradually arriving at clear conceptions of the chief laws of elementary mechanics and of optical instruments. I thus laid the foundation for that interest in physical science and acquaintance with its general principles which have remained with me throughout my life." Or possibly on reading Combe or Swedenborg, or even Humboldt (who wrote frequently on notions of the integration of mind with landscape)? Recall that in 1870 he spoke of "faculties which enable us to transcend time and space" (Wallace 1870c, p. 359), but at that point he was already a convert to spiritualism. In the letter to his brother-in-law Thomas Sims, described in chapter 1, however, he notes, "In my solitude I have pondered much on the incomprehensible subjects of space, eternity, life and death." This was written some four years *before* his study of spiritualism, and seems to support his having had a philosophical interest in the subject going back to at least his early adulthood.

53. At least one modern model of this type has been proposed. See Smith and Derr (2012), Smith (2014a, 2015b), and the web page at http://people.wku.edu/charles.smith/once/writings.htm#2.

54. Wallace (1905b), vol. 2, p. 292.

55. Consider, for example, the possibility that Wallace himself may have had mediumistic abilities, or at least thought he did! As a young man Wallace had developed some considerable skill as a mesmerist, and perhaps in later years he personally experienced contacts—or *thought* he did, at least—with otherworldly entities (e.g., through clairvoyant or out-of-body experiences). The most famous historical example of such a thing is the alleged "personal daimon" (loosely, a "guardian angel") claimed as a companion by Socrates (which Wallace actually discusses in several of his writings). Beyond the mesmerism connection, however, only three (weak) pieces of evidence can be offered in this connection. For one, after his death his son William was quoted as saying, "He was not a sound sleeper, and frequently lay awake during the night, and then it was that he thought out and planned his work. He often told us with keen delight of some new idea or fresh argument which had occurred to him during these waking hours" (Marchant 1916/1975, p. 364). It is under such conditions that hypnogogic experiences might give rise to such notions, real or apparent. A second, and curious, report comes from the journalist Frank Harris (1917, p. 198), who wrote that he once came upon Wallace in a state of reverie, listening, as he stated, to "the music of the spheres"—that is to say, literal music emanating from an otherworldly source. Perhaps Wallace was only having a bit of fun with Harris (or Harris, with his readers), but it is an interesting spectacle to contemplate. As a third item, Wallace once told a friend he thought that "about one person in ten, probably, is a medium" (Allingham and Radford 1907, p. 329). Unfortunately, of course, we can anticipate little real evidence coming to light on these matters.

56. See Smith (1991a, 1992/1999, 2003–2006, 2004a, 2008b, 2015b, 2016b).

‹ 3 ›

Field Study, Collecting, and Systematic Representation

JAMES T. COSTA

While Alfred Russel Wallace was renowned even in his lifetime for the scope and magnitude of his biological collections, his collecting was but a means to an end. He collected to travel rather than traveling to collect, as the income from the sale of his collections back in Britain financed his explorations. To furnish useful data for his natural history research, as well as to fetch a good price, Wallace's specimens had to be thoroughly documented (locality, taxonomic identification), and well curated (labeled, stored or packed). The prodigious number of specimens he collected therefore meant that Wallace had to be supremely well organized: from tagging, preparing, and labeling in the field, to packing and shipping to London. As his livelihood depended on collecting specimens most prized by collectors and museum curators back home, attending to minutiae of morphological variation and geographical locality became critical: this informed identification and classification. The difference between mere intraspecific variation and a new species or subspecies translated into considerable sale price differences. Beyond this, these same data were also central to Wallace's lifelong scientific pursuits.

Wallace's eight years in the Malay Archipelago may have been the "central controlling incident" of his life, as he put it in his autobiography,[1] but the skills and ideas he honed during his earlier four years in Amazonia were formative. The net result of Wallace's combined twelve years in the field was to provide a treasure trove of specimens and observations that he drew upon for decades to come. In this chapter I will first consider Wallace's approach to field collecting as well as the broader cultural context for this endeavor, followed by a consideration of Wallace's approach to taxonomy and systematic organization, and how these relate to his broader scientific and educational interests.

Wallace as a Natural History Collector

Surely very high on the list of things that Wallace was good at was his talent for collecting and organizing facts. It is the mark of a good natural history collector to know the environment well enough to be able to find what one is actually looking for, but this ability alone will not suffice to produce a good scientist. Beyond field skills (and of course these themselves are not easily acquired, nor always easy to put into effect) a scientist must have a grasp of principles of logical arrangement for the things collected. Without such a framework, an understanding of the complexities of nature is impossible, as the primary characteristic of our natural milieu is organized variety. Natural history as a subject of study is most fundamentally a matter of comparisons: how and why observed similarities and dissimilarities are encountered.

It is for this reason that the advances of Linnaeus were so important to the subsequent development of natural history theory, especially biology. An intelligible scientific theory of evolution was not possible before it was realized that, for example, sharks and whales, despite their both being large, streamlined creatures that inhabit the world's oceans, are actually not very closely related to one another. Once this was understood, if only on strictly morphological/physiological grounds, a whole range of questions opened up that could be productively pursued, both by field collectors and by professional "armchair naturalists." Most of the early field collectors were not philosophers or theoreticians; they were either just earning a living, or perhaps exploring new regions in hopes of turning up new opportunities. The few professional biologists and geologists received collected materials, sorting them out and coming to understand that certain things live, or had lived, only in certain places, under certain conditions. But why? The prevailing view before Darwin and Wallace was that there was an underlying plan, and it was the task of "philosophical naturalists" to document and classify the diversity of life in order to recognize that plan.[2]

Wallace and Bates, when they journeyed to the New World, were by no means the first field naturalists who had given attention to the whys of diversity, and not just the whats. But their predecessors had not, on the whole, given much consideration to possible evolutionary explanations. Humboldt, some fifty years earlier, was alert to causalities, but largely of a more immediate sort: for example, the influence climate has on the distribution of organisms. Even Darwin began his expedition on the *Beagle* with no preconceptions of an evolutionary sort. He did not become an evolutionist (transmutationist in the parlance of the time) until March 1837, some months after his return home, and only came to the concept of natural selection a year and a half later.

Still, both Wallace and Darwin later commented that it was an early interest in natural history collecting—notably in collecting beetles—that sent them down their eventual roads toward discovery. In his remarks on receiving the first Darwin-Wallace Medal on 1 July 1908, the fiftieth anniversary of the reading of his and Darwin's papers on evolution by natural selection, Wallace asked, "Why did so many of the greatest intellects fail, while Darwin and myself hit upon the solution of this problem?" Some probably found his answer surprising: "First (and most important, as I believe), in early life both Darwin and myself became ardent beetle-hunters." Wallace explained that the huge species diversity of beetles, their "endless modifications" of shape, size, and coloration, and "innumerable adaptations" to diverse environments could only excite wonder and provoke questions: "It is the constant search for and detection of these often unexpected differences between very similar creatures, that gives such an intellectual charm and fascination to the mere collection of these insects; and when, as in the case of Darwin and myself, the collectors were of a speculative turn of mind, they were constantly led to think upon the 'why' and the 'how' of all this wonderful variety in nature."[3]

In his travels both in Amazonia (1848–1852) and Southeast Asia (1854–1862), Wallace focused mainly on beetles, butterflies, birds, and mammals, but also collected terrestrial mollusks, fish, reptiles, ferns, and other groups both for sale and for his personal collections (figure 3.1). Besides a crash course in bird skinning and preservation techniques with Bates shortly before their departure for Amazonia,[4] the pair's specimen-preparation know-how was largely a matter of "on-the-job training," acquired through trial and error. Wallace's misfortune in losing nearly two years' worth of specimens (along with nearly all of his notebooks and drawings) when his homeward-bound ship, the *Helen*, burned mid-Atlantic, is well known.[5] Eighteen months later he headed east to the Malay Archipelago, where he spent the next eight years crisscrossing the archipelago from one end to the

Figure 3.1 Statue of Wallace with his collecting garb and equipment. Sculpted by artist Anthony Smith, the life-size bronze statue was unveiled at the Natural History Museum, London, in November 2013. Copyright James T. Costa.

other, procuring an astoundingly large and diverse collection that included a prodigious number of new species (see next section). This collecting was, again, a means to an end: "the main object of all my journeys was to obtain specimens of natural history, both for my private collection and to supply duplicates to museums and amateurs," as he explained in the preface to *The Malay Archipelago*.

As a collector Wallace was diligent but ruthless, like all good field collectors of the time—their main piece of field equipment was the gun. (Indeed, the gun remains an important tool for professional bird collectors at research museums.) Wallace rarely expressed remorse for the innumerable specimens he killed; on the contrary, he reported many of his kills dispassionately or, notably in regard to his collections of birds of paradise and orangutans, as adventure narratives. This was certainly by design in *The Malay Archipelago*, where his detailed narratives of stalking and shooting orangs and rare and storied birds helped make that travel memoir a bestseller (it is thus no coincidence that both orangs and birds of paradise are also mentioned in the subtitle of the book). Such accounts, which were initially drafted extensively in Wallace's field journals and "Species Notebook,"[6] perhaps make the most difficult reading for modern sensibilities. Similarly, in other writings Wallace exhibits the emotional dissonance common among collectors who marvel at the beauty of their catch even as they kill the object of their admiration. This may be best expressed in one famous passage in which Wallace described his first sight, and capture, of the birdwing butterfly *Ornithoptera priamus poseidon*:

> I trembled with excitement as I saw it coming majestically towards me, and could hardly believe I had really succeeded in my stroke till I had taken it out of the net and was gazing, lost in admiration, at the velvet black and brilliant green of its wings, seven inches across, its golden body, and crimson breast. It is true I had seen similar insects in cabinets, at home, but it is quite another thing to capture such one's self—to feel it struggling between one's fingers, and to gaze upon its fresh and living beauty, a bright gem shining out amid the silent gloom of a dark and tangled forest. The village of Dobbo held that evening at least one contented man.[7]

Wallace did not view this destructive killing as wanton or wasteful, of course; his collections were in the broadest sense educational, and furnished data on the taxonomy, comparative anatomy, and geographical distribution of species for himself and other philosophical naturalists. Increasingly, much of the trade in birds of paradise was not for scientific study, however, but for fashion: their plumes and sometimes entire birds adorned ladies' hats. Wallace decried this craze, and later in life supported legislation aimed at preventing exploitation of birds for this purpose.[8] The demand (and price) for these birds peaked between 1905 and 1915.[9] Even by 1857 greater birds of paradise had become sufficiently common in the trade that they had declined considerably in value: "Some years ago, two dollars each were paid for these skins," Wallace reported in one paper, "but they have gradually fallen in value, till now there is scarcely any trade in them. I purchased a few in Dobbo at 6*d*. each."[10]

Wallace in the Field

Assistance Local and Global

Wallace was not an "expeditionary traveler," by which I mean he was not part of a sponsored program of exploration in the manner of Captain Cook's voyages of the late eighteenth century or the US Exploring Expedition of 1838–1842—or the voyages of Darwin, Hooker, and Huxley for that matter, all of which were Royal Navy surveying expeditions. Rather, Wallace traveled light, usually without fellow collectors (except for his first year in Amazonia, when he worked with his friend and kindred spirit Henry Walter Bates), but always with hired assistants. His success in the field, particularly during his extended travels in extremely remote and isolated regions, was owed in no small measure to his facility with languages, his keen interest in the native cultures he encountered, and the evident fairness and respect he displayed toward his assistants. These assistants were essential, from trained collectors who traveled with Wallace for long periods, to locals (Malays, Papuans, Chinese, etc.) hired temporarily for particular collecting trips. Wallace had endless problems with most of his assistants, however, reflecting the uneasy nature of the "master-servant" relationship and very different cultural priorities. From Wallace's point of view, his hired hands were often unreliable—disappearing without notice at any point during an expedition—as well as careless, lazy, and in some cases prone to gambling and theft. The reasons are not as simple as reflecting a Western/Eastern cultural divide or an expression of underlying racism; a Malay youth named Ali became Wallace's most long-serving and trusted assistant, while Wallace's first assistant, Charles Allen, who traveled with him from England, fell into the "careless" category.

The son of a London carpenter, Allen worked with Wallace for his first two years in the East (1854–1856), but he departed, having found that the life of a field collector did not suit him. Wallace later rehired Allen to collect independently. Early on, however, Wallace found his fellow countryman exasperatingly sloppy at insect pinning and bird stuffing when they were together, "this after 12 months' constant practice and constant teaching!" he complained to his sister Fanny. In this letter Wallace perhaps only half-jokingly described the qualities he needed in a field assistant:

> let me know his character, as regards *neatness* and *perseverance* in doing any thing he is set about . . . whether he is quiet or boisterous, forward, or sly, talkative or silent, sensible or frivolous, delicate or strong. Ask him whether he can live on rice and salt fish for a week on an occasion—Whether he can do without wine or beer and sometimes without tea, coffee or sugar—Whether he can sleep on a board . . . likes the hottest weather in England . . . is too delicate to skin a stinking animal . . . can walk 20 miles a day . . . can work for there is sometimes as hard work in collecting as in any thing. . . . Can he draw (not copy), can he speak French? Does he write a good hand? Can he make anything—Can he saw a piece of board straight?[11]

In contrast to Allen, Ali traveled with Wallace for most of his time in the archipelago. Wallace came to trust Ali completely: he was organized, careful, expert at shooting, skinning and sailing, and adept at training other assistants and helping Wallace negotiate with locals. Ali even saved Wallace's life, nursing him during severe illnesses. Wallace later had nothing but praise for the young Malay, and when he left the East, he gave him as a parting gift his guns, ammunition, tools, and money, making Ali a wealthy young man. Forty-five years later, a Harvard zoologist collecting on Ternate was approached by an elderly local who introduced himself as "Ali Wallace"—the ultimate compliment to Wallace by Ali was the adoption of his name.[12]

Although Wallace was a lone traveler in one sense, he was very much a part of the Victorian collective enterprise of fieldwork. As historian of science Jane Camerini has compellingly shown, the interpersonal relationships and communal nature of the scientific collecting enterprise were central to Wallace's success in the field.[13] If hired local assistance was crucial for the success of a field collector like Wallace in remote locales, success was equally owed to the network of colonial expatriates living and working around the world, from the formal and informal social norms that opened doors to fellow Europeans regardless of class or wealth status back home, to the colonial administrative institutions and transportation and communication networks, and to capable assistance on the home front. As Camerini put it in one excellent overview of the subject, Wallace "could not have started on his career as a tropical collector had there not been in place a heterogeneous society of collectors, agents, naturalists, editors and publishers, and Europeans settled sparsely in Brazil and much more extensively in the East Indian archipelago. The networks of colonial culture, in all their religious, economic, and military complexity, were as necessary to Wallace's scientific achievements as were his scientific forebears."[14] The first line of assistance on the home front was his capable and well-connected London agent, Samuel Stevens, who insured and superintended the sale of Wallace's specimens, kept the scientific community abreast of Wallace's explorations by periodically publishing extracts of Wallace's letters, and diligently tended to Wallace's requests for funds, books, and supplies.[15] Wallace was very fortunate to have had such a competent, honest, and genuinely good person handling his affairs at home. He also benefited from the global mercantile and naval network of which Victorian Britain was so central a part. Wallace's travels to Amazonia and Southeast Asia were facilitated by commercial ships; the second trip, arranged by the Royal Geographical Society, was supposed to have been with the Royal Navy, but he ended up going on a Peninsular & Oriental (P&O) steamer. He depended on steamships as well as canoes and praus to travel and collect more locally in these regions. Regional networks of steamers (often traders contracted to carry mail) making regular circuits were also crucial for Wallace's ability to ship letters and specimen consignments home, and frequently, if irregularly, receive journals, magazines, newspapers, and personal letters. The multistage mail route between England and the remote Malay Archipelago gives some insight into this network: Wallace's specimens and correspondence would have been carried by two to three Dutch steamship companies between the islands, Java, Singapore, Sri Lanka, and the Gulf of Suez, then camel train and boat to Alexandria, across the Mediterranean by steamship to Southampton (via Malta and Gibraltar), and finally train to London.[16] Thus, despite his forays into terra incognita, for most of his years in the tropics Wallace remained very much in touch with events going on back home, even participating in correspondence of a technical nature. Many of his contributions to these exchanges found their way into professional journals of the period. Beyond these, moreover, for a couple of years at the beginning of his Malay Archipelago expedition Wallace apparently acted as a traveling correspondent for the *Literary Gazette*, a London-based review.[17] What kind of arrangements there were are unknown, but perhaps he exchanged his services for provision of free reading materials.

Vagaries of Collecting

As for collecting itself, Wallace's activities were dictated in part by the weather. Collecting was sporadic in the rainy season, and he used his downtime to work on his collections, read, write, and reflect, as he remarked in his autobiography, where he described the circumstances under which he produced his famous "Sarawak law" paper: "It was written during the wet season, while I was staying in a little house at the mouth of the Sarawak river, at the foot of the Santubong mountain. . . . [D]uring the evenings and wet days I had nothing to do but to look over my books and ponder over the problem which was rarely absent from my thoughts."[18] Under fair conditions his collecting kept him very busy. While in Singapore soon after arriving in the East, he provided a brief account of his daily collecting routine to his mother:

> I will tell you how my day is now occupied. Get up at half-past five. Bath and coffee. Sit down to arrange and put away my insects of the day before, and set them safe out to dry. Charles [Allen] mending nets, filling pincushions, and getting ready for the day. Breakfast at eight. Out to the jungle at nine. . . . Then we wander till two or three, generally returning with about 50 or 60 beetles, some very rare and beautiful. Bathe, change clothes, and sit down to kill and pin insects. Charles ditto with flies, bugs, and wasps; I do not trust him yet with beetles. Dinner at four. Then to work again till six. Coffee. Read. If very numerous, work at insects till eight or nine. Then to bed.[19]

The following year, in Borneo, he described his routine again in a letter published in the *Zoologist*, this time giving more detail about his collecting paraphernalia:

> To give English entomologists some idea of the collecting here, I will give a sketch of one good day's work. Till breakfast I am occupied tick-

> eting and noting the captures of the previous day, examining boxes for ants, putting out drying-boxes and setting the insects of any caught by lamp-light. About 10 o'clock I am ready to start. My equipment is, a bag-net, large collecting-box hung by a strap over my shoulder, a pair of pliers for Hymenoptera, two bottles with spirits, one large and wide-mouthed for average Coleoptera, &c., the other very small for minute and active insects, which are often lost by attempting to drop them into a large mouthed bottle. These bottles are carried in pockets in my hunting-shirt, and are attached by strings round my neck; the corks are each secured to the bottle by a short string. The morning is fine, and thus equipped I first walk to some dead trees close to the house frequented by Buprestidæ.[20]

That day Wallace and his assistant caught "94 beetles, 51 different species, 23 of which are new to my collection. . . . I have been out five hours, and consider this a very good day's work."

Note the comment in the passage above about "examining boxes for ants." Like all tropical field collectors, Wallace waged an incessant battle to keep pests and scavengers from destroying his specimens, foremost among them ants:

> They are great enemies to any dead animal matter, especially insects and small birds. In drying the specimens of insects we procured, we found it necessary to hang up the boxes containing them to the roof of the verandah; but even then a party got possession by descending the string, as we caught them in the act, and found that in a few hours they had destroyed several fine insects. We were then informed that the Andiroba oil of the country (Brazil), which is very bitter, would keep them away, and by well soaking the suspending string we have since been free from their incursions.[21]

In New Guinea he despaired over a superabundant stinging black ant that "immediately took possession of my house, building a large nest in the roof, and forming papery tunnels down almost every post. They swarmed on my table as I was at work setting out my insects, carrying them off from under my very nose, and even tearing them from the cards on which they were gummed if I left them for an instant."[22] In an account of his cramped accommodations in Lombok, Wallace related how "ants swarmed in every part of it, and dogs, cats and fowls entered it at pleasure," continuing, "My principal piece of furniture was a box, which served me as a dining-table, a seat while skinning birds, and as the receptacle of the birds when skinned and dried. To keep them free from ants we borrowed, with some difficulty, an old bench, the four legs of which, being placed in cocoa-nut shells filled with water, kept us tolerably free from these pests."[23]

Other pests were even more difficult to combat. "Lean and hungry dogs" in the Aru Islands were his greatest enemies, he said, constantly trying to make off with his bird specimens: "If my boys left the bird they were skinning for an instant, it was sure to be carried off. Every thing eatable had to be hung up to the roof, to be out of their reach. . . . Every night, as soon as I was in bed, I could hear them searching about for what they could devour, under my table, and all about my boxes and baskets, keeping me in a state of suspense till morning, lest something of value might incautiously have been left within their reach."[24] Blowflies were at times as voracious as the dogs: in New Guinea they "settled in swarms on my bird skins when first put out to dry, filling their plumage with masses of eggs, which, if neglected, the next day produced maggots. They would get under the wings or under the body where it rested on the drying-board, sometimes actually raising it up half an inch by the mass of eggs deposited in a few hours."[25]

Illness was another perennial concern in the field. Wallace clearly had a robust constitution to survive the rigors of living for extended periods under the most basic conditions, relying on self-medication and, on occasion, the ministrations of his assistants or locals. He nearly died on several occasions in both Amazonia and the Far East, and on at least two occasions suffered personal losses to disease: Wallace's younger brother Herbert, who had come to Amazonia to assist him but did not take to tropical fieldwork, succumbed to yellow fever on his way home. And in New Guinea, where Wallace and his assistants were afflicted with fever and dysentery, he lost one of his assistants to disease. At times fever or ulcerated insect bites prevented Wallace from collecting, something he found especially frustrating. "Wounds or sores in the feet are especially difficult to heal in hot climates, and I therefore dreaded them more than any other illness" he wrote in *The Malay Archipelago*. He was philosophical about these setbacks, wryly commenting on the swarms of mosquitoes and sand flies that seemed "bent upon revenging my long-continued persecution of their race," but lamenting that while the stings and bites of the teeming insects could be suffered uncomplainingly, "to be kept prisoner by them in so rich and unexplored a country, where rare and beautiful creatures are to be met with in every forest ramble . . . is a punishment too severe for a naturalist to pass over in silence."[26]

Despite these setbacks and others caused by sometimes unreliable hired hands, robbery, shipwrecks, and threats of pirates, Wallace (aided by his more capable assistants) was a remarkably productive field collector. By his own account his travels in the East alone yielded over 125,000 specimens: nearly 110,000 insects, 7,500 mollusks, 8,050 bird skins, and 410 mammals and reptiles,[27] including as many as 5,000 species new to science in the estimate of entomologist and Wallace scholar George Beccaloni.[28] The Natural History Museum in London (originally the British Museum, Natural History division) became the single largest repository of Wallace's vertebrate collections and insects, with the second-largest collection of his insects being at the University Museum at Oxford.[29]

An Embarrassment of Riches: Collecting and Organizing Wallace's Collections

Wallace's collections, both those sold through Stevens and those intended for private study, yielded innumerable taxonomic and systematic publications. Wallace's interest was never merely descriptive, however. During his first four years in the East, Wallace had planned a book-length exposition on evolution,[30] an undertaking that would have had to wait until his return to England, as he commented to Darwin in 1857: "The mere statement & illustration of the theory in [the 1855 Sarawak law] paper is of course but preliminary to an attempt at a detailed proof of it, the plan of which I have arranged, & in part written, but which of course requires much [research in British] libraries & collections."[31] As part of this research program he intended to study and describe his extensive private collections. He abandoned his book plan once he learned that Darwin was already doing this, however, and soon after he arrived home in 1862 he realized that he did not have the time necessary to thoroughly analyze his vast collections, which he estimated to consist of approximately 3,000 bird skins (comprising approximately 1,000 species) and some 20,000 beetles and butterflies (consisting of about 7,000 species). Of the approximately 5,000 new species he had collected, Wallace named just 295 of them himself, including 120 butterflies in nine publications, 70 scarabaeid beetles in one large monograph, and 105 birds in eleven publications.[32] He farmed out large portions of his remaining collection to specialists for description, as he related in *The Malay Archipelago*:

> Nearly two thousand of my Coleoptera, and many hundreds of my butterflies, have been already described by various eminent naturalists, British and foreign; but a much larger number remains undescribed. Among those to whom science is most indebted for this laborious work, I must name Mr. F. P. Pascoe, late President of the Entomological Society of London, who has almost completed the classification and description of my large collection of Longicorn beetles (now in his possession), comprising more than a thousand species, of which at least nine hundred were previously undescribed, and new to European cabinets.
>
> The remaining orders of insects, comprising probably more than two thousand species, are in the collection of Mr. William Wilson Saunders, who has caused the larger portion of them to be described by good entomologists. The Hymenoptera alone amounted to more than nine hundred species, among which were two hundred and eighty different kinds of ants, of which two hundred were new.[33]

The most extensive treatment of Wallace's insects was produced by the English entomologist Francis Polkinghorne Pascoe, whose *Longicornia Malayana*, published between 1864 and 1869, was a monumental treatise on Wallace's long-horned beetles and their relatives. Around 1867 Wallace decided to sell most of his remaining insect and bird collection, retaining "only a few boxes of duplicates to serve as mementoes of the exquisite or fantastic organisms which I had procured during my eight years' wanderings."[34] Most of these specimens now form the Wallace Collection in the Natural History Museum, London.

In addition to specimen collections, Wallace often recorded natural history notes and other observations on selected species, and drew illustrative sketches. His surviving sketches of Amazonian palms, fishes of the Rio Negro, and ethnological artifacts reveal considerable artistic talent.[35] He produced a number of drawings during his travels in Southeast Asia as well, including some in color. Several of these were used as the basis for woodcut illustrations in *The Malay Archipelago*, along with ten expertly executed maps. It is worth noting, too, that Wallace often sketched in his field notebooks. The Species Notebook of 1855–1859, for example, includes sketches of birds of paradise and quite a few insects, including over four pages of detailed drawings of beetle mouthparts accompanying a series of preserved beetle wings, perhaps serving as a sort of field guide and to help understand the relationships between them.[36]

Labeling and Nomenclatural Issues

The talent that Wallace had for organizing is nowhere more apparent than in his orderly documentation of his specimen collections—a crucially important skill, to be sure, when one's livelihood depends on accurate record keeping as well as identification. As George Beccaloni has pointed out, Wallace was unusual for his time in that he put distinctive locality labels on every specimen he collected, a practice dating at least to his time in the Amazon. This practice reflects Wallace's conviction that recording precise locality data was critical to accurately delineating the geographical distribution of species, which in turn was critical for understanding the geological, climatic, and evolutionary factors that determine the origin and dispersal of species. Wallace chastised his fellow naturalists in his 1852 paper "On the Monkeys of the Amazon" for providing inadequate locality information (see chapter 10), and a decade later waxed eloquent on the central importance of such data in an oft-quoted passage from his paper "On the Physical Geography of the Malay Archipelago."[37]

Wallace's use of distinctive labels facilitates the identification of specimens as his (except where his labels were replaced by later owners). Insect specimens were given locality labels (about 8 mm in diameter), while the birds and mammals had rectangular labels (marking with a red stripe the ones he intended to keep for his personal study).[38] Wallace often used preprinted labels on his vertebrate specimens. He also experimented with label designs (see figure 3.2); in the Species Notebook he drew several models, ultimately coming up with one giving only the specific epithet, noting "Best

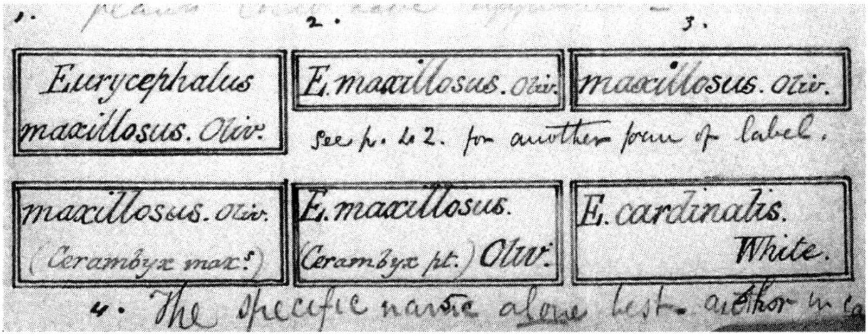

Figure 3.2 Detail from Wallace's Species Notebook, showing a set of draft cabinet labels. He settled on a design with the species name alone, relegating the author to an accompanying catalogue. From Linnean Society manuscript 180, p. 36. By permission, Linnean Society of London.

form . . . This form gives the shortest label, which is important where there are many small species & mostly single specimens."[39]

Economy of space was an overriding concern in the field. Wallace kept meticulous records of his collections in a series of field notebooks and species registers, which were generally written tête-bêche ("head to toe"), with entries made from both ends of the notebook inverted with respect to one another, effectively creating two notebooks in one in order to economize on space. Most of Wallace's notebooks are dedicated to collections, unsurprisingly, except for the Species Notebook, with its extensive natural history observations, memoranda, and evolutionary speculations (its reverse or "verso" side contains the homemade field guide previously mentioned, however, as well as extensive tables of daily collecting results).

A common problem that Wallace confronted as a field naturalist whose livelihood as well as scientific interests hinged on accurate identification of specimens was the lack of universally agreed-upon species identifications and naming rules. The problem stemmed from multiple causes. Then, as now, population variation can introduce confusion: is a given novel variant a new species, a subspecies, or mere variety or even single variant individual? "Lumpers" tend to err on the conservative side in interpreting variation, while "splitters" are prone to see novel species more often in such cases. More problematically in Wallace's day, naturalists often did not know when a species had already been described and so tended to mint new, redundant, names. (And sometimes they *did* know but for reasons of nationalism or self-advancement bestowed the name of their choice regardless.) Introducing further confusion, different sexes of dimorphic species were sometimes mistakenly named as different species, an error all too easy to make when different specimens are caught by different people in different places. By the mid-nineteenth century there had been two centuries of freewheeling taxonomic practice, and naturalists and scientific societies were only just making the first tentative steps toward adopting standards and rules.[40]

Perhaps the single biggest headache for Wallace was keeping track of taxonomic synonyms, wherein different naturalists give different scientific names to the same organism. Synonyms proliferated in the absence of internationally recognized codes for naming species. For example, the emerald fruit beetle *Cetonia aruginosa* described by Drury was the same species as Scopoli's *Scarabaeus speciosissimus*, Olivier's *Cetonia aurata*, Fabricius's *Cetonia fastuosa*, and Herbst's *Cetonia speciosissimus*. Which name should prevail? Without any universally accepted rules such as priority for the first describer, the name that should be consistently applied was a matter of opinion (and naturalists' opinions often diverged). As a result it was more or less standard practice in the mid-nineteenth century (and still is today) to list all known names every time a species was catalogued. In his Species Notebook Wallace voiced great frustration with the proliferation of taxonomic synonyms and the burdens this created.[41] Ever practical minded, he devoted several pages to remedies, including (1) proposals to stop the proliferation of synonyms; (2) a suggestion to publish "synonymical catalogues," compendia of synonyms to serve as reference works, obviating the need for naturalists to review synonyms time after time; and (3) the idea of forming international committees to oversee publications designated for new-species descriptions.[42] One entry in the Species Notebook hints at Wallace's practical plan for organizing insect specimens, pairing the brief label design mentioned previously with a reference catalogue of synonyms: "is not the [species] name alone sufficient & best for a cabinet, leaving authority and synonymes [*sic*] to be had by reference to the catalogue. I am inclined to think so."[43]

Wallace was ahead of his time with his creative solutions, and while not all of his proposals were practicable, some have, in effect, been realized today: for example, international arbiters for nomenclatural rules (such as the International Commission on Zoological Nomenclature), and open-access taxonomic catalogues and databases that include lists of taxonomic synonyms and their authors. The modern Code of Zoological Nomenclature traces its ancestry to the rules and recommendations developed by a committee convened by the British Association for the Advancement of Science (BAAS). Chaired by geologist and ornithologist Hugh Strickland (1811–1853), the committee issued what became known as the "Stricklandian Code" in 1842.[44] This became a standing committee on rules of nomenclature, and in 1863 Wallace was appointed to serve on a review committee set up to revise the 1842 rules. The committee was charged with "power to add to their number, to report on the changes, *if any*, which they may consider it desirable to make in the Rules of Nomenclature drawn up at the instance of the Association by Mr. H. E. Strickland and others, with power to reprint these Rules, and to correspond with foreign naturalists and others on the best means of ensuring their general adoption." The members had their work cut out for them, noting in their report at the 1865 BAAS meeting that

> [s]ince the time that Mr. H. E. Strickland's Rules and Recommendations were printed in the Reports of the British Association, zoological nomenclature has not been improved. Whether it is from the rules and recom-

> mendations not being sufficiently well known, or from an idea that no one has any right to interfere with or make rules for others, many gentlemen appear to cast them away, and do not recognize them at all, while others accept or reject just what pleases themselves; in consequence many very objectionable names have been given, and a very base coinage and spurious combinations have been going on.[45]

Wallace was a diligent committee member, both via correspondence with fellow members and at meetings. The 1865 BAAS meeting report noted that "Mr. Wallace had brought with him a written memorandum containing notes of what he thought could be altered or modified with advantage."[46] Not all of his recommendations were accepted, however. Although Wallace and others strongly opposed Strickland's rule stating that "[a] name whose meaning is glaringly false may be changed," they were not successful in getting it deleted from the code.[47]

Wallace also took an active interest in communicating the newly developed code, recognizing that nothing would improve if naturalists did not know about it. Accordingly, he asked for copies to distribute, and urged his friend (and committee member) Alfred Newton to do the same:

> Have you heard from Sir W. [Jardine] about the copies of the "Nomenclatural Rules." They should be sent to all really working naturalists if any good is to be done. At least 50 copies should be sent to the Secretaries of the Linnean & Zoological Societies for distribution, or no good will be done. We want a general expression of opinion & several I have spoken to know nothing about it. Will you as a personal friend of Sir W. J. & a member of the Committee write & ask to have the residue of the copies printed sent to London for distribution. I know at least a dozen working Entomologists & Conchologists who ought to have them. They should be sent liberally abroad.[48]

A final point worth making in regard to Wallace's collecting is his tendency, rather unusual for the time, to collect *series* of selected birds and insects, not simply to have multiple salable specimens but also to illustrate geographical, developmental, or sexual variation. This reflects Wallace's overriding biogeographical and evolutionary interests. For example, besides species registers tallying his collecting successes, his Species Notebook includes compilations of butterfly species listing which islands they occur on.[49] Wallace aimed to be thorough, collecting both sexes of every possible species in certain groups, giving him a big-picture view of the distribution of certain groups in all of their diversity, in turn giving insight into evolutionary history—precisely the goal that Wallace articulated to his friend Henry Walter Bates years before back in England: "I begin to feel rather dissatisfied with a mere local collection—little is to be learnt by it. . . . I sh[oul]d like to take some one family, to study thoroughly—principally with a view to the theory of the origin of species. . . . By that means I am strongly of [the] opinion that some definite results might be arrived at."[50] Perhaps the best-known example of a masterful treatment by Wallace that was made possible by such data is found in his famous paper on the Malayan papilionid butterflies.[51] Here Wallace presented a striking analysis of sex-limited mimicry (see chapter 4 for a discussion of Wallace's work on this and related phenomena) as well as geographical variation.[52] For example, certain swallowtail species in the archipelago were found to be tailless, while their close continental relatives to the west sported the well-developed tails that the family is known (and named) for. Within the archipelago itself, he found that on the island of Celebes (Sulawesi) the butterflies of at least three families exhibited a curious wing morphology, with wings abruptly curved and pointed (falcate), while close relatives of surrounding islands lacked this trait.[53] Series were also helpful in correcting erroneous assumptions based on limited knowledge of some groups. For example, in New Guinea Wallace reported that he was "much pleased to obtain a fine series of a large fruit-pigeon with a protuberance on the bill . . . and to ascertain that this was not, as had been hitherto supposed, a sexual character, but was found equally in male and female birds."[54]

Wallace on Classification and Biological Systematics

Wallace's immense collections yielded abundant material for classification and analysis at home, as we have seen. Their extent—and his deep knowledge of the comparative anatomy of the groups he collected most intensively, insects and birds—enabled Wallace to authoritatively engage questions of principles of classification and representation of relationships among taxa, both of which were much debated at the time. Perhaps the only point of agreement among naturalists was the desirability of a "natural" system of classification—that is, an orderly arrangement of taxa that reflects, as far as is practicable, actual or true relationships. (This was pursued by many in a nonevolutionary context, in terms of the "plan" of creation, while naturalists such as Wallace, Darwin, and Bates were inclined to an evolutionary interpretation of the natural system.) True relationships were to be based on *affinities*, as opposed to *analogies*, mere similarities in external appearance and behavior. Strickland defined these terms succinctly in an influential 1840 paper,[55] in which he also argued strongly against any a priori symmetrical system of classification—specifically the so-called "Quinarian" or "Circular" system of entomologist William Sharp Macleay (1792–1865).[56] Strickland found this system completely unnatural and at odds with an accurate system based on affinities, necessitating as it does combining or separating taxa simply to preserve symmetry. Though he recognized that relationships were best represented by ramifying branches of variable length,[57] Strickland's interpretation was not evolutionary (transmutational, in the language of his day); he simply urged mapping out affinities and representing them in their truest form—a form that was, however, suggestive of transmutation to Wallace.

Wallace's approach to classification was complex, indicative of a still-unresolved tension between those who wish to "systematize" strictly on the basis of evolutionary relationships (i.e., a genealogical "family tree" approach) and those who looked more toward ecological relationships (i.e., in regard to the way ecological space is organized). He strongly agreed with Strickland, however, and promoted his method of representing taxonomic affinities. Indeed, Wallace's grappling with these issues dates to his scientifically formative years in the early 1840s, likely in direct response to reading Strickland's 1840 paper. In the first volume of his autobiography, Wallace describes his general approach as follows (in discussing an early Mechanics' Institute lecture he gave in Neath):

> I devoted a large portion of my lecture to the question of classification in general, showed that *any* classification, however artificial, was better than none, and that Linnaeus made a great advance when he substituted generic and specific names for the short Latin descriptions of species before used, and by classifying all known plants by means of a few well-marked and easily observed characters. I then showed how and why this classification was only occasionally, and as it were accidentally, a natural one; that in a vast number of cases it grouped together plants which were essentially unlike each other; and that for all purposes, except the naming of species, it was both useless and inconvenient. I then showed what the natural system of classification really was, what it aimed at, and the much greater interest it gave to the study of botany. I explained the principles on which the various natural orders were founded, and showed how

> often they gave us a clue to the properties of large groups of species, and enabled us to detect real affinities under very diverse external forms.
>
> I concluded by passing in review some of the best marked orders as illustrating these various features. . . . Its chief interest to me now is, that it shows my early bent towards classification, not the highly elaborate type that seeks to divide and subdivide under different headings with technical names, rendering the whole scheme difficult to comprehend, and being in most cases a hindrance rather than an aid to the learner, but a simple and intelligible classification which recognizes and defines all great natural groups, and does not needlessly multiply them on account of minute technical differences. . . . It is this attraction to classification, not as a metaphysically complete system, but as an aid to the comprehension of a subject, which is, I think, one of the chief causes of the success of my books, in almost all of which I have aimed at a simple and intelligible rather than a strictly logical arrangement of the subject-matter.[58]

Among other things Wallace admired in Strickland were the geologist's ideas for visual representation of taxonomic affinities. Strickland drew an analogy with mapping—geographical representation—something that surely appealed to Wallace, whose facility with surveying and mapping was well developed by then. Wallace's first attempt to apply this program came with his "Attempts at a Natural Arrangement of Birds," written in the field and published in 1856. In this paper Wallace explicitly adopted Strickland's treelike diagrams, but with his own twist. Rather than having all groups branching from extant taxa, Wallace allowed for "empty" nodes—essentially creating unrooted trees—signifying extinct ancestral groups (common ancestors, in modern terms), giving his diagrams an implicit evolutionary interpretation.[59] He also made use of branch length to denote the amount of difference between the groups (see figure 3.3). Wallace urged that "in every systematic work each tribe and family should be illustrated by some such diagram, without which it is often impossible to tell whether two families follow each other because the author thinks them allied, or merely because the exigencies of a consecutive series compels him so to place them."[60]

One intriguing Species Notebook entry, likely made around the time he was working on the birds paper, reads, "Systems of nature, compared to fragments of dissected map or picture or mosaic.—approximation of fragments shew that all gaps have been filled up."[61] It is not clear whether this has an ecological meaning, relating perhaps to the modern niche concept, or is referring to evolutionary history, the "gaps" in modern attempts at a

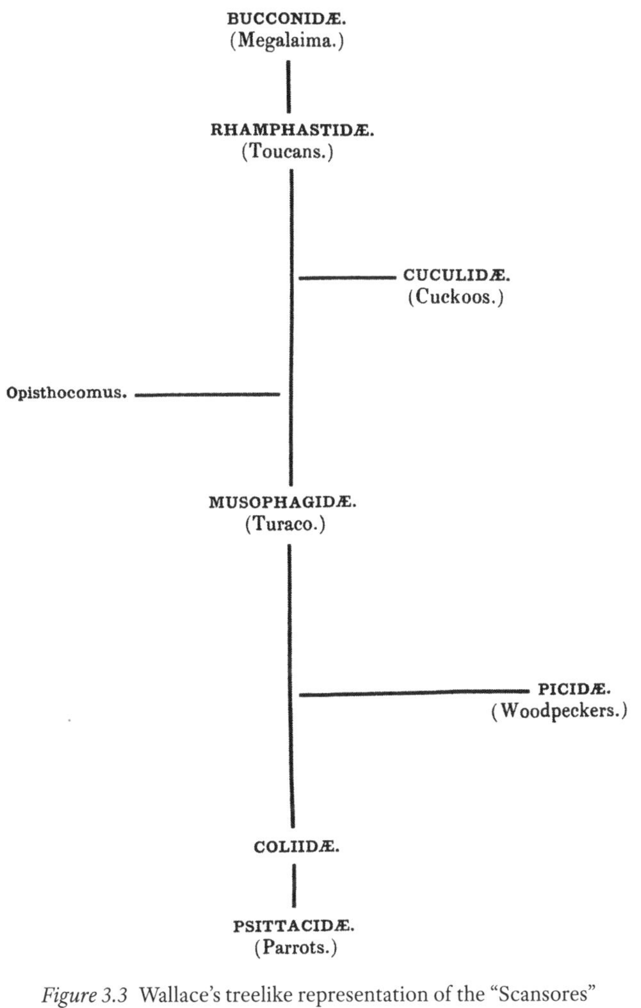

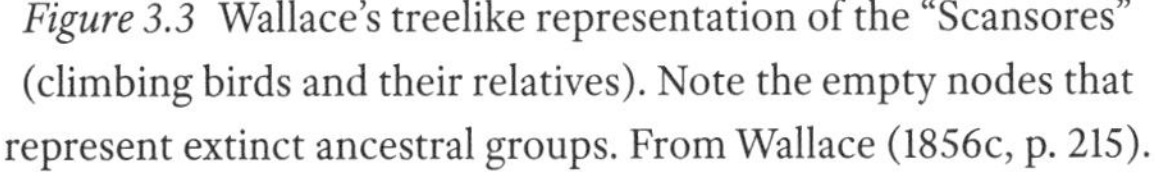
Figure 3.3 Wallace's treelike representation of the "Scansores" (climbing birds and their relatives). Note the empty nodes that represent extinct ancestral groups. From Wallace (1856c, p. 215).

natural arrangement of taxa being filled by extinct species, and the "fragments" the remaining extant ones. The latter interpretation is consistent with some passages from "Attempts at a Natural Arrangement of Birds," for example where Wallace states that "all gaps between species, genera, or larger groups are the result of . . . extinction" during former epochs, and that "the gaps shown [in his diagram] have been all filled up by genera and families forming a natural transition from one of our groups to the other." In criticizing the quinarian system of classification, Wallace notes that "whole orders must have existed of which we are absolutely in ignorance," in which case the circular or quinarian system would be "quite imperceptible in the mere fragment [of remaining species] we have an acquaintance with."[62]

Other aspects of Wallace's thinking on affinities and analogies are found in the Species Notebook as well, such as entries on the significance of affinities for informing classification. Wallace dedicated eight pages to a discussion of relationships within different mammalian, avian, and invertebrate groups, the evolutionary import of which is evident in his frequent use of such terms as "transition" and "modification," and the idea that transitional linking forms are to be found basally, not apically, in the branches of evolutionary trees.[63] Wallace further commented on the use of embryology to inform classification, citing the work of the distinguished Swiss-American paleontologist, glaciologist, and comparative anatomist Louis Agassiz (1807–1873). Agassiz was famously hostile to the evolutionary ideas of Wallace and Darwin, but in his field notebook Wallace clearly saw support for transmutation in Agassiz's ideas; as he put it years later in *Contributions to the Theory of Natural Selection*: "Agassiz . . . insists strongly that the more ancient animals resemble the embryonic forms of existing species; but as the embryos of distinct groups are known to resemble each other more than the adult animals . . . this is the same as saying that the ancient animals are exactly what, on Darwin's theory [and his own!], the ancestors of existing animals ought to be; and this, it must be remembered, is the evidence of one of the strongest opponents of the theory of natural selection."[64]

Extensions of Wallace's Approach to Systematics

Wallace extended his fascination with systematics to a number of subjects outside the domain of biological description per se. Later we will see related efforts in his dealings with biogeographical regions and vaccination statistics; for the moment, however, we can consider some educational ramifications. As indicated earlier, Wallace was sensitive to the idea that subject matter as presented in a book should adhere to "simple and intelligible" arrangements of the information conveyed. In this thinking he followed the lead of Humboldt, whose book chapters typically identified and flowed around core themes (perhaps reflecting Humboldt's views on the growing, but incomplete, nature of knowledge), eschewing simple chronological organization.[65] In Wallace's case, he follows suit most obviously in the organization of his 1869 book, *The Malay Archipelago*. In its preface he writes,

> I visited some islands two or three times at distant intervals, and in some cases had to make the same voyage four times over. A chronological arrangement would have puzzled my readers. They would never have known where they were; and my frequent references to the groups of islands, classed in accordance with the peculiarities of their animal productions and of their human inhabitants, would have been hardly intelligible. I have adopted, therefore, a geographical, zoological, and ethnological arrangement, passing from island to island in what seems the most natural succession, while I transgress the order in which I myself visited them as little as possible.[66]

Similar thoughts underlie the organization of two other important books by Wallace, *The Geographical Distribution of Animals* (1876) and *Island Life* (1880). But it is another aspect of the layout of the first of these two works that is most interesting. The importance of the former to the development of the field of biogeography in general is discussed in chapter 10, but it also made its mark in a different way. In an effort to organize great amounts of distributional data into an easily accessible whole, Wallace again followed Humboldtian models in compiling an extensive set of tables and maps that aggregated species-level data to the family level (there being far too many species to deal with individually in a single work), and geographical data into six faunal regions each divided into four subregions. With this device it became far easier to discern general patterns. But he additionally employed another device of generalization that would find great favor among readers as well as visitors to museums and zoological parks over the next 150 years.

In an effort to give readers some idea of the combined look of landscapes and animals typical of each of the twenty-four subregions he treated, Wallace had a series of plates made up that depicted the landscape of each populated with representative flora and fauna mingled in semirealistic arrangement. This became one of the main inspirations for the "faunal diorama" displays that were a mainstay of twentieth-century natural history museums.[67] Beyond this, and in more recent years, some zoological parks have taken the idea further by placing together living animals to form "biome" exhibits. Wallace himself advocated a plan for the Epping Forest preserve that would have produced "biome"-like assemblages of geographically representative tree species, but this came to naught when his application for its superintendence failed in 1878.

Meanwhile, Wallace was advancing some original thoughts on other aspects of museum design and accessibility. His first essay on the subject, which appeared in 1869, featured a number of suggestions on design and focus that he thought would make them more effective institutions, "best fitted to interest, instruct, and elevate the middle and lower classes, and the young. It is more in accordance with their tastes and sympathies, as shown by the universal fondness for flowers and birds, and the great interest excited by new or strange animals." To this end Wallace argued that public museums "should contain a series of objects to illustrate all the sciences which treat of the earth, nature, and man. These are—1, Geography and Geology; 2, Mineralogy; 3, Botany; 4, Zoology; 5, Ethnology."[68] In related writings his recommendations ranged from location (in the center of a park or public garden) and illumination (well-lit, naturally) to specimens (a limited number, in good condition) and design of exhibit space (low ceilings, and a limited number of side-gallery displays against walls). But Wallace's most urgent recommendations related to his central passion: geographical distribution. In the late 1880s Wallace published several essays and letters concerning what he viewed as the superior arrangement of the museums he'd visited in the United States during his lecture tour of 1886–1887.[69] He was especially impressed with the geographical arrangement at Harvard's Museum of Comparative Zoology (MCZ), which he first visited in November 1886. His diary records his impressions of this visit: "splendid arrangement—(1) Gen. Typical collection of Animal [Kingdom] in one moderate room. (2) Gen. illustrative series of each class or sub-kingdom—Mammalia &c. Birds &c. each in one room—(3) Geog[raphical] arrangement illustrating faunas of N. America, S. America, Europe—Asia, Africa, Australia—each in one room—Excellent!!!"[70] Wallace was also impressed with the way specimens were didactically arranged in the MCZ, particularly in a manner illustrative of evolutionary principles. The museum's founder, Louis Agassiz, was a staunch opponent of Wallace and Darwin's evolutionary ideas, but his son Alexander, who became director of the museum after his father's death in 1873, was more accepting. The irony that exhibits at the MCZ lucidly illustrated principles so strongly opposed by its founder, while exhibits in the country where evolution by natural selection was first discovered paled in comparison, was not lost on Wallace: "It is surely an anomaly that the naturalist who was most opposed to the theory of evolution should be the first to arrange his museum in such a way as best to illustrate that theory, while in the land of Darwin no step has been taken to escape from the monotonous routine of one great systematic series of crowded specimens arranged in lofty halls and palatial galleries, which may excite wonder, but which are calculated to teach no definite lesson."[71] In time some of the early-twentieth-century evolutionary exhibits of the MCZ became instructive in another way, as snapshots of obsolete evolutionary iconography and hypotheses (e.g., fossil horses arranged in a linear series reminiscent of early ideas of orthogenesis). Many of those familiar with the MCZ regarded it as something of a "museum of a museum," albeit not in a pejorative way. Yet its biogeographical arrangement of specimens was highly innovative at the time of its founding, and a feature that Wallace greatly admired.

⁂

Wallace's approaches to field study, collecting, and systematic representation (both in two-dimensional written form and the three-dimensional form of museum exhibits) were closely interrelated. The common thread running through these seemingly disparate interests is geographical distribution (and Wallace's overarching interest in species origins). Whether collecting series of specimens of varied locales to better understand the extent and distribution of species and their variants, urging fellow naturalists to provide precise locality data for their specimens, using paper cut-out cards to experiment with representing the relationships of taxa, or arguing for new and illustrative approaches to specimen display and arrangement in museums, Wallace's firm foundation in field study profoundly informed his lifelong interest in understanding and communicating evolutionary and geographical relationships of species.

Acknowledgments

I thank Charles Smith and George Beccaloni for helpful editorial comments on earlier versions of this chapter, Andrew Berry and George Beccaloni for many enlightening conversations about Wallace, and Lynda Brooks and Elaine Charwat at the Linnean Society of London for their kind assistance in facilitating my study of Wallace's Species Notebook. My Wallace studies were greatly facilitated by a yearlong scholarly leave as fellow of the Wissenschaftskolleg zu Berlin, with additional support from Western Carolina University and Highlands Biological Station.

Notes

1. Wallace (1905b), vol. 1, p. 336.

2. See Raby (1996), introduction and chapter 1; and Asma (2001), chapters 4 and 5.

3. Wallace (1909a), pp. 7–9.

4. Wallace's and Bates's agent Samuel Stevens was instrumental in tutoring the two on specimen preservation techniques and equipping them. They also consulted the preservation chapters in Swainson (1835); see Slotten (2004), pp. 41–45.

5. See Wallace (1852a).

6. Wallace kept four field journals in Southeast Asia (Linnean Society manuscripts 178a–178d). His "Species Notebook" (Linnean Society manuscript 180; Costa 2013b) was initially used for this purpose, from Wallace's arrival in Singapore in April 1855 until June 1856.

7. Wallace (1869c), p. 434.

8. Sir John Lubbock (Lord Avebury), an MP who was a friend and neighbor of Darwin's and an accomplished entomologist, was active in the movement to enact laws to limit the use of bird plumage in the millinary trade and other ornamental purposes. In a 1908 letter to Lubbock, Wallace wrote, "Allow me to wish every success to your Bill for preserving beautiful birds from destruction. To stop the *import* is the only way—short of the still more drastic method of heavily *fining* every one who wears feathers in public, with *imprisonment* for a second offence. But we are not yet ripe for that" (quoted in Hutchinson 1914, vol. 2, p. 258). See also Doughty (1975).

9. See Swadling (1996).

10. Wallace (1857c), p. 414.

11. Wallace letter to Frances (Fanny) Sims, 25 June 1855 (WCP359).

12. See Rookmaaker and van Wyhe (2012) and van Wyhe and Drawhorn (2015) for accounts of Allen and Ali, respectively.

13. Camerini (1996, 1997).

14. Camerini (1996), p. 64.

15. See Stevenson (2009) for an informative account of Stevens as agent and naturalist.

16. See, for example, van Wyhe and Rookmaaker (2012), pp. 251–252, for the possible route of one of Wallace's letter packets.

17. Smith (2008a); see Wallace (1854a, 1854b, 1854c, 1855a, 1855d).

18. Wallace (1905b), vol. 1, p. 354.

19. Wallace letter to Mary Ann Wallace, 28 May 1854 (WCP354; Marchant 1916b, vol. 1, p. 49).

20. Wallace (1855c), p. 4805.

21. Wallace (1853b), pp. 13–14.

22. Wallace (1869c), p. 169.

23. Ibid., p. 169.

24. Ibid., pp. 470–471.

25. Ibid., p. 514.

26. Ibid., p. 465–466.

27. Ibid., p. viii.

28. See http://wallacefund.info/wallaces-specimens.

29. For information on Wallace's vertebrate collections, see Sharpe (1906), Baker (2001), Steinheimer (2003, pp. 179–180), and Cranbrook et al. (2005). A summary of Wallace's insect collection at Oxford is given by A. Z. Smith (1986), appendix B, pp. 157–158.

30. That Wallace was planning to write a book arguing for species change (and wrote out parts of this future work in his Species Notebook, Linnean Society manuscript 180) was first pointed out by historian of science H. Lewis McKinney (1972b, pp. 27–43), and subsequently discussed at length by Costa (2013b, 2014).

31. Wallace letter to Darwin, 27 September 1857 (DCP2145, WCP4080).

32. See http://wallacefund.info/wallaces-specimens.

33. Wallace (1869c), pp. v–vi.

34. Wallace (1905b), vol. 1, p. 404.

35. Wallace's palm drawings (Linnean Society manuscript 182) illustrate his book *Palm Trees of the Amazon and Their Uses* (Wallace 1853c); many of these are reproduced in Knapp (1999), which also includes a number of Wallace's fish and ethnological drawings. Raby (2001) also reproduces several of Wallace's drawings. See, finally, the "Picture Galleries" link on the Wallace Fund website (http://wallacefund.info/) for color scans of Wallace's drawings and paintings.

36. See Costa (2013b) for a facsimile reproduction and annotated transcription of the Species Notebook (Linnean Society manuscript 180).

37. Wallace (1863b). Portions of his remarks are discussed in chapter 11.

38. Wallace letter to Samuel Stevens, 21 August 1856 (WCP1703): "My private collection of Birds are mostly in boxes separate & need not be opened: among the others, all which have a red stripe on the tickets, are private" (emphasis in the original).

39. See Costa (2013b), pp. 100, 112 (Species Notebook, pp. 36, 42).

40. The International Commission on Zoological Nomenclature, founded in 1896, develops, refines, and applies the International Code of Zoological Nomenclature (http://iczn.org), including rules on synonymy and priority. The current code grew out of recommendations first issued in 1842 (see below).

41. Wallace variously described synonymy and the problems it created as a "disgrace to Natural History," a "source of error & perplexity," a "blot upon our science," and an "absurdity," among other things; see Costa (2013b), pp. 166, 280, 284 (Species Notebook, pp. 69, 126, 128).

42. See Costa (2013c) for an overview of Wallace's creative proposals for combating synonymy from the Species Notebook, and his later engagement with the scientific community over this and related issues.

43. Costa (2013b), p. 112 (Species Notebook, p. 42).

44. The committee's report was published as Strickland et al. (1842).

45. British Association for the Advancement for Science (1866), pp. 26, 27.

46. Ibid., p. 27. Wallace also made several recommendations for refining rules to fellow committee member William Jardine (e.g., WCP4193, WCP4194, and WCP4195).

47. McOuat (1996), p. 514.

48. Wallace letter to W. Jardine, 9 December 1863 (WCP3535); Wallace letter to A. Newton, 22 November 1863 (WCP4004); emphasis in the original.

49. See Costa (2013b), pp. 522–530, for Wallace's species compilations for the butterfly families Morphidae and Satyridae.

50. Wallace letter to Henry Walter Bates, 11 October 1847 (WCP348).

51. Wallace (1865a).

52. Wallace's explanation for the evolution of sexual dimorphism, in which change occurs in female rather than male phenotype, has been supported by molecular-phylogenetic studies of female-limited mimicry in papilionids (Kunte 2008); see chapter 5.

53. Wallace speculated that the falcate form gave those Celebes species greater speed and maneuverability, enabling them to better evade predators—or at least predators past: "there seems no unusual abundance of insectivorous birds to render this necessary; and as we can not believe that such a curious peculiarity is without meaning, it seems probable that it is the result of a former condition of things" (Wallace 1869c, p. 288). Wallace further discussed this phenomenon in *The Malay Archipelago* (1869c, pp. 286–289).

54. Wallace (1869c), p. 539.

55. Strickland (1840b, p. 185) wrote that affinity "consists in those *essential* and *important* resemblances which determine the place of a species in the natural system, while . . . [analogy] expresses those *unessential* and (so to speak) *accidental* resemblances which sometimes occur between distantly allied species without influencing their position in the system." See also Strickland (1840a). The term "homology" was coined two years later by British paleontologist and comparative anatomist Richard Owen (1804–1892).

56. The history of visual representations and metaphors for taxonomic relationships is a rich field of study; see O'Hara (1991) for a lucid treatment of efforts to visually represent the natural system in the nineteenth century, and Archibald (2014) for a comprehensive history of visual metaphors of biological order.

57. "The natural system may, perhaps, be most truly compared to an irregularly branching tree," Strickland (1840b, p. 190) wrote. And, just as the form of a tree can be represented by sketching its branches, and so on, "so may the natural system be drawn on a map."

58. Wallace (1905b), vol. 1, pp. 200–201.

59. Ornithologist Robert J. O'Hara (1987, 1991, 1993) discusses Wallace's approach to "systematic generalization" in relation to his adoption of Strickland's diagrams. O'Hara (1987) argues that "Wallace's systematic work, particularly his advocacy of Strickland's mapmaking approach, was a central part of the evolutionary argument he

was building in the 1850's, and was at least as important as his often-cited 1855 work in biogeography."

60. Wallace (1856c), p. 207.

61. Costa (2013b), p. 132 (Species Notebook, p. 52).

62. Wallace (1856c), p. 206. Camerini (1993) discusses Wallace's map metaphor in this passage, arguing that it served to simultaneously represent his ideas about evolutionary origins and geographical distributional patterns.

63. Costa (2013b), pp. 180–197 (Species Notebook, pp. 76–84).

64. Wallace (1870b), p. 301.

65. Kutzinski and Ette (2011, 2012).

66. Wallace (1869c), p. vi.

67. Wallace did not actually invent the diorama, as discussed in Voss and Sarkar (2003), but as many museum directors knew and were admirers of this work (e.g., Henry Fairfield Osborn at the American Museum of Natural History), the link is inescapable.

68. Wallace (1869b), p. 245. This essay was reprinted in Wallace (1900a), vol. 2, pp. 1–16.

69. See Wallace (1887a, 1887b, 1887c, 1887d).

70. A. R. Wallace *North American Journal* (Linnean Society manuscript 177), pp. 8–9; emphasis in the original. See Smith and Derr (2013, p. 21) for an annotated transcription.

71. Wallace (1887a), p. 359.

‹ 4 ›

Wallace, Darwin, and Natural Selection

JAMES T. COSTA

As the codiscoverers of the theory of evolution (transmutation) by natural selection, Darwin's and Wallace's respective views on the details of the evolutionary process have naturally been of great interest to historians. The two believed they had derived precisely the same theory, but on closer inspection their views as of 1858–1859 differed in significant ways. In the decades following the first public reading of their papers on the subject in July 1858, and the publication of *On the Origin of Species* in November 1859, the two came to disagree profoundly on some subjects, to hold subtly differing but related views on others, and to reinforce and complement one another on still others. Darwin, who had the advantage of having thought deeply about transmutation and natural selection for over two decades before Wallace's independent discovery set into motion the events of 1858–1859, came to appreciate Wallace's depth of insight even as the two sparred over aspects of evolutionary process and interpretation. If T. H. Huxley was "Darwin's bulldog," Wallace might be considered "Darwin's retriever," with his uncanny ability to bring clarity in his discussions and debates with Darwin, even as he defended and made Darwin more understandable to others.

In this chapter I take a broadly chronological approach in exploring the key aspects of Wallace's evolutionary thinking, beginning with the points of similarity and difference between his and Darwin's respective paths to transmutation and their formulation of natural selection. Most of the chapter is dedicated to an overview of the varied evolutionary topics of interest to Wallace in the post-*Origin* decades, notably (but not solely) those that he and Darwin discussed and debated.

Wallace and Darwin: Early Views and Paths to 1858–1859

From the time of their earliest engagement with transmutationism, Darwin and Wallace differed in their approach and even receptivity to this idea, which most of their contemporaries considered incorrect at best, and heretical at worst. Darwin was something of a reluctant evolutionist at first, little questioning the received orthodoxy of species immutability through his university years and the five-year *Beagle* voyage that followed. It is well known that he converted to the idea of transmutation some five months after his return from the *Beagle* voyage in October 1836, largely on the weight of such evidence as the relationships between fossil and extant species in South America and between Galápagos endemics and allied species of the South American mainland.[1] Thus did geography and paleontology play key roles in convincing Darwin of transmutation. It was another year and a half before he discovered the principle of natural selection.[2] Wallace, fourteen years Darwin's junior, was a teenager in grammar school at that time.

Darwin's conversion to transmutationism was to determine the course of his scientific career, but early on he showed no indication of heterodoxy. The early-career Darwin worked hard to make his mark in British scientific society; the many specimens he collected on the *Beagle* voyage were being described by leading naturalists, and he was active in the learned scientific societies, reading papers, contributing to discussion, and accepting leadership positions such as secretary of the Geological Society. In short order he also produced several well-received books: three geological works and a travel narrative later reissued as the best-selling *Voyage of the Beagle*. Other than hints given in the latter work, Darwin kept his transmutational speculations largely private, eventually sharing them with a few select friends between the mid-1840s and mid-1850s. In private, however, he steadily pursued a good many lines of evidence to support the theory, synthesizing information from the scientific literature, querying innumerable colleagues and correspondents, and conducting numerous experiments in his study, greenhouse, garden, and surrounding woodlands.

Darwin's approach can be described as a pursuit of *consilience*—accumulating evidence in support of transmutation by natural selection along many, seemingly unrelated, lines that mutually reinforce one another.[3] He did so for some twenty years until his hand was forced by Wallace's discovery of natural selection in 1858. This marked the beginning of Wallace's full understanding of transmutation (one that included a mechanism, natural selection), in contrast with Darwin, who had been gathering evidence for the full theory for two decades, and along the way had come to grasp such special cases as sexual and family-level selection. Wallace's path of exploration of the subject thus differed in a key respect from Darwin's, beyond his later start: while Darwin investigated lines of evidence in support of transmutation *and* his hypothesized mechanism of natural selection, Wallace explored many of the same lines of evidence for transmutation, but without knowledge of a mechanism.[4]

Unlike Darwin, Wallace was an eager convert to transmutation. By his own account Wallace became a transmutationist in 1845, upon reading the anonymously published *Vestiges of the Natural History of Creation*. As he related to his friend Henry Walter Bates, he regarded the transmutationism of *Vestiges* "an ingenious hypothesis strongly supported by some striking facts and analogies but which remains to be proved by more facts & the additional light which future researches may throw upon the subject," a hypothesis that "furnishes both an incitement to the collection of facts & an object to which to apply them when collected."[5] Three years later he and Bates realized their ambitious plan to travel to Amazonia, paying their way with exotic specimens sold by their agent back in London as a means to pursuing the question of species origins.[6]

Wallace's approach to the species question was initially geographical (see chapter 10), in the belief that minutely mapping the geographical distribution of species and varieties would shed light on the process by which new species arise. In Amazonia his working hypothesis posited a geological connection as well. He thought that geologically young areas contain more recently arisen species, and we may surmise that he believed there to be a relationship between the two.[7] It is not known to what extent Wallace might have developed this idea, however, as nearly all of his notebooks and other materials were lost when his homeward-bound ship burned in the Atlantic.[8] What is clear is that two years later, soon after arriving in Southeast Asia to continue his explorations, Wallace described related geographical and paleontological patterns of species distribution in what many historians consider to be a thinly veiled argument for transmutation: "Every species has come into existence coincident both in space and time with a pre-existing closely allied species."[9] Thus, Wallace converged on the same significant intersection of biogeography and geology as did Darwin some years prior.

Wallace and Darwin in 1858: The Action of Natural Selection

At the Linnean Society of London meeting on 1 July 1858, Wallace's essay announcing his discovery of natural selection ("On the Tendency of Varieties to Depart Indefinitely from the Original Type") followed two unpublished pieces on the same subject from Darwin; their contributions were later published collectively under the title "On the Tendency of Species to Form Varieties; and on the Perpetuation of Varieties and Species by Natural Means of Selection."[10] I do not need to discuss the Wallace-Darwin priority issue and related arguments here;[11] rather, I briefly summarize the similarities and differences in the views of Darwin and Wallace as expressed in their 1858 papers to provide insight into their respective takes on the evolutionary process as of that date.[12]

Darwin's and Wallace's belief that they had hit upon precisely the same theory was based on their similarly expressed general formulations of natural selection.[13] In their respective papers, both Darwin and Wallace relate the following key ideas: (1) the ubiquity of the struggle for existence, (2) the illusion of balance or harmony in nature, (3) the abundance of variation or variability, (4) the natural tendency of populations to increase exponentially, and (5) natural checks on this exponential increase.[14] Darwin dubbed this process natural selection, but Wallace did not give a name. Beyond this core process, both naturalists further discussed a role for geological or environmental change as responsible for changing selection pressure over time, and described the outcome of long-continued selection (over geological time) in terms of a branching or treelike pattern of divergence. Despite these general similarities, it has periodically been asserted that Wallace's concept of the action of natural selection differed from that of Darwin's,[15] stemming from differences in which certain ingredients for selection were conceptualized.

Varieties and Variants, Groups and Individuals

At its most basic the selection dynamic stems from abundant and heritable variation plus strong population pressure and struggle; with these criteria fulfilled, natural selection of individuals carrying variations that confer superior survivorship and reproductive potential follows. Some historians have held that in 1858 Wallace thought more in terms of whole varieties (groups) than individual variations per se.[16] If so, this would suggest that

Wallace's selection dynamic pertained to varieties or "races" wholesale and not the individual-based model we see in Darwin's formulation, leading some to go so far as to assert that Wallace does not deserve credit as codiscoverer of natural selection.[17] In my view this charge has no merit.

Confusion stems from Wallace's interchangeable use of terms: variation, varieties, variety, and so on. At that time the word "variety" often referred to distinctive groups (akin to domestic breeds) as well as variant individuals and minute variations themselves (discussed later in this chapter),[18] and there are passages in Wallace's Ternate essay reflecting all of these meanings.[19] The essay was couched largely in terms of varieties in the group or "breed" context, but this reflects its overarching aim: to refute geologist Charles Lyell's claims that species cannot change, a position largely based on the assertion that domestic varieties (again, the sense of "breeds") can vary only so much, and inevitably revert to the parental form when they run wild.[20] Hence Wallace's title for the essay: "On the Tendency of Varieties to Depart Indefinitely from the Original Type." Wallace clarified his view of selection acting on individuals when the Ternate essay was reprinted in *Contributions to the Theory of Natural Selection* (1870), where he inserted a footnote explaining that "[domestic varieties] will vary, and the variations which tend to adapt them to the wild state, and therefore approximate them to wild animals, will be preserved. *Those individuals which do not vary sufficiently will perish.*"[21] He later offered further clarification to E. B. Poulton, an Oxford-based evolutionary biologist (and champion of Wallace's and Darwin's views) who had written Wallace on the subject. Wallace replied that he "used the term 'varieties' because 'varieties' were alone recognised at that time, individual variability being ignored or thought of *no importance*. My 'varieties' therefore included 'individual variations.'"[22]

Still, Wallace was not always clear on how he saw natural selection acting on individuals versus groups, or at least could seem off the mark by modern standards. An example of this is found in Wallace's discussion of the tree-boring instinct of bark beetles he observed in the Aru Islands and New Guinea. He observed beetles attracted to a certain tree species that produces copious amounts of sticky latex, killing the beetles in large numbers as they attempted to bore into the tree.[23] Wallace opined that beetle species that find the tree attractive might be doomed to extinction, while those that are repelled by the tree will survive.[24] A more nuanced view reflecting a more subtle understanding of natural selection acting on individual variation would suggest that, as *individuals* of a species with the fatal attraction are killed, naturally arising variant individuals that are less and less attracted would survive and propagate, such that the species would evolve aversion to the deadly tree over time.

Hard versus Soft Selection

A related aspect of selection as conceived by Darwin and Wallace concerns the concept of "soft" versus "hard" selection, terms introduced in 1968 by evolutionary geneticist Bruce Wallace.[25] Soft selection can be thought of as density-dependent selection, where fitness varies as a function of population density and frequency of a trait. Competition is an example of soft selection: the severity of competition varies as a function of number of competitors per unit area. Hard selection, in contrast, is largely independent of population density, consisting of selection pressures imposed by more or less strict environmental factors—nutritional or environmental needs, capacity to escape or repel predators, and so on. Historians have noted that Darwin appeared to incorporate both hard and soft selection in his model for selection, while Wallace seemed to focus on hard selection only.

In their 1858 papers both Darwin and Wallace commented on selection pressure stemming from abiotic environmental conditions (e.g., intense cold, or desert conditions). They also referred to such biotic factors as predation, or how organisms procure food. But Darwin emphasized competition as not merely another class of biotic "soft selection" pressures, elevating this to center stage in a way that Wallace did not. Darwin saw competition as the most pervasive and important mode of action of selection, even driving evolutionary diversification (a creative process, generating new species). Wallace, in contrast, did not provide a role for competition in his formulation of natural selection in the Ternate essay.[26] In Darwin's view, then, different forms of selection have creative or eliminative roles in evolution, while at that point Wallace addressed the latter role only. This important difference in these naturalists' concepts of selection is discussed in the next section.

Principle of Divergence

Perhaps the single most significant difference between Darwin's and Wallace's concepts of selection pertains to the process that Darwin dubbed the "principle of divergence." By the late 1850s this became a key aspect of the process of transmutation in Darwin's thinking, taking center stage as the very driver of evolutionary diversification. As he described it in a letter to Asa Gray, the American botanist,

> Another principle, which may be called the principle of divergence, plays, I believe, an important part in the origin of species. . . . it follows, I think . . . that the varying offspring of each species will try (only few will succeed) to seize on as many and as diverse places in the economy of nature as possible. Each new variety or species, when formed, will generally take the place of, and thus exterminate its less well-fitted parent. This I believe to be the origin of the classification and affinities of organic beings at all times; for organic beings always *seem* to branch and sub-branch like the limbs of a tree from a common trunk, the flourishing and diverging twigs destroying the less vigorous—the dead and lost branches rudely representing extinct genera and families.[27]

In Darwin's divergence model, the ability of individuals to "seize on as many and as diverse places in the economy of nature as possible" stems from direct competition between closely related variant individuals, varieties, species, and so on. A great many scholars have analyzed and commented upon Darwin's principle of divergence, including the role of competition and the likely origins of the concept in political economy and the so-called "physiological division of labor,"[28] but space constraints preclude a comprehensive discussion here. Suffice it to say that Darwin came to believe that natural selection alone, simply as a process of eliminating less-fit individuals from the population, was inadequate for generating the diversity of life. Accordingly, his divergence principle was the actual engine of speciation, a process where competition afforded a creative role for selection. Wallace, in contrast, did not have a model resembling Darwin's divergence principle in any way. This is not to say he did not appreciate divergence per se, in the sense of the branching and rebranching of lineages over time. He, like Darwin, certainly held that concept, but this is not to be confused with the competition-based *principle* of divergence that Darwin conceived.[29]

Note that Darwin's model of divergence as the primary means of speciation led him to downplay the importance of isolation (allopatric speciation, in modern terms) in favor of speciation in continuous geographical areas (sympatric speciation). Wallace, on the other hand, was consistently wed to an allopatric model of speciation. This difference is discussed later in the chapter, so here I will simply note that insofar as divergence (speciation) goes hand in hand with competitive exclusion and niche partitioning (both modern ecological concepts that have their origin in Darwin's thinking[30]), Darwin's principle of divergence model necessarily entails close interaction of incipient varieties and species. This view of speciation is at odds with modern evolutionary thinking, which holds that allopatric speciation is of far greater importance than sympatric speciation. Darwin's de-emphasis of isolation in speciation was to create problems for him later, when the German naturalist Moritz Wagner (1813–1887) took up this issue and claimed that Darwin had not in fact discovered the true mode of species origins.[31]

Domestication Analogy

Another aspect of the evolutionary thinking of Darwin and Wallace circa 1858 worthy of notice here pertains to domestic varieties. Darwin saw the domestication process as analogous to natural selection in the wild. Indeed, Darwin's close study of agricultural improvement manuals in the months following his conversion to transmutation in 1837 reflects his conviction that domestic breeds (development of which constitutes a limited form of transmutation) somehow held the key to understanding transmutation in nature.[32] Years later, he opened the *Origin* with an extended discussion of domestication, including the results of his special study of pigeon breeds undertaken to establish both the abundance of natural variation in virtually all points of structure, and the object lesson these birds provide for thinking in terms of genealogical (evolutionary) relationships. To Darwin, domestic breeds are analogs of varieties in nature, and his overarching argument was that varieties are but incipient species.

Wallace did not comment on domestication per se in the Ternate essay, and it has sometimes been asserted that he did not accept the domestication-as-analogy argument (and therefore did not understand artificial selection). This misperception stems from two sources. The first is the way that Wallace set up the central argument of the Ternate essay. As noted, this essay was largely aimed at Charles Lyell, whose anti-transmutation views were well known. In one key argument Lyell observed that domestic breeds or varieties tend to revert to the parental form when they run wild, as they soon interbreed. Lyell believed that, regardless of degree to which a breed has been developed, varieties have a limited capacity for change and, if they are released into nature, are ephemeral. Wallace argued for a dynamic (selection) where varieties "depart indefinitely" from the parental type, and in setting up his argument he dismissed Lyell's assertion about reversion of domestic varieties as irrelevant. Of course they would revert, Wallace argued, but this has no bearing on species and varieties in nature because domestic varieties are unnatural to begin with—they are not "analogous to or even identical with those of domestic animals." Thus Wallace put domestic varieties aside in the essay, and proceeded with an argument showing the conditions under which varieties in *nature* do indeed exhibit sustained change, or divergence, from their parental form.

Wallace's statement that domestic varieties are not *analogous* to wild ones would seem, superficially, to put him at odds with Darwin, who developed the domestication-as-analogy concept. This is an incorrect reading, however. It is understandable but erroneous to extrapolate from Wallace's argument to an assumption that he did not see domestic varieties as in any way informative. This is evident from his private Species Notebook, which includes an explicit argument for domestic breeds as evidence for transmutation: "In a few lines Lyell passes over the varieties of the Dog & says there is no transmutation—Is not the change of one original animal to two such different animals as the Greyhound & the bulldog a transmutation?—Is there more essential difference between the ass the giraffe & the zebra than between these two varieties of dogs[?]"[33] Moreover, in the Species Notebook Wallace also gives a hypothetical scenario for the continued change and divergence of species using a domesticated animal as an example (spaniels).[34]

Contributing to the misreading of Wallace on this point is a statement he made later, in the preface to his 1889 book *Darwinism*: "It has always been considered a weakness in Darwin's work that he based his theory, primarily, on the evidence of variation in domesticated animals and cultivated plants. I have endeavoured to secure a firm foundation for the theory in the variations of organisms in a state of nature."[35] Taken by itself, Wallace's statement seems, again, to reject the notion that domestication can provide insights into selection and transmutation in nature. Read in context, however, it is clear that Wallace was referring to what he perceived as Darwin's reliance on domestication as a key line of evidence for transmutation. He concurred fully with Darwin that artificial selection is powerful, and that domestic breeds can serve as a potent example of how long-continued selection can lead to striking modifications.[36] But artificial selection is just that—artificial—and so Wallace felt that a stronger foundation for transmutation is detailed evidence for variation, struggle, and selection in nature.

Special Cases of Natural Selection

Between late 1838, when Darwin first conceived of natural selection, and 1858, when his ideas were made public for the first time, Darwin thought deeply about the theory. Along the way he came to appreciate special cases or nuances in how natural selection works under certain circumstances. He dubbed one special case sexual selection, while he described but did not name the second, which we call family-level selection. In contrast to Darwin's twenty years of rumination on natural selection before the theory went public, Wallace had six months at best. It is well known that he wrote up the theory immediately after he thought of it in a malarial fever in February 1858. Mailing it off to Darwin, the next he heard (in mid-fall 1858) was that the paper already had been publicly read, and published. Wallace was thrilled on one hand, but also mildly annoyed that the manuscript he had hastily written was read without the benefit of his reviewing and refining it.[37] In any case Wallace certainly did not have time to reflect on the theory to any significant degree, and if additional nuanced aspects of selection had occurred to him at that time, he did not record them.

Wallace and Darwin: Key Evolutionary Themes of the 1860s–1880s

In the post-*Origin* decade Wallace and Darwin had a steady correspondence about evolutionary ideas, and Wallace visited Darwin at Down House on more than one occasion after he returned to England in 1862. During his eight years traversing the Malay Archipelago and collecting a multitude of specimens, he also published a succession of insightful papers, mainly on collecting, ornithology, entomology, and biogeography. Papers on these themes continued after his return, but transmutation now more explicitly informed his analysis of some topics. For example, in his 1863 paper on hummingbirds Wallace argued that key morphological traits provide "unmistakeable evidences of a common ancestry" between hummingbirds and swifts, an overtly evolutionary way to frame the relationship of these bird groups. In that same year Wallace also published his first spirited defense of Darwin—refuting Samuel Haughton's criticism of Darwin's scenario for how natural selection gave rise to the cell-building behavior of bees—and read a paper on geographical distribution, referring to the "admirable chapters on this topic in the 'Origin of Species'" in which "Mr Darwin has given us a theory as simple as it is comprehensive."[38]

As Wallace emerged as an eloquent and fierce defender of his and Darwin's ideas—ideas that were already becoming known as "Darwinian," something Wallace contributed to[39]—it also became apparent that the two originators of the theory of evolution by natural selection did not always see things in precisely the same light, including the utility of the term "natural selection" itself.[40] While Wallace continued in his characteristically magnanimous fashion to defer to Darwin in regard to priority and the depth and breadth of work on the subject, it is equally characteristic that he collegially debated Darwin on topics on which they disagreed. In this section I will highlight some of the key areas of agreement and disagreement between Wallace and Darwin through the 1860s.

Variation and Varieties

The degree and distribution of intra- and interspecific variation had long been considered of central importance to defining species and varieties. The inherent complexities of different types of variation, how it is phenotypically expressed, and its heritability, together with inconsistent and imprecise use of such terms as variety, breed, and race, created confusion among naturalists. The English naturalist Edward Blyth expressed the issue succinctly in his 1835 attempt at classifying animal varieties: "The term 'variety' is understood to signify a departure from the acknowledged type of a species, either in structure, in size, or in colour; but is vague in the degree of being alike used to denote the slightest individual variation, and the most dissimilar breeds which have originated from one common stock. The term is, however, quite inapplicable to an animal in any state of periodical change natural to the species to which it belongs. Varieties require some classification."[41] Wallace took the discussion of varieties and variation to a new level by introducing a geographical component. He opened his masterful paper "On the Phenomena of Variation and Geographical Distribution as Illustrated by the Papilionidae of the Malayan Region" (first read in 1864, published in 1865) with a succinct description of the different forms of variation that are often confounded: "I shall proceed to consider these under the heads of—1st, simple variability; 2nd, polymorphism; 3rd, local forms; 4th, coexisting varieties; 5th, races or subspecies; and 6th, true species."[42] Evolutionary biologist James Mallet pointed out that this was likely the first attempt in the post-*Origin* period to distinguish between, and classify, geographical and nongeographical forms often lumped under the term "variety."[43] In so doing Wallace advanced and helped clarify a crucial aspect of Darwin's theory largely neglected in the *Origin*: namely, exactly what *are* species?

Definition of Species

In the *Origin* Darwin argued repeatedly that varieties are incipient species, and he made much of what he labeled "doubtful forms"—forms with unclear taxonomic status, the kind of thing one might expect from dynamically evolving populations.[44] But it was a common criticism that he failed to explicitly define what a species is, a serious issue for a book that claimed to be on the origin of species.[45] Darwin had a de facto species definition that was based on morphology. That is, species were diagnosed on the basis of morphological characters, and while there was inherent ambiguity in this, he believed that with careful study even highly variable and complex groups of species could be so delineated.[46]

While naturalists of the time could not agree on what constitutes a species based on morphology in particular cases, most did have a sort of "rule of thumb" generic species concept. They adhered to the view that different species simply could not interbreed—a species concept based on "reproductive isolation," an idea that has been around for a very long time, greatly predating the so-called biological species concept so often misattributed to German-American evolutionary biologist Ernst Mayr.[47] In essence, the members of "good" species can interbreed, while members of separate species were intersterile. Darwin was sharply critical of the idea that hybrid sterility was a universal rule, and his chapter on hybridism in the *Origin* is an extended argument against this view. He realized that nature was far more subtle, and messy, than such overgeneralizations would have naturalists believe. He discussed at great length exceptions to this so-called rule: there are cases aplenty of seemingly "good," commonly recognized, distinct species that do interbreed, and others where some members of the very same species cannot. Then there are the cases of asymmetrical hybrid fertility, where a cross consisting of males of species A and females of species B yields fertile offspring, while the inverse cross of females of A and males of B does not.

So while Darwin did not doubt that at a certain level hybrid sterility is real and does say something about the distinctness of entities that cannot interbreed, he looked to morphology to diagnose species and varieties. Wallace concurred. Indeed, in his 1865 paper Wallace gave a succinct definition of species virtually identical with the modern biological species concept:

"Species are merely those strongly marked races or local forms which, when in contact, do not intermix, and when inhabiting distinct areas are generally believed to have had a separate origin, and to be incapable of producing a fertile hybrid offspring." But he immediately pointed out the flaw in this definition:

> But as the test of hybridity cannot be applied in one case in ten thousand, and even if it could be applied, would prove nothing, since it is founded on an assumption of the very question to be decided—and as the test of separate origin is in every case inapplicable—and as, further, the test of non-intermixture is useless, except in those rare cases where the most closely allied species are found inhabiting the same area, it will be evident that we have no means whatever of distinguishing so-called "true species" from the several modes of variation here pointed out, and into which they so often pass by an insensible gradation.[48]

Wallace's criticism is insightful: a definition based simply on lack of interbreeding is useless since it can rarely be put to the test, and in any case it is somewhat tautological in *defining* species in terms of the mechanism posited to *generate* new species. (What are species? Entities that cannot interbreed. How do they form? Lack of interbreeding.) Wallace thus reinforced Darwin's operational species definition based on morphology, a pragmatic if problematic means of diagnosing species that is used to this day,[49] and underscored Darwin's view that distinct species are but "well-marked varieties" connected by "intermediate gradations," often in intermediate geographical areas.[50]

Wallace was led to engage the problem of defining species through his treatment of the swallowtail butterflies (Papilionidae) of the Australasian region. In the context of archipelagos where more or less distinct forms occur on different islands, there are no "intermediate" geographical areas. How, then, are species and varieties distinguished? This problem led Wallace to suggest that the *degree* of difference between any two forms should be down-weighted in favor of considering only *permanent* differences:

> The rule, therefore, I have endeavoured to adopt is, that when the difference between two forms inhabiting separate areas seems quite constant, when it can be defined in words, and when it is not confined to a single peculiarity only, I have considered such forms to be species. When, however, the individuals of each locality vary among themselves, so as to cause

> the distinctions between the two forms to become inconsiderable and indefinite, or where the differences, though constant, are confined to one particular only, such as size, tint, or a single point of difference in marking or in outline, I class one of the forms as a variety of the other.[51]

Species, then, were defined by Wallace partially in geographical terms: they consist of morphologically similar groups of individuals that coexist spatially (overlapping ranges) and lack intermediates. Sometimes, however, interfertility trumps morphology, as we will see in the next two sections.

Selection, Hybrid Sterility, and Speciation

An important related issue regarding the process by which new species arise pertains to hybrid sterility: at what point do two diverging populations become independent, on their own evolutionary trajectory, and what role if any does selection play in that process? This was perhaps the first subject on which Wallace's and Darwin's rather different perspectives on how selection operates became obvious. The issue centered on the disinclination or inability of diverging varieties to interbreed (with one another, or their parent species) in the absence of a physical barrier. Free interbreeding would homogenize them, preventing divergence. But what prevents interbreeding of closely allied forms living in immediate proximity? Darwin was not sure, but envisioned that some barrier, perhaps physiological, arose as a by-product of incipient divergence. He first suggested this idea in his 1868 treatise *Variation of Animals and Plants under Domestication*. In keeping with his usual model of gradual evolution, Darwin argued that progressive reduction of fertility could not arise *directly* through selection, because those individuals with reduced reproductive success would be quickly displaced by those that were fully interfertile. No advantage to individuals was gained by having fewer offspring, even if beneficial to the species, so selection could not play a role.[52] Wallace disagreed, holding that selection could accelerate divergence by disfavoring hybrids. In his model the physiological mechanism preventing interbreeding arose directly through the action of selection, and he was puzzled why Darwin did not concur: "I do not see your objection to *sterility* between allied species having been aided by natural selection. It appears to me that,—given a differentiation of a species into two forms each of which was adapted to a special sphere of existence,—every slight degree of sterility would be a positive *advantage*, not to the *individuals* who were sterile, but to *each form*."[53]

The issue became quite confused when Darwin's acolyte George Romanes later proposed a theory he termed "physiological selection,"[54] downplaying the role of natural selection in species origins and proposing a mechanism of "physiological variation" whereby some individuals are suddenly rendered incapable of intercrossing with sister varieties or parent forms, leading to sympatric speciation. The idea of natural variation in the degree of interfertility was acceptable to Wallace, but he viewed Romanes's physiological selection model as weak at best. The physiological selection idea arose at a time (in the decade following Darwin's death in 1882) when selection was under siege, with neo-Lamarckian and other non-Darwinian mechanisms of evolution increasingly in vogue. Romanes's model was in that vein, holding that intersterile varieties arose via sudden mutation and that natural selection à la Darwin and Wallace thus played no role in the origin of species.[55] An experimental physiologist, Romanes had become Darwin's assistant in the great naturalist's later years, collaborating with Darwin on the snap-trap mechanism of Venus flytraps. Now he appeared to be the intellectual heir apparent as the leading Darwinian, except that—in Wallace's view—he threatened to undermine the Darwin-Wallace theory of evolution by natural selection.

Wallace published a devastating critique of Romanes's physiological selection hypothesis in the 1 September 1886 issue of the *Fortnightly Review*. Entitled "Romanes versus Darwin: An Episode in the History of the Evolution Theory," this essay systematically torpedoed Romanes's "three cardinal difficulties" with natural selection, namely (1) the claim that domesticated varieties of a species tend to be fully interfertile despite often striking outward differences, while natural species are intersterile despite little outward differences; (2) the swamping effect of intercrossing of varieties, preventing differentiation into species; and (3) the preponderance of apparently useless variations and traits, which cannot have arisen via natural selection. Wallace then took on physiological selection directly, a theory, he said, that "sounds feasible when not closely examined; but . . . really slurs over insuperable difficulties, and when viewed in the light of the known facts of variation and natural selection it will be seen that the supposed results could not follow."[56] Other leading evolutionists agreed with Wallace, and physiological selection was discarded, though Romanes fought back with increasingly bitter and acrimonious attacks on Wallace and others until his death in 1894.[57]

In chapter 7 of his 1889 book *Darwinism*, Wallace further discussed the central problem of swamping of novel variants and prevention of divergence.[58] To use modern terms, he held that sympatry at a large scale resolves into allopatry on a highly localized or microgeographical scale, following Darwin's lead. There are two important aspects to this idea that represent separate but related lines of modern research and debate. One, which harkens back to Wallace's 1865 Malayan papilionid butterfly paper mentioned earlier (and discussed further below), is the question of whether the speciation process can be seen as a series of discrete stages, where incipient varieties become increasingly divergent owing to micro- or macrogeographical isolation and/or ecological adaptation (i.e., the formation of so-called "ecotypes").[59] This view embraces species formation as a dynamic, multi-staged process,[60] focusing attention on mechanisms underlying speciation without becoming bogged down in the particulars of this or that species concept. Another is the question of whether selection can promote speciation by reducing the capacity for (or frequency of) hybridization between varieties. In theory this would occur if the offspring of hybrids suffered a reduction in viability or fitness compared to the offspring of "pure" parental forms. Postzygotic fitness reduction could lead to prezygotic avoidance of mating, perhaps through recognition signals (which could take the form of visual, olfactory, and/or behavioral cues). Now termed "reinforcement" or, appropriately, the "Wallace effect," this continues to be an important, albeit controversial, line of investigation in evolutionary biology today.[61]

It should be pointed out, finally, that Wallace always saw a role for spatial separation or isolation in the speciation process. In addition to scenarios like that discussed in this section (microgeographical-scale isolation), in other writings Wallace articulated a process essentially identical with the modern theory of peripatric speciation. Generally attributed to Ernst Mayr,[62] peripatric speciation is a process whereby the initial stages of divergence of subpopulations occur on the margins (periphery) of the main range of a species, perhaps assisted by gradients in environmental conditions. As the subpopulation becomes better adapted to the conditions of the marginal area, it subsequently expands and may displace the parent species altogether. Anticipating Mayr by ninety years, Wallace described the process we term peripatric speciation in his 1880 book *Island Life*: "It is now very easy to understand how, from such a variable species, one or more new species may arise. The peculiar physical or organic conditions that render one part of the area [range] better adapted to an extreme form [variety or subpopulation] may become intensified, and the most extreme variations thus having the advantage, they will multiply at the expense of the rest."[63]

Mimicry: Sex-Limited and Other Forms

Another important phenomenon pertaining to the nature of species and varieties and the action of selection is sex-limited mimicry, long of great interest to naturalists. This is also a sizable subject; this chapter gives only a general overview of Wallace's thinking on this and other forms of mimicry.[64] Sex-limited mimicry is characterized by strong sexual dimorphism, where (usually) females are completely unlike the opposite sex in appearance, and indeed convergent on the coloration patterns of other (unpalatable) species altogether that occur in the same area. This phenomenon is found in numerous swallowtail species worldwide, including the common eastern North American tiger swallowtail (*Pterourus glaucus*). Wallace's treatment focused on the group he knew best: the Australasian papilionids.

Based on their interpretations of morphology and coloration, many naturalists held that different morphs simply represent distinct species, and a good deal of taxonomic confusion (in the modern view) stemmed from different species descriptions and names applied to what are now recognized merely different sexes of the same species. In his 1865 papilionid paper Wallace was the first person to recognize such females for what they are—conspecifics of the males despite their differences in appearance—describing the wonder of the phenomenon with a striking human analogy.[65] He rested his case for conspecific relationship of the sexual morphs on reproductive competence: dimorphic males and females both arise from eggs laid by a single mother, and he observed the dimorphic sexes *in copulo*.[66] Wallace was thus able to combine geographical and reproductive factors in his definition of species.

Wallace also recognized the significance of the divergent female papilionids in terms of their imitation of other species. It should be pointed out that Wallace did not originate the mimicry concept, which is attributed to his friend and Amazon traveling companion, Henry Walter Bates.[67] In November 1861 Bates delivered a striking paper on *Heliconius* butterflies of the Amazon at the Linnean Society, providing the first overtly evolutionary explanation for mimicry, a term he coined.[68] In the phenomenon that later became known as Batesian mimicry, selection could act on vulnerable, palatable butterflies, favoring the evolution of wing coloration patterns converging on those of noxious, well-defended (and so unpalatable) species in the same area. This was a special type of "protective resemblance," some forms of which had long been appreciated by naturalists (such as the strong resemblance of some syrphid flies to stinging wasps, thereby gaining protection thanks to the wasps' fearsome reputation). Bates's paper was published the following year, and Darwin and Wallace were delighted with his analysis. Darwin applauded him, writing, "I have just finished after several reads your Paper. In my opinion it is one of the most remarkable & admirable papers I ever read in my life. The mimetic cases are truly marvellous & you connect excellently a host of analogous facts. . . . You have most clearly stated & solved a wonderful problem."[69] Coincidentally, Wallace sent Bates a congratulatory letter on the very same day as Darwin.[70] Two years later, in his own analysis of female-limited mimicry in papilionid butterflies, Wallace saw the same selection dynamic at work: "I am disposed to believe that we have here a case of mimicry, brought about by the same causes which Mr. Bates has so well explained in his account of Heliconidae, and which thus led to the singular exuberance of polymorphic forms in this and allied groups of the genus *Papilio*."[71] Later in this paper he devoted a section to the subject, listing and discussing fifteen species of papilionids, the species they mimic, and their geographic range (figure 4.1).

Wallace followed up three years later with a celebrated review entitled "Mimicry, and Other Protective Resemblances among Animals,"[72] in which he presented three "mimicry laws" and explored other kinds of mimicry, such as what is now termed "aggressive" mimicry (e.g., wolf in sheep's clothing, and resemblance to dangerous animals[73]). Wallace further discussed Batesian mimicry, to use the modern term, in his 1869 book *The Malay Archipelago*. After discussing camouflage,[74] he introduced the form of protective resemblance "which has been happily termed 'mimicry' by Mr. Bates, who first discovered the object of these curious external imitations of one insect by another belonging to a distinct genus or family, and sometimes even to a distinct order."[75] Wallace even applied the concept to birds that occur on the island of Buru: rather timid orioles of the genus *Mimeta*, which he argued mimic the "plentiful, and very pugnacious" honey-suckers of the genus *Tropidorhynchus*.[76]

In Bates's concept of mimicry, the palatable species (mimic) evolves to resemble an unpalatable species (model). Neither Bates nor anyone else considered a scenario with *two* unpalatable species, however. In 1878 German naturalist Fritz Müller was the first to do just that, describing a system where two or more equally well-defended species mutually reinforce one another by converging on the same coloration pattern. Fittingly, this is termed Müllerian mimicry today. Wallace immediately embraced and expanded upon Müller's idea, considering both unequal degrees of unpalatability and unequal abundance between the joint mimics. In an 1882 re-

Figure 4.1 Plate 1 from a paper on the Malayan Papilionidae, representing forms of *Papilio memnon* from Borneo, Java, and Sumatra. From Wallace (1865a).

view of the subject, he discussed the significance of "various degrees of protection," stating that "there is however yet another cause which may have led to mimicry in these cases, and one which does not appear to have been discussed by Dr. Müller."[77] In typical Wallace style he then attempted a synthesis, in effect suggesting that common genetic background of related and partially protected groups (members of the same genus or family, for example) facilitates convergence on a common coloration pattern. This in turn favors the evolution of mimicry of this group by other, palatable, species in the same area: "the advantage of sharing in this partial protection has led species of altogether unprotected and much persecuted groups to gain some protection by mimicking them, whenever their general form, habits, and style of coloration offered a suitable groundwork for variation to act upon." "If these views are correct," he concluded, "we shall have the satisfaction of knowing that all cases of mimicry are explicable by one general principle; and it seems strange to me now that I should not have seen how readily the principle is applicable to these abnormal cases."[78] (In the modern view, however, Wallace may have overstated things in his attempt to generalize Müllerian mimicry as a widespread, virtually universal phenomenon.)

Although Wallace and his circle viewed the different forms of mimicry as among the best evidence for evolution by natural selection, soon after Wallace's death some biologists used mimicry to make precisely the opposite argument. The classic Darwinian explanation for Batesian and other forms of mimicry is accepted today, and many students of evolutionary biology are surprised to learn that this was not always the case. Soon after the turn of the twentieth century natural selection was largely rejected as a significant evolutionary process, and mimicry was, ironically, cited as evidence. This view culminated in R. C. Punnett's 1915 book *Mimicry in Butterflies*, rejecting the Darwin-Wallace model and championing instead the mutationism of Hugo De Vries and other early Mendelians. Punnett advanced the alternative view that uniformity of genetic background would increase the likelihood that a suite of variations giving just the right mimetic effect could arise suddenly, as a genetic sport.[79]

Dichromatism and Sexual Selection

Sex-limited mimicry highlights a related evolutionary issue for Wallace and Darwin: sexual dimorphism, specifically in the form of dichromatism. This takes the form of differential coloration of the sexes, typically where one sex (usually male) is brightly colored while the other (typically female) is drab or camouflaged. Darwin became increasingly interested in this phenomenon through the 1860s as he worked to develop his theory of sexual selection, contemplating a book-length treatment. This culminated in *The Descent of Man* (1871), much of which was dedicated to a survey of secondary sexual characters and sexual dimorphism in the animal kingdom, complementing one of his key hypotheses concerning human racial diversification via sexual selection. Darwin's working hypothesis was that dimorphic groups were ancestrally monomorphic. Accordingly, in the case of dichromatic birds and butterflies, he held that both sexes were ancestrally drab in coloration. Males diverged from the females over time through the action of sexual selection (specifically female choice) to become brightly colored.

Many striking cases of sexual dichromatism are found in birds and butterflies, and Wallace's rich collections from the Malay Archipelago furnished abundant examples. If there was one species that drew Darwin and Wallace into a more in-depth comparison of their views, it was likely the southeast Asian butterfly *Diadema* (now *Hypolimnas*) *anomala*, family Nymphalidae. This species exhibits a reversal of the usual pattern of male/female dichromatism, with cryptic brown males and bright metallic-blue females. Wallace realized that the females mimicked a species of *Euploea* (family Danaidae).[80] Discussing this butterfly with Darwin by letter, Wallace declared that he "sometimes doubt[ed] whether sexual selection has acted to produce the colours of male butterflies." He went on to state that he thought that it was merely "advantageous for the females to have less brilliant colours, & that colour has been produced merely because in the process of infinite variation all colours in turn were produced."[81] In other words, male color may be simply some physiological by-product, not necessarily functional in any way, while less brilliant colors may be selected for in females. Thus ensued a years-long debate between Wallace and Darwin over the relative importance of natural versus sexual selection and the nature of inheritance in the evolution of dichromatism.[82]

In the immediate post-*Origin* years Wallace had accepted Darwin's views on sexual selection,[83] but on reflection he saw roles for both sexual and natural selection, especially in the sex-limited mimicry form of dichromatism. This is because such mimicry is a form of protective coloration, which by definition stems from the action of natural rather than sexual selection. He generalized his view into the idea that selection would act to promote crypsis in the more vulnerable sex. In an 1867 letter to Darwin, Wallace gave a succinct summary of this view: "I have been writing a popular article on

'Mimicry' & allied phenomena in which I have tried to bring together all the groups of facts which bear upon the subject. While doing this I have become more than ever convinced of the powerful effect of 'protective resemblances' in determining and regulating the development of colour." He went on to explain that while pondering why in some brightly colored birds the two sexes are similarly colored, he was led to a simple explanation: sexual selection operated on *both* sexes, which meant that in cases of dimorphism something checked the action of sexual selection in the less conspicuously colored sex. That something was natural selection via predation pressure: "Hence the law,—that where birds [nest] in holes in the ground, or in holes in trees, or build covered nests, the females will generally be as gaily coloured as the males."[84] He published a comprehensive treatment of this phenomenon the following year. "A Theory of Birds' Nests" was a tour de force that underscored his disagreement with Darwin. In a footnote Wallace commented that "[i]n his 'Origin of Species,' fourth edition, p. 241, Mr. Darwin recognises the necessity for protection as sometimes being a cause of the obscure colours of female birds; but he does not seem to consider it so very important an agent in modifying colour as I am disposed to do. . . . I impute the difference, in the great majority of cases, to the greater or less need of protection in the female sex in these groups of animals."[85]

Wallace thus argued that sexual selection should produce bright coloration equally in both sexes, but in the case of incubating females that construct exposed nests, natural selection counteracts sexual selection to produce drab coloration that better blends with the surroundings. This may seem contrary to modern views of bright coloration arising through female choice; was Wallace suggesting that *male* choice operates as well? Yes.[86] In this regard there are two notable aspects to Wallace's argument. One pertains to the nature and inheritance of variation. Wallace held that new variations conferring coloration were inherited (and initially expressed) equally in both sexes. Selective preference for more brightly colored partners could lead to each sex becoming more conspicuously colored. He further argued that in cases where females are exposed to danger, natural selection trumps sexual selection on the female, *and*, more questionably, results in a change in heritability of the coloration traits such that they become sex-limited (with males inheriting and expressing the traits, but not the females).

The two were initially in broad agreement. Darwin accepted Wallace's views for the most part,[87] although he thought Wallace pushed protective coloration too far, while Wallace reiterated his qualified agreement with

Darwin.[88] But the occasion of Wallace's "Theory of Birds' Nests" paper in 1868 brought the debate to a new level. Darwin was convinced that sex-limited mimicry proved that sex-limited inheritance of new variations was likely, as opposed to Wallace's view that the two sexes inherited such variations equally. In the case of female-limited mimicry, if mimetic coloration is transmitted to sons as well as daughters, the fact that sons do not express this coloration means that something must prevent it—that must be natural selection. But why would selection prevent mimicry coloration in males? He declared that he could see no reason for this: "the variations leading to beauty must *often* have occurred in the males alone, and been transmitted to that sex alone. Thus I should account in many cases for the greater beauty of the male over the female, without the need of the protective principle."[89]

Darwin and Wallace never came to an agreement over sex-limited inheritance.[90] The difference in their views came down to the degree of importance they ascribed to different aspects of the phenomenon. Did sex-limited expression of traits occur from the start through strict sex-limited inheritance, or was it the end result of selection acting on traits otherwise inherited by each sex equally? How frequent and important are "exceptions" to the rule of bright male coloration in species with concealed nests? Wallace acknowledged a few exceptions (e.g., brilliantly colored sunbirds, family Nectariniidae) but downplayed their importance, while Darwin felt that exceptions were far more numerous and therefore far more problematic for Wallace's position. Wallace answered Darwin's letter with a long and far-ranging letter of his own that included, among other things, a particularly incisive criticism of Darwin's position:[91] if sex-limited inheritance was the norm, shouldn't we find cases where non-nest-incubating males are camouflaged while the females are brightly colored? In instances where males sit on the nest, camouflage is expected. But if sex-limited inheritance occurs, and selection is not so important, as Darwin argued, then at least *occasional* cases of cryptically colored nonincubating males should be found. However, this did not appear to be the case. Darwin concurred, though he tried to come up with an alternative explanation.

For a time Wallace accepted sex-limited inheritance as the explanation for sex-limited mimicry, and Darwin came to acknowledge that natural selection favoring protective coloration was more common than he previously thought. He wrote Wallace on 30 April 1868, "[W]e shall never convince each other. I sometimes marvel how truth progresses, so difficult is it for one man to convince another unless his mind is vacant. Nevertheless, I myself to a certain extent contradict my own remarks; for I believe *far more* in the importance of protection than I did before reading your articles."[92] Darwin's comment that he and Wallace "would never convince each other" proved prophetic. As he thought more and more deeply about the problem, he found himself diverging further and further from his younger colleague, to his distress.[93] Wallace marshaled a lengthy, fifteen-proposition argument reiterating his case for clearly adaptive coloration stemming from natural selection, and novel color variants initially being inherited equally by the sexes. But Darwin held fast to his belief that in species with female-specific coloration, the colors had likely been inherited only by daughters to begin with. He took the argument further by turning Wallace's model of the relationship between bright-colored females and concealed nests on its head. As he put it later in *Descent of Man* (1871), "[I]n most cases as the females were gradually rendered more and more brilliant from partaking of the colours of the male, they were gradually led to change their instincts (supposing that they originally built open nests), and to seek protection by building domed or concealed nests."[94] Wallace viewed it the other way around, believing that coloration could change more readily than instincts: for birds having the instinct of concealed nest building, selection favored the acquisition of bright coloration in the female (presumably through male choice).

The two ultimately agreed to disagree: "I grieve to differ from you," Darwin wrote, "and it actually terrifies me, and makes me constantly distrust myself. I fear we shall never quite understand each other."[95] For his part, Wallace responded that he was "sorry to find that our differences of opinion on this point is a source of anxiety to you. Pray do not let it be so. The truth will come out at last, and our difference may be the means of setting others to work who may set us both right. After all, this question is only an episode (though an important one) in the great question of the origin of species, and whether you or I are right will not at all affect the main doctrine—that is one comfort."[96] This sentiment was echoed in another letter by Wallace a decade later, after the two had grown even more widely apart in their views. Darwin reignited their debate in *Descent of Man* a few years later. Wallace disagreed with his treatment of the subject, especially in regard to what he felt was a misrepresentation of his views. Darwin seems to have steadfastly ignored Wallace's argument that females did not have to be "specially modified" for protective coloration, but merely be *prevented* by natural selection from changing from cryptic coloration to bright coloration. Darwin portrayed Wallace as arguing that females were "specially modified," a subtle but important difference. He apologized to Wallace and clarified how his statement should have been worded, yet in the second edition of *Descent of Man* his account remained unchanged.[97]

Darwin ultimately rejected Wallace's theory of dichromatism, namely that cryptic coloration (or coloration in the less conspicuous sex) evolved by natural selection for protection. Instead, he invoked sexual selection and the sex-limited inheritance of coloration traits. Wallace always held that color variants were initially inherited equally by the two sexes, and somehow were converted to sex-limited expression in some lineages, while Darwin never budged from his position that sex-limited inheritance from the start was more likely since he did not think it was possible for selection to convert the traits from equal- to unisexual inheritance. They were at their widest disagreement at the debate's end in the late 1870s, when Wallace rejected female choice altogether. He had long intimated that he did not believe female choice could operate in insects, though he initially accepted it for birds. But in the mid-1870s Wallace concluded that mere "taste" or whim on the part of females for this or that odd trait was unlikely to be sustained over generations, while Darwin argued that aesthetic tastes of females played a central role.

The great argus pheasant (*Argusianus argus*) became the exemplar of his argument: the perfectly round eyespots of the male's wing feathers originated, in Darwin's view, eons ago as the merest suggestion of a spot. That females would somehow develop a taste for those early splotches and consistently selected for ever-bigger and more symmetrical spots over innumerable generations seemed incredible to Wallace. He could see how female choice could act in general, yet he saw difficulties when you zoomed in on particular cases. Thus while Wallace could refer (in his 1876 *Geographical Distribution of Animals*) to the "remarkable confirmation of Mr. Darwin's views" seen in the beautiful and elaborate tail feathers of pheasants (see figure 4.2), "developed in the male bird for the purpose of attractive display,"[98] just a short time later he published an article in which he rejected sexual selection as an explanation for such traits, arguing that coloration and similar secondary sexual traits were simply a physiological by-product.[99] This became known as Wallace's "vigor theory," developed most fully in his 1878 book *Tropical Nature*. While bright coloration was selected against in females, Wallace argued, such coloration had no such limitation in males, and its expression thus reflected male vigor and activity.

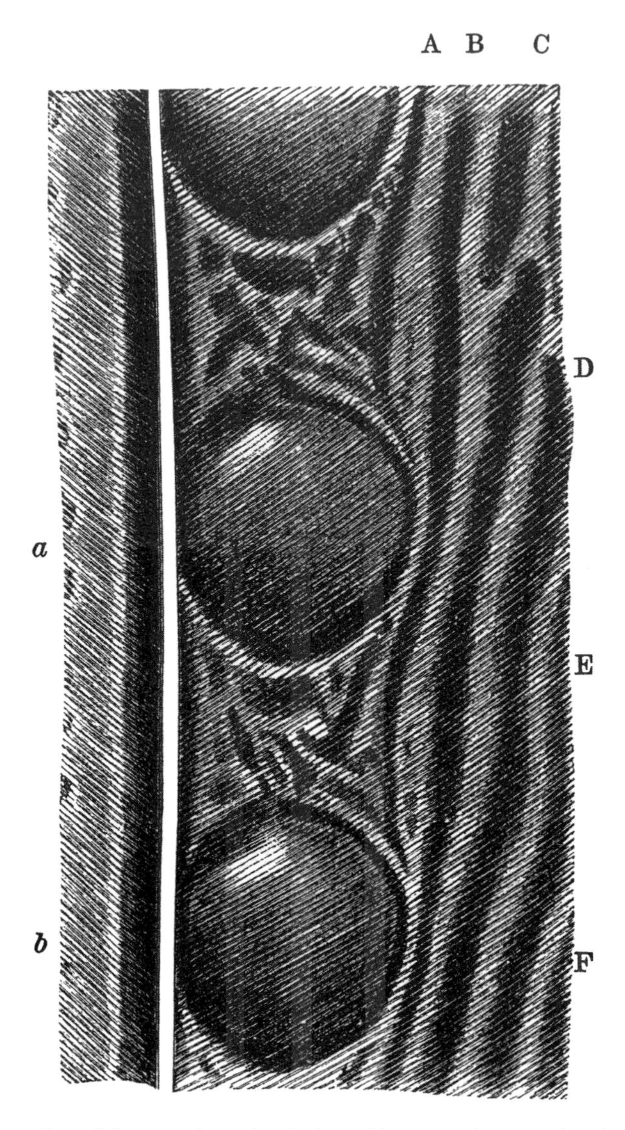

Figure 4.2 Part of the secondary wing feather of the argus pheasant, showing ocelli. From Darwin (1871), vol. 2, fig. 56.

In the modern view, both Wallace and Darwin were correct. Female "tastes" (mate preferences) may be whimsical, but they can be sustained insofar as they increase fitness, of course. And beyond increasing reproductive success merely due to aesthetic selection of mates (today's "sexy sons" model), the aesthetic traits may themselves signal something about the health and robustness of the males. This is the idea behind such modern theories of sexual selection as the "handicap principle," "honest signalling," and "good genes,"[100] where the quality of traits catching female's eyes (bright coloration, for example) are affected by nutrition, parasite load, pathogens, etc. The brighter the males, the healthier. A related idea, the handicap principle, posits that elaborate ornaments (think peacocks' tails) and bright coloration put the males bearing them at a disadvantage, expending more energy to lug around the ornament or slowing them down or making them more conspicuous to predators. By virtue of the fact that they live and compete robustly for mates, they must be especially healthy to have overcome such handicaps. In modern sexual selection theory this idea is generally attributed to Israeli evolutionary biologist Amotz Zahavi, who dubbed the theory the "handicap principle" in the mid-1970s,[101] but Wallace is the originator of the concept. Darwin, however, was disconcerted but largely unmoved by Wallace's new position: "I am very sorry that you have given up sexual selection. I am not at all shaken, and stick to my colours like a true Briton," he wrote, and followed up by saying, "[Y]ou will not be surprised that I differ altogether from you about sexual colours. That the tail of the peacock and his elaborate display of it should be due merely to the vigour, activity, and vitality of the male is to me as utterly incredible as my views are to you."[102]

Historian Malcolm Kottler concluded, in his elaborate analysis of the Darwin-Wallace debate over dichromatism and related issues, that neither Darwin nor Wallace won their long debate, there being truth in both of their positions. For example, in mimetic systems such as *Papilio* butterflies there is evidence that it is the female, responding to natural selection for protective mimetic coloration, that deviates from an ancestral monomorphic condition, while in other systems it is the male that changes—or there is evidence of repeated change in both directions.[103] And a century of research confirms the reality of sexual selection via both male competition and female choice in a multitude of avian, arthropod, and mammalian groups.[104] I leave the final word on Darwin's and Wallace's complex and ongoing debate over sexual selection to Oxford evolutionary biologist Richard Dawkins, who noted simply that "the dichotomy, between beauty for beauty's sake on the one hand, on the Darwin side, and beauty as a badge of utilitarian usefulness on the other side, which is Wallace's view . . . lasts right into the twentieth century."[105] (And, I would add, the twenty-first.)

Aposematism

Bates's elucidation of the evolution of mimicry as a form of protective coloration—imitation of brightly colored distasteful butterflies in the same area—was viewed as a triumph of the Darwin-Wallace concept of natural selection. Curiously, although Bates proposed that the species being mimicked were protected by a substance rendering them distasteful to predators,[106] the idea of the butterflies' bright coloration as an *advertisement* of unpalatability apparently did not occur to Bates or Darwin. Darwin tended to view bright coloration in terms of sexual selection, especially female choice (discussed above), so he was puzzled by the vivid, contrasting colors found in some caterpillars. As the juvenile (larval) stage of Lepidoptera, sexual selection would seem to be inapplicable to caterpillars. He consulted Bates on the matter, who referred him to Wallace: "On Monday evening I called on Bates," Darwin wrote to Wallace, "and put a difficulty before him, which he could not answer, and, as on some former similar occasion, his first suggestion was, 'You had better ask Wallace.' My difficulty is, why are caterpillars sometimes so beautifully and artistically coloured? Seeing that many are coloured to escape danger, I can hardly attribute their bright colour in other cases to mere physical conditions." In the same letter Darwin hinted that he thought sexual selection must have something to do with it: "If anyone objected to male butterflies having been made beautiful by sexual selection, and asked why should they not have been made beautiful as well as their caterpillars, what would you answer? I could not answer, but should maintain my ground."[107]

Wallace suggested that bright colors were advertisements of the caterpillars' unpalatability. He pointed out that it was well known that many caterpillars are protected by crypsis, blending in with their background, while others deter predators with hairs or spines. He continued,

> Now supposing that others, not hairy, are protected by a disagreeable taste or odour, it would be a positive advantage to them never to be mistaken for any of the palatable catterpillars, because a slight wound such as would be caused by a peck of a bird's bill almost always I believe kills a growing catterpillar. Any gaudy & conspicuous colour therefore, that would plainly distinguish them from the brown & green eatable catterpillars, would enable birds to recognise them easily as a kind not fit for food, & thus they would escape seizure which is as bad as being eaten.[108]

Darwin was impressed: "Bates was quite right, you are the man to apply to in a difficulty. I never heard any thing more ingenious than your suggestion & I hope you may be able to prove it true."[109] In the last part of that sentence Darwin alluded to Wallace's suggestion that the idea could be experimentally tested. Wallace followed up, first with a presentation at the 4 March 1867 meeting of the Entomological Society of London and then with an open letter to the editor of the *Field*, asking readers to provide observations on caterpillars eaten or rejected by birds.[110] He suggested two sets of experiments:

> Those who keep insectivorous birds, such as thrushes, robins, or any of the warblers (or any other that will eat caterpillars), may offer them all the kinds they can obtain, and carefully note (1) which they eat, (2) which they refuse to touch, and (3) which they seize but reject . . . Those who do not keep birds, but have a garden much frequented by birds, may put all the caterpillars they can find in a soup plate or other vessel, which must be placed in a larger vessel of water, so that the creatures cannot escape, and then after a few hours note which have been taken and which left.[111]

This nineteenth-century version of crowd sourcing was often used to good effect by Darwin, but Wallace was disappointed that he received but one response.[112] He then asked his friend John Jenner Weir, who kept an aviary, to assist. The obliging Jenner Weir confirmed Wallace's hypothesis, and read two papers on the subject at the Entomological Society of London.[113]

The technical term for warning coloration is "aposematism," coined by E. B. Poulton in his 1890 book *The Colours of Animals*, from the Greek word for signs or signals, *sematos*. A vast body of empirical and theoretical work on the ecology and evolution of aposematism has developed since, including phylogenetic treatments and extensions of the concept into the realms of acoustic aposematism and amplification of aposematic signals through gregariousness.[114]

Human Cognitive Evolution

Perhaps the most serious disagreement between Darwin and Wallace arose in the late 1860s over the evolution of human consciousness and cognition. Wallace intimated to Darwin that he would be exploring "some limitations to the power of natural selection" in a forthcoming article, and said he was "afraid that Huxley & perhaps yourself will think them weak and unphilosophical." An apprehensive Darwin replied that he was "intensely curious" about Wallace's article, and hoped he had not "murdered too completely your own and my child."[115] Wallace's paper in the *Quarterly Review*, mainly a review of the tenth edition of Lyell's *Principles of Geology*, was as shocking as Darwin feared.[116] Wallace argued that the rich complexities of the human mind could not have arisen through gradual evolution by natural selection, since even so-called savages possessed capacities for things they never even dreamed of, abilities they never exercised such as those leading to complex mathematics or musical compositions. Wallace was being "more Darwinian than Darwin" in his strict gradualism, arguing that every stage must be expressed and have its use for selection to act upon it and favor its further improvement. According to the chauvinism of the day, the "savage" races clearly represented a lower evolutionary stage of humanity, so how could they possess the same rich cognitive abilities as Europeans, latent abilities that would never be expressed? Wallace came to see "preternatural" agency of some kind (see chapter 2) guiding human mental evolution (and related attributes such the organs of speech, manual dexterity, etc.) once human physical evolution had reached a certain point via natural selection.

Wallace's bombshell of a paper has been seen by many scholars as an about-face on a crucial and sensitive topic. Certainly Darwin, Huxley, Hooker, and their circle thought so. But Charles Smith has suggested that this was no about-face; rather, Wallace long had reservations about human cognition, and it may only have been his embrace of spiritualism in the mid-1860s that led him at last to crystallize and express thinking on humans.[117] In any case, certainly no position espoused by Wallace did more damage to his reputation in scientific circles than his public endorsement of spiritualism coupled with his declaration of the inadequacy of natural selection to account for the complexities of human consciousness.[118]

Instinct and Experience

One final subject of great interest to both Darwin and Wallace considered here is the nature and evolution of instinct. Darwin recognized the difference between instinctive behavior and what he termed habit (learned behavior in modern terms), and of course held that, like any trait, heritability and inherent variability determined whether natural selection could act upon behavior, and perhaps by definition saw innate, instinctive behaviors as being shaped by selection. Emphasizing continuity in the animal kingdom, in Darwin's view "higher" animals like vertebrates, including humans, differ in *degree* and not *kind* from less complex organisms such as invertebrates. Thus, in the *Origin*, he first set up highly complex (and instinctive) behaviors—for example, hexagonal cell building in honeybees, obligate slave-making in certain ants, and obligate parasitism in cuckoos—as instinctive behaviors commonly and erroneously claimed to be too complex to have arisen through gradual evolution. Darwin then argued in each case for continuity linking these complex forms with related species, showing the gradual steps by which the highly complex expressions of behavior could have arisen from simpler antecedents.[119]

Wallace broadly agreed with Darwin regarding continuity in the organic world, and complex adaptive behaviors arising from less complex ancestral forms. He also delved more deeply into the subject of instinct, however, first attempting to clarify the question by offering a definition of instinct and then critiquing claims of instinct in nature (and in humans). This topic had long interested Wallace, as evidenced by some forty pages of his Species Notebook bearing entries on the subject.[120] His discussions in the notebook focused largely on bees' cells, birds' nests, and putative instinctual behaviors in humans.

The construction of honeycombs by bees, with its seeming mathematical precision, was long cited by natural theologians as evidence of a divine intelligence at work. In chapter 7 of the *Origin*, Darwin argued for the gradual acquisition of the cell-building instinct in bees, pointing to the simpler cell and comb types constructed by "collateral relatives" of honeybees, the bumblebees and *Melipona* bees. Darwin was attacked by the Rev. Samuel Haughton for suggesting that natural selection could be responsible for the cell-building instinct of bees, among other complex behaviors, prompting an immediate and spirited response from Wallace in the *Annals and Magazine of Natural History*.[121] Wallace was in full agreement with Darwin about the agency of natural selection in giving rise in stepwise fashion to such wonderful instincts as cell-building behavior, a view he long held. Thus in his Species Notebook of the 1850s Wallace noted that "in investigating instinct we proceed by degrees, from the easy & near to the difficult & remote." He commented on how comparative anatomists like Owen or Huxley gained insight into morphological homologies by analyzing structural similarities and differences through "tracing modifications step by step," in a series, not by simply comparing seemingly disparate forms at the ends of a morphological spectrum. So too, Wallace suggested, should instincts be compared step-by-step across a series of species in order to gain insight into how the most complex expressions might have derived from the less complex.[122]

More incisively, in the Species Notebook Wallace also defined instinct in terms of learned behavior: "The performance of a complicated act absolutely without previous instruction or knowledge of it." He also suggested how to experimentally test for instinct:

> it is said & repeated, that birds & insects build nests gather & store food & provide for their future wants without any instructions from their fellows & without even knowing that such acts have been performed by others. This however is assumed. It has never been tried. Bees have never been carefully secluded from the larva state, & then loosed in an enclosure where they could not communicate with other bees. Birds reared from the egg in confinement have not been shewn to make the same nest as their fellows.[123]

He penned a similar comment in another notebook, his Insect Register.[124] His point is that for all we know, learning and experience may be behind the apparently "instinctive" behaviors of birds and bees. These speculations immediately precede several pages of Species Notebook entries in which Wallace questioned whether humans have any instinctive behavior. His interest was perhaps aroused because supposed uncanny "instinctive" abilities were sometimes cited as evidence of the special creation of humans. Wallace later developed an argument against this position in an essay entitled "On Instinct in Man and Animals" in the 1870 collection *Contributions to the Theory of Natural Selection*. Along the way he critiqued the standard example of human instinctive behavior of his day, namely suckling behavior by infants, and the putative ability of Indians to find their way through "trackless wilderness" without aids of any kind. In the latter case he argued that experience is involved, learning to read the lay of the land they call home, concluding, "It appears to me, therefore, that to call in the aid of a new and mysterious power to account for savages being able to do that which, under similar conditions, we could almost all of us perform, although perhaps less perfectly, is almost ludicrously unnecessary."[125]

In this vein, while not denying the possibility of instinctive behaviors, Wallace became increasingly critical of unexamined *claims* of instinct. Thus in one letter to the editor in answer to a critic of his 1870 essay, Wallace insisted that the evidence customarily given in support of claims of instinct in animals in fact "do not prove instinct," and furthermore, quoting his 1870 essay, that "experiments on instinct have not been sufficiently carried on, and I conclude, not that there is no such thing as instinct, but that it should not be accepted as proved in any particular case 'until all other possible modes of explanation have been exhausted.'" "I am open to conviction by facts," he declared, and cited an intriguing series of experiments conducted by the naturalist Douglas Spalding on newly hatched chicks and ducklings, which suggested that several behaviors appeared to be expressed without prior experience. "But," he concluded, "we have as yet no experiments to show that the exceedingly complex actions involved in the higher instincts can be so performed."[126] In the same year Wallace published a largely favorable review of a book by Belgian astronomer Jean-Charles Houzeau, comparing the mental faculties of animals in relation to humans. Wallace approvingly described Houzeau's take on the difficulties of ascribing many behaviors in the animal world to "blind instinct," and his assertion that, while the "faculty of invention" may be more developed in humans than other animals, "these differences . . . ought not to blind us to the existence of the faculty in various degrees of development among many animals."[127]

These were to remain Wallace's themes concerning instinct: acknowledging that instinct could occur, but critical of the shoddy reasoning behind the putative examples advanced by naturalists. One class of examples often cited was homing behavior in dogs—we might call these the "incredible journey" type of case: "Now the power many animals possess to find their way back over a road they have travelled blindfolded (shut up in a basket inside a coach for example) has generally been considered to be an undoubted case of true instinct."[128] Wallace then advanced an alternative hypothesis: perhaps the acute sense of smell of the confined dog permits it to take note of the successive odors along the way, impressing upon its mind "a series of images as distinct and prominent as those we should receive by the sense of light." He urged readers to attempt experiments like those of Spalding to help settle the question: "The animals' previous history must be known and recorded; a sufficient number of experiments, at various distances and under different conditions, must be made, and a person of intelligence and activity must keep the animal in sight, and note down its every action till it arrives home. If this is done I feel sure that a satisfactory theory will soon be arrived at, and much, if not all the mystery that now attaches to this class of facts be removed."[129]

I conclude this discussion with one of Wallace's last commentaries on the matter, published in a long essay-review entitled "The Problem of Instinct." In it Wallace made a remarkable suggestion. The essay largely consisted of a review of the 1896 book *Habit and Instinct*, by C. Lloyd Morgan. It was evident that in the twenty-five years since his earliest essays on the subject, not much had changed. He pointed once again to the paucity of experimental efforts aimed at testing for instinctive behaviors: "There is probably no subject in the whole range of biology, the study of which has been so universally neglected as Instinct," he declared in the opening sentence. But Wallace applauded Morgan's efforts at experimentation as an exception. Morgan had replicated Spalding's experiments (finding that many did not stand up to scrutiny), and conducted a great many of his own. Among other "instinctive" phenomena Morgan reviewed nest-building behavior, migratory behavior, and of course the homing instinct of dogs and cats. Wallace again described the experiment testing for the latter that he had first pointed out in 1873 ("A moderate number of such experiments would settle the question of instinct or sense-observation, and it is to be hoped, now that a more intelligent interest is taken in the subject, such experiments will be made."[130]) Near the conclusion of his essay Wallace took up the question of whether acquired behaviors could become "instinctive" (innate).

The idea of inheritance of acquired characters, an essentially Lamarckian process, had a long and problematic history. Even Darwin became increasingly Lamarckian over time, but Wallace was steadfastly critical of the idea. However, he saw in Morgan's book a way for natural selection to act upon acquired traits: "Acquired modification thus helps on congenital change by giving time for the necessary variations in many directions to be selected, and we have here another answer to the supposed difficulty as to the necessity of many coincident variations in order to bring about any effective advance of the organism."[131] Unbeknownst to Wallace, a year before his essay appeared an American psychologist named James Mark Baldwin elaborated a similar process in a paper entitled "A New Factor in Evolution." Now termed the "Baldwin effect," coined by the noted evolutionary biologist George Gaylord Simpson,[132] this is a mechanism by which traits (including behavioral) acquired by plasticity or habit, may, if beneficial, become genetically based if mutations underlying the trait arising by chance over time are selectively favored—a process of genetic assimilation. Philosopher Daniel Dennett described the process lucidly: "Thanks to the Baldwin effect, species can be said to pretest the efficacy of particular different designs by phenotypic (individual) exploration of the space of nearby possibilities. If a particularly winning setting is thereby discovered, this discovery will *create* a new selection pressure: organisms that are closer in the adaptive landscape to that discovery will have a clear advantage over those more distant."[133] Although at times controversial, the Baldwin effect is now largely accepted by evolutionary biologists, confirmed by both theoretical and empirical studies.[134] Rightly ascribed to Baldwin, who developed the theory, we may nonetheless add this phenomenon to the long list of evolutionary insights conceived by Wallace over the course of his long and intellectually productive life.

Conclusion

The range of subjects addressed in this chapter gives an indication of the richness of Wallace's evolutionary ideas and contributions over many decades—and this is apart from his contributions to evolutionary biogeography, treated in chapter 10. Some of his ideas were ahead of their time, and in general many have stood the test of time. Others have not. Collectively, they are a testament to Wallace's depth and breadth of thinking on evolutionary subjects. Even where deemed incorrect by modern standards, Wallace served to enliven and stimulate debate and discussion, itself a contribution to the scientific community. And as defender of his and Darwin's central contribution—evolution by natural selection—he had no equal, despite his reservations over the applicability of the theory to the human mind.

Needless to say, the overview in this chapter is not an exhaustive treatment, and more could be said about a number of lesser evolutionary topics of interest to Wallace. Just one that bears mention here relates to Wallace's long-standing attention to basic adaptive relationships. These enthralled Darwin too, who was perhaps at his most lyrical when writing of such exquisite adaptations as the fringe of hairs lining the legs of water beetles or the graceful plumes of wind-dispersed seeds. Wallace too was struck by the intricateness and beauty of adaptation, and while he generally saw structure and function in terms of adaptive utility, he was not one to blindly ascribe adaptive function to every trait. One early indication of this is found in Wallace's 1853 book, *A Narrative of Travels on the Amazon and Rio Negro*: "In all works on Natural History, we constantly find details of the marvellous adaptation of animals to their food, their habits, and the localities in which they are found. But naturalists are now beginning to look beyond this, and to see that there must be some other principle regulating the infinitely varied forms of animal life."[135] A few years later Wallace gave another telling discussion of adaptive utility. Writing about the canine teeth of orangutans, he notes that these great apes are vegetarian, yet have large canines that are not even used defensively. "Do you mean to assert, then, some of my readers will indignantly ask, that this animal, or any animal, is provided with organs which are of no use to it? Yes, we reply, we do mean to assert that many animals are provided with organs and appendages which serve no material or physical purpose." He continued,

> We conceive it to be a most erroneous, a most contracted view of the organic world, to believe that every part of an animal or of a plant exists solely for some material and physical use to the individual—to believe that all the beauty, all the infinite combinations and changes of form and structure should have the sole purpose and end of enabling each animal to support its existence—to believe, in fact, that we know the one sole end and purpose of every modification that exists in organic beings, and to refuse to recognize the possibility of there being any other. Naturalists are too apt to *imagine*, when they cannot *discover*, a use for everything in nature. . . . The separate species of which the organic world consists being parts of a whole, we must suppose some dependence of each upon all; some general design which has determined the details, quite independently of individual necessities.[136]

Darwin, and modern evolutionary biologists, would broadly agree that not all traits are adaptive (though in the case of the orang's canines, Darwin would point to sexual selection). But even as Wallace cautioned against a reflexively adaptationist view of traits, he also knew adaptation when he saw it.

An excellent example of this is the case of the comet orchid of Madagascar and its pollinator, an episode that once again involved Wallace coming to Darwin's defense. This orchid, *Angraecum sesquipedale*, sports large, white star-shaped flowers and very long nectary spurs (the species epithet *sesquipedale* means "foot and a half" in reference to the spur). Its pollinator was unknown, but judging from other long-spurred orchids Darwin guessed there must be a moth in Madagascar with an exceedingly long proboscis to do the job. In his 1862 orchid book, Darwin wrote of this species, "It is . . . surprising that any insect should be able to reach the nectar: our English sphinxes have probosces as long as their bodies: but in Madagascar there must be moths with probosces capable of extension to a length of between ten and eleven inches!"[137] Darwin was thinking in terms of co-evolution of plant and pollinator; related species have shorter spurs, and one that long must have evolved with a pollinator capable of reaching the nectar at the bottom. Five years later the antievolutionist Duke of Argyll attacked Darwin's explanation of the long nectar spur in his book *The Reign of Law*. Wallace rallied with a critical review of the duke's book and a counterargument: "the laws of multiplication, variation, and survival of the fittest, already referred to, would under certain conditions necessarily lead to the production of this extraordinary nectary."[138]

Wallace was so convinced that Darwin's co-evolutionary hunch was correct that he commissioned an illustration of a hypothetical sphinx moth visiting comet orchids to accompany his review (figure 4.3), and even identified possible candidates for the mystery moth (suggesting a species of *Macrosila*). "That such a moth exists in Madagascar may be safely predicted; and naturalists who visit that island should search for it with as much confidence as astronomers searched for the planet Neptune—and they will be equally successful!" he wrote.[139] In 1903, a population of *Macrosila morganii*, now *Xanthopan morganii*, was indeed discovered in Madagascar with a proboscis of just the right length, and was subsequently confirmed visiting *Angraecum sesquipedale* flowers. It was duly named subspecies *praedicta* in honor of Wallace and Darwin's prediction of its existence—a fitting memorial to both the successful prediction and their acute understanding of the process of evolution by natural selection.

Acknowledgments

I am grateful to Charles Smith for his helpful editorial comments on earlier manuscript versions of this chapter, and for providing a high-resolution version of figure 4.3. Special thanks to Andrew Berry and George Beccaloni for numerous stimulating conversations on Wallace's life and thought, and Lynda Brooks and Elaine Charwat at the Linnean Society of London for their kind assistance in facilitating my study of Wallace's Species Notebook. I am further grateful to the Linnean Society for permission to publish pages from the Species Notebook here. My Wallace studies were greatly facilitated by a year-long scholarly leave as fellow of the Wissenschaftskolleg zu Berlin, with additional support from Western Carolina University and Highlands Biological Station.

Figure 4.3 Artist's rendition of a long-proboscis sphinx moth that Wallace predicted would be found, based on the theory of natural selection. From Wallace (1867f).

Notes

1. Darwin was clear about these joint sources of inspiration. In a journal entry made in the summer of 1837 he wrote, "In July opened first note Book on 'transmutation of species'.—Had been greatly struck from about month of previous March on character of S. American fossils & species on Galapagos Archipelago—These facts origin (especially latter) of all my views" (CUL-DAR158, p. 13, Cambridge University Library). Twenty-two years later, in the introduction to *On the Origin of Species*, he wrote, "When on board H.M.S. 'Beagle' . . . I was much struck with certain facts in the distribution of the inhabitants of South America, and in the geological relations of the present and past inhabitants of that continent" (Darwin 1859, p. 1).

2. See Browne (1995) for the finest detailed account of Darwin's process of discovery during and after the *Beagle* voyage, and Costa (2009b) for an overview.

3. The term "consilience" was coined by the polymath Cambridge scholar William Whewell in his *Philosophy of the Inductive Sciences* (1847, p. 65), a founding treatise on the philosophy of science. Consilience is defined as a situation where different and apparently unrelated sets of observations combine and reinforce one another: "the cases in which inductions from classes of facts altogether different have thus jumped together, belong only to the best established theories which the history of science contains. . . . I will take the liberty of describing it by a particular phrase; and will term it the *Consilience of Inductions*."

4. See Costa (2015) for a discussion of Wallace's consilient approach.

5. Wallace letter to Henry Walter Bates, 28 December 1845 (WCP346).

6. In 1847 Wallace had written to Bates (WCP348) that he had begun "to feel rather dissatisfied with a mere local collection; little is to be learnt by it. I [should] like to take some one family to study thoroughly, principally with a view to the origin of species. By that means I am strongly of the opinion that some definite results might be arrived at." Recently, historian John van Wyhe has argued that Wallace and Bates did not embark upon their travels in order to pursue the question of species origins, a view that has found little support among historians of evolutionary biology. See van Wyhe (2013, 2014), Costa and Beccaloni (2014), and Smith (2015a).

7. For example, in his paper "On the Habits of the Butterflies of the Amazon Valley," read at the Entomological Society of London soon after his return home and published in 1854, Wallace (1854a, p. 258) noted that *Heliconia* butterflies "are exceedingly productive in closely allied species and varieties of the most interesting description, and often having a very limited range; and as there is every reason to believe that the banks of the lower Amazon are among the most recently formed parts of South America, we may fairly regard those insects, which are peculiar to that district, as among the youngest of species, the latest in the long series of modifications which the forms of animal life have undergone."

8. Wallace's poignant if journalistic account of this tragedy was published in the *Zoologist* (Wallace 1852a). See also the letter written by Wallace to his friend Richard

Spruce from the ship that rescued him and the *Helen*'s crew, dated 19 September 1853 (WCP349).

9. This is the "Sarawak law"; Wallace (1855b), p. 186. Contrary to the view of most historians, van Wyhe (2016) argues that the Sarawak law paper does not represent any kind of statement on transmutation by Wallace.

10. Darwin and Wallace (1858).

11. This literature is extensive; see chapter 6 in Costa (2014) for a review.

12. Elsewhere I have closely compared the papers: see Costa (2014), chapter 5.

13. Upon reading Wallace's manuscript, Darwin wrote to Lyell that he "never saw a more striking coincidence. . . . If Wallace had my M.S. sketch written out in 1842 he could not have made a better short abstract! Even his terms now stand as Heads of my Chapters" (Darwin letter to Charles Lyell, 18 June 1858 [DCP2285]). Lyell and Hooker wrote to the Linnean Society that the two had "independently and unknown to one another, conceived the same very ingenious theory."

14. See Costa (2014), pp. 214–217.

15. See, for example, the treatments of Bowler (1976), Bulmer (2005), and especially Kottler (1985).

16. For example, Bowler (1976) and Ruse (2013).

17. For example, Bock (2009). Paleontologist H. F. Osborn was perhaps the first to make this claim, in his 1894 book *From the Greeks to Darwin*.

18. Pointed out by Mayr (1982), and discussed by Kottler (1985).

19. See Costa (2014), pp. 219–220.

20. See Costa (2013b).

21. Wallace (1870b), p. 40; emphasis added.

22. Poulton (1896), p. 80; emphasis in the original.

23. In one paper Wallace (1860b, p. 219) wrote that he found on a felled latex-producing tree "dozens of a species of Scolytidae, with their abdomens protruding from the holes they had bored, but all dead. With a remarkable deficiency both of *instinct* and *reason,* the little creatures had dug their own graves, and were all glued fast by the hardening of the milky sap. In a few days more there were hundreds so killed; indeed it appeared as if not one escaped. It seems evident, therefore, that this tree could not have been the proper food of this species, or the right place to deposit its eggs. I have since observed exactly the same occurrence in another locality." (emphasis in the original).

24. Wallace first recorded his thoughts on these beetles in the Species Notebook (Costa 2013b, p. 422), and later included the account essentially unaltered in *The Malay Archipelago* (Wallace 1869c, p. 481). See discussion in Costa (2014), p. 220.

25. See the retrospectives on this concept by B. Wallace (1975) and Reznick (2016).

26. See Costa (2014), pp. 222–225.

27. Darwin letter to Asa Gray, 5 September 1857 (DCP2136); also published in Darwin and Wallace (1858), pp. 52–53.

28. For example, Limoges (1968), Browne (1980), Schweber (1980), Ospovat (1981), Kohn (1985), Kottler (1985), and Tammone (1995). Darwin dedicated considerable space to the divergence principle in *On the Origin of Species* (see Costa 2009a, pp. 111–126).

29. Treated in some detail in Costa (2014), pp. 225–227.

30. See Stauffer (1960) for a discussion of Darwin's ecological thinking.

31. His commitment to the principle of divergence led Darwin to emphasize ecological and "partial" isolation in the formation of new species, which would be consistent with speciation in continuous geographical areas. In 1868 Moritz Wagner wrote Darwin arguing that complete isolation is essential for the formation of new species; he was not alone in pointing out that any variants that arise in incipient subpopulations would be blended together in large continuous areas, preventing evolutionary change. More or less complete separation or isolation was necessary to permit the formation of new varieties and thus species—hence, Wagner's model was called the "separation theory." Darwin inserted the following statement in the fifth *Origin* edition (1869): "Moritz Wagner has lately published an interesting essay [on the importance of isolation], and has shown that the service rendered by isolation in preventing crosses between newly formed varieties is probably greater even than I have supposed," but also that he could "by no means agree with this naturalist, that migration and isolation are necessary for the formation of new species" (Peckham 1959, p. 196). Although he maintained that he never dropped isolation altogether, spatial isolation clearly took on secondary importance in Darwin's thinking. See Sulloway (1979) for a comprehensive discussion of Darwin's changing view of isolation in speciation.

32. Discussed in Browne (2003), pp. 409–410; and Costa (2009b), pp. 888–889. See early treatments by Wood (1973) and Bartley (1992).

33. Costa (2013b), p. 110 (Species Notebook [Linnean Society manuscript 180)], p. 4; emphasis in the original).

34. In the Species Notebook (pp. 39–40; Costa 2013b, pp. 106–108) Wallace writes, "In fact what positive evidence have we that species only vary within certain limits? Let us suppose that every variety of the Dog but one was to become extinct & that one say the spaniel, to be gradual spread over the whole world, subjected to every variety of climate & food, & domesticated by every variety of the human race. Have we any reason for supposing that in the course of ages a new series of varieties quite distinct from any now existing would not be developed,—& then should the same process be repeated & one of these varieties farthest removed from the original, again be spread over the earth & be subjected to the same variety of conditions, does it not seem probable that again new varieties would be produced, & have we any evidence to show that at length a check would be placed on any further change & ever after the species remain perfectly invariable under any circumstances whatever. Those who advocate variation within definite limits must suppose so, though the only ground for their opinion is that the varieties which have been produced under the influence of man have certain limits."

35. Wallace (1889a), p. vi.

36. Chapter 4 of *Darwinism* (Wallace 1889a), entitled "Variation of Domesticated Animals and Cultivated Plants," draws heavily upon Darwin's "remarkable volumes" (as Wallace put it), *The Variation of Animals and Plants under Domestication* (1868). In the opening paragraph Wallace sets forth his thesis, discussing how variations in domestic breeds "have been increased and accumulated by artificial selection, since we are thereby better enabled to understand the action of natural selection."

37. See Beccaloni (2008) for Wallace's later corrections and annotations to his personal copy of the published version of the Ternate essay.

38. Wallace (1863a, p. 8491; 1863c; 1864a, p. 1).

39. See "Coda" in Costa (2014), pp. 265–299.

40. Darwin received criticism from several quarters over his term "natural selection," including from Wallace. The term is problematic in that "selection" implies conscious action by a selector, personifying nature. By the third *Origin* edition (1861, p. 85), Darwin was defending the term: "In the literal sense of the word, no doubt, natural selection is a misnomer; but who ever objected to chemists speaking of the elective affinities of the various elements?—and yet an acid cannot strictly be said to elect the base with which it will in preference combine. . . . It has been said that I speak of natural selection as an active power or Deity; but who objects to an author speaking of the attraction of gravity as ruling the movements of the planets? Every one knows what is meant by such metaphorical expressions. . . . So again it is difficult to avoid personifying the word Nature; but I mean by Nature, only the aggregate action and product of many natural laws." Wallace thought "survival of the fittest" was a superior phrase, adopting it from philosopher Herbert Spencer, and went so far as to strike "natural selection" through much of his copy of the *Origin* and to write "survival of the fittest" beside each. Darwin reluctantly followed suit, and beginning with the fifth *Origin* edition (1869, p. 72), he added Spencer's phrase after defining "natural selection": "But the expression often used by Mr. Herbert Spencer of the Survival of the Fittest is more accurate, and is sometimes equally convenient." See discussion by Paul (1988).

41. Blyth (1835), pp. 40–41.

42. Wallace (1865a), p. 5.

43. Mallet (2004), p. 447.

44. Darwin discussed "doubtful species" in chapter 2 of the *Origin*, making much of the disagreement among naturalists over species identifications for many plants. See Costa (2009a, pp. 47–52) for Darwin's discussion with annotations.

45. Darwin struggled with species definitions, and reviewed the literature to better understand how his fellow naturalists defined species. In one letter to Joseph Hooker he wrote, "I have just been comparing definitions of species, & stating briefly how systematic naturalists work out their subject. . . . It is really laughable to see what different ideas are prominent in various naturalists minds, when they speak of 'species' in some resemblance is everything & descent of little weight—in some resemblance seems to go for nothing & Creation the reigning idea—in some descent the key—in some ste-

rility an unfailing test, with others not worth a farthing. It all comes, I believe, from trying to define the undefinable" (Darwin letter to Joseph D. Hooker, 24 December 1856 [DCP2022]).

46. In the *Origin,* Darwin gave a hypothetical scenario where a "young naturalist" attempts to delineate species of a group. The young naturalist is initially struck by differences between groups and names many species, but with closer study comes to appreciate the similarities and intergradations that link them, realizing that some of the entities named as separate species should in fact be joined. I speculate (Costa 2009a, pp. 50–51) that Darwin was referring to his own experience studying the taxonomy of barnacles.

47. The "biological species concept" (which defines species as "actually or potentially interbreeding groups of individuals") is often attributed to Mayr (1942), who did much to popularize the interbreeding concept after the modern synthesis but did not originate it. See Templeton (1989), Mallet (2008b), and K. de Queiroz (2005, 2007), for overviews and history of species concepts.

48. Wallace (1865a), p. 12.

49. See Mallet (2004, 2008b, 2009) for insightful discussions of Wallace's role in the development of the modern biological species concept. I am indebted to Jim Mallet's admirable work in this area for my own abbreviated account here.

50. Costa (2009a), p. 485.

51. Wallace (1865a), p. 4.

52. See Darwin (1868), vol. 2, pp. 185–189 for his discussion of sterility and selection.

53. Wallace letter to Darwin, 24 February 1868 (DCP5922).

54. Romanes (1886).

55. Even Darwin became somewhat confused (and confusing) on this point, incorporating use and disuse and other neo-Lamarckian processes into later editions of the *Origin* in response to critics. See Costa (2009a, pp. 491–495) for an overview of successive changes in the *Origin*'s editions, and Bowler (2003, pp. 160, 236–237) for a discussion of Darwin and neo-Lamarckian processes.

56. Wallace (1886b), p. 312.

57. See chapter 17 in Slotten (2004) for an excellent account of this episode, and Romanes's fraught and duplicitous relationship with Wallace.

58. See Wallace (1889a), pp. 173–179, "The Influence of Natural Selection upon Sterility and Infertility."

59. Lowry (2012) reviews the ecotype concept in relation to speciation theory.

60. Mallet et al. (2007), Mallet (2008a).

61. See, for example, Servedio and Noor (2003), Servedio (2004), Ortiz-Barrientos et al. (2004), Johnson (2008), and Matute and Ortiz-Barrientos (2014) for treatments of reinforcement theory and the Wallace effect. See also the case study of Silvertown et al. (2005) and the commentary on this paper by Ollerton (2005).

62. Mayr (1970), Provine (2004).

63. Wallace (1880b), p. 59.

64. See reviews by Ruxton et al. (2004), Caro, Hill, et al. (2008), Caro, Merilaita, et al. (2008), Forbes (2009), and especially Caro (2017).

65. “The phenomena of *dimorphism* and *polymorphism* may be well illustrated by supposing that a blue-eyed, flaxen-haired Saxon man had two wives, one a black-haired, red-skinned Indian squaw, the other a woolly-headed, sooty-skinned negress—and that instead of the children being mulattoes of brown or dusky tints, mingling the separate characteristics of their parents in varying degrees, all the boys should be pure Saxon boys like their father, while the girls should altogether resemble their mothers. This would be thought a sufficiently wonderful fact; yet the phenomena here brought forward as existing in the insect-world are still more extraordinary; for each mother is capable not only of producing male offspring like the father, and female like herself, but also of producing other females exactly like her fellow-wife, and altogether differing from herself” (Wallace 1865a, p. 10).

66. Wallace observed that “[w]e have here, therefore, distinct species, local forms, polymorphism, and simple variability, which seem to me to be distinct phenomena, but which have been hitherto all classed together as varieties. I may mention that the fact of these distinct forms being one species is doubly proved. The males, the tailed and tailless females, have all been bred from a single group of the larvæ, by Messrs. Payen and Bocarmé, in Java, and I myself captured in Sumatra a male *P. Memnon*, L., and a tailed female *P. Achates*, Cr., ‘in copulâ’” (Wallace 1865a, p. 6).

67. In fact, Wallace puzzled over the phenomenon but did not hit upon an explanation. In a postscript to a letter to Darwin dated December 1860 (WCP4079), he commented that “‘Natural Selection’ explains <u>almost</u> everything in Nature, but there is one class of phenomena I cannot bring under it,—the repetition of the forms & colours of animals in distinct groups, but the two always occurring in the same country & generally on the <u>very same spot</u>. These are most striking in insects, & I am constantly meeting with fresh instances” (emphasis in the original).

68. Bates (1862).

69. Darwin letter to Henry Walter Bates, 20 November 1862 (DCP3816).

70. Wallace letter to Henry Walter Bates, 20 November 1862 (WCP4668).

71. Wallace (1865a), p. 8.

72. Wallace (1867e).

73. An interesting form of aggressive mimicry is found in Neotropical lepidopteran caterpillars and pupae bearing eyespots, often coupled with a startle display; see Janzen et al. (2010).

74. Camouflage was found in the remarkable brush-footed butterflies of the genus *Kallima* that Wallace found in Sumatra, which bear a striking resemblance to dead leaves. See Caro, Meriaita, et al. (2008), Caro (2017), and chapter 5 of this volume for treatments of Wallace on cryptic coloration.

75. Wallace (1869c), p. 404.

76. Wallace (1869c), pp. 404–405.

77. Wallace (1882a), p. 87.

78. Ibid., p. 87.

79. See Kimler (1983) for an overview of the early Mendelians' view of mimicry and their arguments with naturalists such as Poulton who championed the Darwin-Wallace-Bates school.

80. Wallace read a paper on these butterflies at the 27 August 1866 meeting of the British Association for the Advancement of Science in Nottingham (Wallace 1866b).

81. Wallace letter to Darwin, 24 February 1867 (DCP5416); emphasis in the original.

82. See Kottler (1980) and Cronin (1991, chapters 6 to 9) for the most comprehensive treatments of the debate; also Baker and Parker (1979), Badyaev and Hill (2003), Caro, Hill, et al. (2008, pp. 157–159), Milam (2010, chapter 1), Allen et al. (2011), and Caro (2017) for further discussions of Wallace's and Darwin's views on dichromatism.

83. In his paper on Malayan papilionids, Wallace (1870b, pp. 155–156) cited Darwin and sexual selection explicitly: "The fact of the two sexes of one species differing very considerably is so common, that it attracted but little attention till Mr. Darwin showed how it could in many cases be explained by the principle of sexual selection."

84. Wallace to Darwin, 26 April 1867 (WCP4087); emphasis in the original.

85. Wallace (1868a), pp. 84–85.

86. Male choice is a separate but related issue over which Wallace and Darwin disagreed. Wallace believed that male choice occurred just as commonly as female choice; in other words, *mutual* sexual selection was the norm except in cases where natural selection opposed it. Darwin believed that male choice was rare (humans constituted one exception), anticipating in general form the modern theory that suggests that male investment in the form of energetically inexpensive sperm and the fact that males tend to seek out mates means that they are adapted to be the less choosy sex. Darwin put forth this argument in a letter dated 30 April 1868 (DCP6146), and succeeded in convincing Wallace of this point.

87. For example, in a letter written 5 May 1867 (DCP5528), Darwin commented to Wallace that he had "long recognized how much clearer & deeper your insight into matters is than mine," and outlined ideas on expression of coloration and the action of selection much in line with Wallace's own thinking: among other things, Darwin held that "characters may again arise in either sex and be transmitted to both sexes, either in an equal or unequal degree. In this latter case I have supposed that the survival of the fittest has come in to play with female birds, & kept the female dull-coloured."

88. For example, Wallace wrote Darwin (24 February 1868 [DCP5922]) that he thought it "probable that the development of many or perhaps all of these characters may be aided by sexual selection or be wholly due to it,—but I think there is abundant evidence to prove that the *cause* of colour and fine plumage . . . acts almost *equally* on both sexes *as the rule*; and that when these are wholly absent in the female, their absence has been produced by the need for protection, or by their being otherwise incompatible with the females' parental duties."

89. Darwin letter to Wallace, 15 April 1868 (DCP6121).

90. The modern genetic understanding is that both sex-limited inheritance and equal-sex inheritance occur; indeed, both processes have been shown to contribute to dimorphic coloration patterns in different lineages of nymphalid butterflies of the genera *Bicyclus* and *Junonia* (Oliver and Monteiro 2011).

91. Wallace letter to Darwin, 28 April 1868 (DCP6144).

92. Darwin letter to Wallace, 30 April 1868 (DCP6146); emphasis in the original.

93. In his 28 April 1868 letter (DCP6144) to Wallace, Darwin wrote, "[P]ray do not suppose that because I differ to a certain extent, I do not thoroughly admire your several papers and your admirable generalisation on birds' nests." The following August (DCP6322), he commented that "[i]n truth, it has vexed me much to find that the further I get on, the more I differ from you about the females being dull-coloured for protection . . . this has *much* decreased the pleasure of my work" (emphasis in the original). One month later (16 September 1868 [DCP6368]), Darwin wrote to Wallace that he "will be pleased to hear that I am undergoing severe distress about protection and sexual selection: this morning I oscillated with joy towards you; this evening I have swung back to the old position, out of which I fear I shall never get."

94. Darwin (1871), vol. 2, p. 171.

95. Darwin letter to Wallace, 23 September 1868 (DCP6386).

96. Wallace letter to Darwin, 4 October 1867 (DCP6408).

97. In a 27 January 1871 letter (DCP7460), Wallace protested to Darwin, "But my view is, and I thought I had made it clear, that the female has (in most cases) been simply *prevented* from acquiring the gay tints of the male (even when there was a tendency for her to inherit it) because it was hurtful;—and, that when protection is not needed, gay colours are so generally acquired by *both sexes* as to show, that inheritance by *both sexes* of colour variations is the most usual, when not prevented from acting by nat. selection" (emphasis in the original).

98. Wallace (1876a), vol. 1, p. 340.

99. Wallace (1877c).

100. Cronin (1991).

101. Zahavi (1975).

102. Darwin letters to Wallace, 17 June 1876 (DCP10538) and 31 August 1877 (DCP11121).

103. For example, Burns (1998), Ord and Stuart-Fox (2006), Kunte (2008), Oliver and Monteiro (2011), Prum (2010).

104. Helena Cronin's 1991 *The Ant and the Peacock* (1991) remains the best overview of the central issues in sexual selection. See also Eberhard (1996), Choe and Crespi (1997), Avise and Ayala (2009, chapters 9–12), and Hoquet (2015).

105. Richard Dawkins, Oxford University Press blog commenting on Helena Cronin's *The Ant and the Peacock*, http://blog.oup.com/2008/07/dawkins_cronin/.

106. Bates (1862), pp. 502–515.

107. Darwin letter to Wallace, 23 February 1867 (DCP5415).

108. Wallace letter to Darwin, 24 February 1867 (DCP5416).

109. Darwin letter to Wallace, 26 February 1867 (DCP5420).

110. Wallace's presentation on bright colors in caterpillar larvae, and the follow-up discussion, were published in the *Journal of Proceedings of the Entomological Society of London* (Wallace 1867b).

111. Wallace (1867c).

112. See Wallace (1870b), p. 119.

113. Jenner Weir (1869, 1870).

114. For example, Mappes et al. (2005), Marples et al. (2005), Ruxton and Sherratt (2006), Blount et al. (2009), and Stevens and Ruxton (2012). See also chapters 5–9 in Ruxton et al. (2004).

115. Wallace letter to Darwin, 24 March 1869 (DCP6681); Darwin letter to Wallace, 27 March 1869 (DCP6684).

116. Wallace (1869d). Darwin wrote to Wallace (14 April 1869 [DCP6706]), "If you had not told me I should have thought [Wallace's argument about humans] had been added by some one else. As you expected I differ grievously from you, & I am very sorry for it. I can see no necessity for calling in an additional & proximate cause in regard to Man."

117. See Fichman (2001), Smith (2008b, 2008c), and Moore (2008).

118. See treatments by Kottler (1974), Schwartz (1984), Fichman (2001; 2004, pp. 150–157), Slotten (2004, pp. 268–270), Nelson (2008), and Gross (2010).

119. The apparent problem of these "extreme" instinctive behaviors are discussed in chapter 7 of the *Origin*; see Costa (2009a), pp. 207–244, for an annotated discussion of Darwin's argument.

120. Costa (2014), appendix 3, pp. 292–293.

121. Wallace (1863c).

122. Species Notebook (Linnean Society manuscript 180), p. 167; Costa (2013b), p. 362.

123. Species Notebook (Linnean Society manuscript 180), p. 168; Costa (2013b), p. 364.

124. On the back endleaf of his Insect Register (manuscript WCP4767), Wallace wrote, "Instinct. Can a single case be shewn of an animal performing any complex act no part of which it has ever seen performed—? or without having seen the result. For example. Will any bird . . . bred from the egg, build a nest of the same form the same material & in the same situation, as others of its species; if turned for instance into a greenhouse or a glass covered water garden, with choice of materials & situations. This is a test wh[ich] will decide the question of birds building by instinct. I say under such circumstances it will not build a nest having the true characters of that of its species. May 1861" (emphasis in the original).

125. Wallace (1870b), p. 210. The argument about navigation in Indians is also found on pp. 170–172 of the Species Notebook (see Costa 2013b, pp. 368–373).

126. Wallace (1872b), p. 70.

127. Wallace (1872c), p. 470.
128. Wallace (1873a), p. 303.
129. Wallace (1873b), p. 66.
130. Wallace (1897a), p. 167.
131. Ibid., p. 168.
132. Simpson (1953).
133. Dennett (2003), p. 69.
134. Weber and Depew (2003), Badyaev (2009).
135. Wallace (1853b), pp. 83–84.
136. Wallace (1856b), pp. 30–31.
137. Darwin (1862), p. 198.
138. Wallace (1867f), p. 475.
139. Ibid., p. 477.

‹ 5 ›

Wallace on the Colors of Animals: Defense against Predators

HANNAH M. ROWLAND AND ELEANOR DRINKWATER

> The varied ways in which the colouring and form of animals serves for their protection, their strange disguises as vegetable or mineral substances, their wonderful mimicry of other beings, offer an almost unworked and inexhaustible field of discovery for the zoologist, and will assuredly throw much light on the laws and conditions which have resulted in the wonderful variety of colour, shade, and marking which constitutes one of the most pleasing characteristics of the animal world, but the immediate causes of which it has hitherto been most difficult to explain.
>
> —WALLACE, "Mimicry, and Other Protective Resemblances among Animals"

Wallace has been variously described as "the greatest field biologist of the nineteenth century,"[1] the "other discoverer of natural selection,"[2] as well as a "prolific and successful writer."[3] He wrote extensively about animal and plant coloration,[4] and so has also been described as "the father of the field of animal coloration."[5]

This description is fitting, because Wallace was the first scientist to recognize that the conspicuous color patterns of prey could serve as warning signals to would-be predators that prey possessed chemical defenses.[6] He was also among the first scientists to suggest how camouflage and mimicry evolved, as well as providing the first conceptual framework for how to classify the colors of animal and plants.[7]

Wallace's color categories have guided how the field has developed, and his appraisal of the evolutionary drivers of color still informs research on protective colors today.[8] This chapter highlights advances in Wallace-related color research made since the last major review was published.[9] We follow Wallace's original categories, looking first at protective colors, and then at warning colors and mimicry. The topics of mimicry of sexual colors (the appearances that Wallace referred to as typical colors), and the attractive colors of plants, are discussed in chapter 4.

Wallace's Universal Rule of Color

In discussing the battle that goes on between predators and prey, Wallace described "an almost universal rule" of animal colors: concealment from enemies and from prey. So before speaking of the different defensive colors that have evolved in animals, we start with Wallace's ideas on the selective pressure that drives their evolution: attack by predators, and concealment from prey.

In an 1879 essay titled "The Protective Colours of Animals," Wallace emphasized the importance of predators in creating the diversity of animal defenses and prey color.[10] Wallace suggested that insectivorous animals, especially birds, were the agents of natural selection on insect prey. But he needed evidence to back this up. After writing to the editor of *The Field* in March 1867 to request records of bird attacks on caterpillars,[11] Wallace was able to report that birds, frogs, lizards, and spiders ate green and brown caterpillars presented to them. From this, he surmised that predators selected for concealment strategies in their prey, either through color matching or through prey's remaining motionless. This idea of Wallace's, that visually guided predators were responsible for the many different types of antipredator color patterns, has been generally accepted.[12]

Wallace also discussed how prey affect the colors of their predators.[13] In describing the color of hawks, buzzards (buteos), and owls, he wrote, "[A]ny deviation from those tints best adapted to conceal a carnivorous animal would render the pursuit of its prey much more difficult, would place it at a disadvantage among its fellows, and in a time of scarcity would probably cause it to starve to death."[14] In contrast to the wealth of research on predators affecting the color of their prey, comparatively less effort has been given to understanding how prey affect the color properties of their predators. In 2016, however, support was obtained for Wallace's hypothesis that prey affect how raptor color patterns evolve. In a South African study, dark morph (black) sparrowhawks were found to have significantly higher prey-catching rates at low light than white morphs, which in turn have higher catching rates in brighter conditions.[15]

In both predators and prey, concealing colors have evolved to reduce detectability, and this is where we begin our review of Wallace's ideas on antipredator defenses.

Protective Coloration

In their letters to one other, Darwin and Wallace often discussed why some colors "harmonise animals with their general environment,"[16] while others were bright and conspicuous.[17] In this section we discuss four types of concealment identified by Wallace, and how his ideas have been tested. These types are crypsis through background matching, countershading, crypsis via color change, and special imitation by insects—now known as "masquerade." Finally, we treat the question of why zebras are striped—a topic that has been debated continually since Wallace's time!

Background Matching: Crypsis

Many animals possess body colors and patterns that resemble those in the neighboring background.[18] This didn't go unnoticed by Wallace, and in a letter to Frederick Bates, the brother of Henry Walter Bates, Wallace described the appearance of tiger beetles, Cicindelidae, he had observed during his travels.[19] Beetles living on the beach appeared to match the white sand on which they were discovered, whereas those found more inland matched the color of dark sand. He remarked that the "colour so exactly agrees that it was exact perfectly invisible except for its shadow!" Today we refer to this type of color pattern as background-matching crypsis—the appearance of a prey matches the color, lightness, and pattern of the background on which it rests, which reduces its detectability to predators.[20]

Wallace noted the generality of this close similarity between animals' body color and the many different environments they inhabit.[21] For example, he described how "white prevails among arctic animals; yellow or brown in desert species; while green is only a common color in tropical evergreen forests." These examples of background color resemblance puzzled him—he couldn't understand how these adaptations arose. But with time Wallace was able to hypothesize that prey with colors best adapted to concealment from their enemies would have higher levels of survival.[22]

It is now known that prey that match the color of their backgrounds do survive longer than conspicuously colored prey, just as Wallace predicted.[23] And the colors of animals are often correlated with local environmental conditions.[24] Furthermore, in 2001 the spectral reflectance of the elytra and abdomen of the beetle *Odontocheila nicaraguensis* Bates was measured, and found likely to be almost indistinguishable from the leaf litter to a visual predator such as a bird.[25] While this isn't the species Wallace wrote to Bates about, these results confirm what Wallace had described to Frederick Bates 143 years earlier.

Countershading

Many animals from a variety of different taxonomic groups display a graded change in the color of their body's pigments. This shading is most often characterized by darker backs and lighter bellies.[26] Wallace discussed what we now refer to as "countershading" in the context of marine animals such as whales. He hypothesized that this type of coloring would conceal the animals from predators viewing them from above (they would match the dark sea below), and from predators viewing from beneath them (they would match the light sky above). Following Wallace's idea, alternative hypotheses for countershading were offered by Hugh Cott in 1940.[27] Cott suggested that graded pigmentation could act to counterbalance shadows cast on the body of an animal illuminated from above, such that conspicuous self-shadows were obliterated. Counterbalancing shadows would reduce the capacity of predators to perceive three-dimensional cues of body shape.

Most research that has been conducted since Wallace's time has focused on this self-shadow concealment in terrestrial mammals and insects. These studies have shown that artificial prey with countershading have a greater probability of survival than those not displaying this form of patterning,[28] and comparative analyses focusing on ruminants revealed that the most likely function for countershading is crypsis.[29] However, the mechanisms underlying the concealing function of countershading in aquatic systems, as discussed by Wallace, have remained unclear.

In 2015, research on the countershaded freshwater fish *Melanotaenia australis* gave the first support to Wallace's hypothesis. When the fish were housed in pale or dark environments, they changed their body color to match their visual background by altering the relative ratio of their dorsoventral skin darkness.[30] Rather than concealing their shadows, it appeared that the fish were achieving crypsis through background matching, just as Wallace predicted.

Crypsis through Color Change

Many animals, including insects and mammals, change color during development, apparently to maintain the visual match with their environment.[31] Wallace thought that this kind of color change was "rare and quite exceptional,"[32] and was the first to describe the experimental evidence for color change, later summarized in his *Natural Selection and Tropical Nature*.[33] In this collection of essays he reported a study by T. W. Wood about the chrysalids of cabbage white caterpillars that changed color to match the background on which they rested. Wallace also described experiments by Mrs. Mary Barber on the chrysalis of the African butterfly (*Papilio nireus*) that showed a similar phenomenon.[34]

It is now known that color change is not limited to insects, but also occurs in crabs, shrimps, and fishes. In fact, Wallace himself reported changes of color in the chameleon shrimp (*Mysis chameleon*), which is gray when on sand, and brown or green when among seaweed of these two colors.[35] In recent years, scientists have been researching why animals that live in intertidal habitats have the ability to change color.[36] In two such studies, rock gobies (*Gobius paganellus*) and horned ghost crabs (*Ocypode ceratophthalmus*) placed on different backgrounds have been shown capable of rapid color change, with these changes leading to improvements in camouflage match to the background.[37]

We still do not understand the mechanisms underlying color change. But in discussing the examples of caterpillars and chameleon shrimp, Wallace proposed two mechanisms for color change: "a photographic action of the reflected light" and the effects of diet.[38] By the time he published *Darwinism*, his ideas had solidified into two distinct categories: "In one case the change is caused by reflex action set up by the animal *seeing* the colour to be imitated, and the change produced can be altered or repeated as the animal changes its position. In the other case the change occurs but once, and is probably not due to any conscious or sense action, but to some direct influence on the surface tissues while the creature is undergoing a moult or change to the pupa form."[39] Over 130 years later, in 2008, researchers found that caterpillars of the peppered moth (*Biston betularia*), which resemble the color of the host tree twigs on which they feed[40] probably achieve this

Figure 5.1 Protective morphology and coloration in *Kallima inachis*. From Wallace (1879h), p. 136.

by perceiving their visual environment (just as Wallace predicted).[41] However, there is still much to be discovered about color change, and the exact mechanisms underlying the phenomenon.

Masquerade

A number of organisms appear to mimic inanimate objects such as twigs, leaves, stones, and bird droppings. Today we call this type of visual defense "masquerade," but Wallace described it as "protective imitation."[42] In 1867 he described two butterflies, *Kallima inachis* and *Kallima paralekta,* that mimic the structure of leaves "in every stage of decay" (see figure 5.1). Wallace also noted that, coupled with this physical imitation of a leaf, the behavior of these insects aids their concealment.[43] Wallace wrote at length about invertebrates mimicking twigs and other objects (even bird droppings!),[44] and made an important distinction between two elements of this defense: the physical form and the specific behavior: "We thus have size, colour, form, markings, and habits, all combining together to produce a disguise which may be said to be absolutely perfect."[45]

It was eighty-five years before the first research was performed on masquerade. In a series of experiments carried out in 1952 by De Ruiter, twig-like geometrid caterpillars (*Ennomos alniaria, Biston hirtaria,* and *B. strataria*), and twigs chosen to resemble the larvae, were scattered over the floor of an aviary.[46] The individual jays (*Garrulus glandarius*) that were used as visually hunting predators initially ignored both the twigs and the motionless caterpillars, supporting Wallace's idea that the physical form aided their disguise. Four of eight birds that detected the larvae did so only because the insect moved, which also supported Wallace's idea that concealment was enhanced by behavior.

It took a further forty years after De Ruiter's work for the first evidence to emerge for the protective value of behavior in masquerading prey. In a 2015 study, artificial edible caterpillars were created from pastry dough to resemble bird droppings.[47] The researchers tested whether a bent shape enhanced the artificial caterpillar's resemblance to bird droppings and aided their survival. Birds attacked the straight caterpillars almost three times more often than the bent ones. When the researchers repeated the same experiment with green caterpillars—which do not mimic bird droppings—there was no difference in attack rate between the bent and unbent models. This was the first experimental support of Wallace's idea that behavior could help caterpillars masquerade as inedible objects.

Wallace didn't specifically discuss how masquerade tricks predators. One idea is that the mimicry helps the prey blend in with the wider environment, and reduces their detectability. Another idea is that predators detect the prey, but do not identify them as a prey animal. It is only in the last decade that we have discovered that potential prey in fact benefit from being misidentified as inedible objects by their predators.[48]

Why Are Zebras Striped?

> The zebra is conspicuously striped, and stripes on the open plains of South Africa cannot afford any protection.
>
> —DARWIN, *The Descent of Man,* 1871

> But the zebra is a very swift animal, and, when in herds, by no means void of means of defence. The stripes therefore *may* be of use by enabling stragglers to distinguish their fellows at a distance, and they *may* be even protective when the animal is at rest among herbage—the only time when it would need protective colouring. Until the habits of the zebra have been observed with special reference to this point, it is surely somewhat hasty to declare that the stripes "cannot afford any protection."
>
> —WALLACE, "The Colours of Animals and Plants," 1877

One of the topics that Wallace wrote about has received more interest than most, and has been debated for more than a century: the adaptive significance of zebra stripes.[49] Wallace's ideas conflicted with Darwin's, and Wallace perceived Darwin to have made a "hasty" assumption that stripes "cannot afford any protection" to the zebra. Wallace put forward the idea that zebra striping might have more than one function. His ideas were influenced by the work of William H. Brewer and Francis Galton. Brewer had demonstrated asymmetry among the markings of domestic animals, which caused Wallace to suggest that bilateral asymmetries in stripes might aid recognition among zebras, "especially by the sexes and the young."[50] Wallace's discussions with Galton led him to pose that in the twilight, the stripes merge "together into a gray tint," making them difficult to see. Overall Wallace put forward an elegant dual solution to the question of zebra stripes. During the day when the zebra was alert and with the herd, it could rely on speed for protection, so the stripes could be used for recognition.[51] During the twilight hours, however, when the zebra was resting and therefore more vulnerable to attack from predators, the stripes provided much needed camouflage to the animal.

Many alternative explanations to an antipredator function have been suggested since Wallace's time, including that the stripes change how air flows around a zebra's body to help it stay cool, that stripes protect the zebras from biting insects, and that the stripes are a means of reinforcing social bonds. The results of different studies have often presented conflicting evidence.[52]

To get to the bottom of this question, two groups of investigators have used a method called the comparative approach, in which evidence on various behaviors or appearances as they are found in the real world is collected, and the similarities and differences among them compared.[53] In the first study, information on seven species of striped animals that are closely related to zebras was compiled, and scored for number and intensity of stripes. These patterns were then intercorrelated with geographical data to identify which factors might overlap. The only link discovered was between stripe variation and incidence of blood-sucking flies. In the second study researchers analyzed the environmental factors that might explain geographical variation in striping within a single species, the plains zebra (*Equus quagga* or *Equus burchellii*). They found that greater intensities of intraspecific striping were associated with warmer temperatures and high precipitation.

Together these two pieces of research allow us to dismiss predation as driving striping in equids, and to reject a cryptic function.[54] Both studies point to tabanid or other biting flies (e.g., muscid, simuliid, or mosquito) as being the evolutionary driver of striping in Equidae, most probably because of the diseases that they carry.[55]

Is there any evidence that patterns of contrasting stripes aid predator avoidance? Stripes like the zebra's purportedly degrade an observer's ability to judge the speed and direction of moving prey, and are often postulated as the selective agent leading to repeating patterns on zebra, as well as many fish, snake, and invertebrate species. Contrasting stripes were even used to change the appearance of warships in both world wars.[56] Recently researchers have been using computer-game experiments to test the hypothesis that stripes are used to dodge predators.[57] We call this the "motion dazzle hypothesis," and it suggests predators are confused by stripes and can't understand their movement. The results of these studies have shown that, when moving in groups, targets with stripes parallel to the targets' direction of motion are harder to track than those with more conventional background-matching patterns. It appears that some high-contrast patterns may benefit striped animals that live in groups.[58] Clearly, as Wallace suggested in his statement, there is no simple answer as to why zebras are striped. The debate has continued for more than 145 years.

Warning Colors

The second class of colors that Wallace identified was warning colors.[59] He was the first to recognize that the bright, conspicuous color patterns of prey serve as signals to would-be predators that the prey possess chemical defenses.[60] In fact, Wallace resolved Darwin's problem with determining the function of conspicuous markings of juvenile nonreproducing individuals, which posed a real problem for his theory of sexual selection.[61] In 1889 Wallace wrote, "[T]he animals in question are either the possessors of some deadly weapons, as stings or poison fangs, or they are uneatable, and are thus so disagreeable to the usual enemies of their kind that they are never attacked when their peculiar powers or properties are known. . . . They require some signal or danger-flag which shall serve as a warning to would-be enemies not to attack them, and they have usually obtained this in the form of conspicuous or brilliant colouration."[62]

Wallace's friend and contemporary E. B. Poulton later reported evidence supporting a correlation between bright coloration and acceptability (or lack of) to predators.[63] It was Poulton who originally formulated the idea that these color pattern combinations and structures enhance avoidance learning, or capitalize on innate tendencies of vertebrate predators to avoid warning colors and patterning,[64] and coined the term "aposematism." Poulton proposed a variety of traits[65] that could be classed as aposematic: contrast of a prey's colors with the background, possession of two or more contrasting colors within a prey's pattern, and conspicuous behaviors—including the absence of hiding and escape behaviors and sluggishness.[66]

Wallace's ideas have been extensively followed up with empirical research on frogs,[67] snakes,[68] birds,[69] wasps,[70] Lepidoptera,[71] bioluminescent Coleoptera,[72] and perhaps even plants.[73] We now know that warning colors accelerate predator avoidance learning, and enhance predator memory of which prey to avoid.[74] Stronger, more-visible signals facilitate faster avoidance learning compared to weaker signals, and can make the difference between predators learning and not learning to avoid aposematic prey.[75]

Conspicuous versus Distinctive Aposematic Signals

Within his description of warning coloration Wallace stressed the importance of distinguishing between distinctiveness and conspicuousness of the colors of protected organisms. Of conspicuousness he wrote, "[T]o these creatures it is *useful* to be seen and recognised, the reason being that they have a means of defence which, if known, will prevent their enemies from attacking them, though it is generally not sufficient to save their lives if they are actually attacked."[76] But Wallace recognized that it is not sufficient for a chemically defended animal to only be conspicuous; they must also have distinctive features that predators will pay attention to and remember: "[the defense should be] very distinct from the protective tints of the defenceless animals allied to them."[77] Wallace provided an explanation for both the conspicuousness and distinctiveness of the coloration in protected organisms, thus achieving an important milestone in the field of aposematism.

Most of the research distinguishing signal conspicuousness from signal distinctiveness is theoretical.[78] Signal conspicuousness is thought to increase detection rate, and distinctiveness allows predators to distinguish defended from undefended prey. It is suggested these traits have opposing effects on evolution of warning signals: a more conspicuous warning signal cannot evolve unless it makes the prey more distinctive from palatable prey, reducing mistaken attacks by predators. Compared with the amount of theoretical work, the idea has received little experimental testing. However, in one experiment with humans predating on computer-generated prey, more cryptic warning signals evolved if they made the defended prey distinct from undefended prey.[79] It has been suggested that the inconspicuous but distinct coloration of European vipers supports Wallace's idea.[80] There is much to be done to understand the psychology of predators, and how conspicuousness and distinctiveness affect their attacks on prey.

Müllerian and Batesian Mimicry in Butterflies

It is interesting that, despite Wallace's insights into the evolution of warning signals, the occurrence of mimicry initially puzzled him. In a letter to Darwin, Wallace wrote, "Natural selection explains *almost* everything in nature, but there is one class of phenomena I cannot bring under it,—the repetition of forms and colours of animals in distinct groups, but the two always occurring in the same country and generally on the *very same spot*."[81] It was Wallace's old traveling companion, Henry Walter Bates, who formalized the explanation now bearing his name.[82] Batesian mimicry describes the resemblance of a palatable species to an unpalatable species: the palatable species gains protection from predation due to its perceived unpalatability.

Wallace championed Bates's theory, and on 17 March 1864 he read a paper at the Linnean Society titled "On the Phenomena of Variation and Geographical Distribution as Illustrated by the Papilionidae of the Malayan Region."[83] In the paper Wallace introduced the idea of sex-limited mimicry, where an undefended female butterfly would mimic an unpalatable species, but the male would not. Wallace suggested this phenomenon was due to the differing survival pressures on each sex. In his 1867 paper he pointed out that the female needed to exist longer than the male to deposit her eggs in a place adapted for the development and growth of their larvae.[84] This explanation is still accepted in the modern literature.[85]

In addition to his contribution to Batesian mimicry, Wallace also supported and contributed to the subject of Müllerian mimicry, where two or more unpalatable species benefit from sharing warning colors so as to share the cost of teaching naïve predators.[86] Since his time, additional explanations for the benefits of shared warning signals have been offered, including that selection is driven by experienced rather than naïve predators, and that this kind of mimicry is more likely to evolve in complex (as opposed to simple) prey communities. There has also been much debate about the distinction between Batesian and Müllerian mimicry, especially where some mimics have less concentrated levels of chemical defenses—a phenomenon termed quasi-Batesian mimicry.[87] Wallace, who acknowledged that species with fewer defenses would have a greater need for protection, also discussed this controversial topic. Regardless of the mechanisms that explain it, and the dynamics between different species, a great deal of scientific evidence supports Müller's mechanism, as championed by Wallace.

Polymorphisms in Mimicry

In his famous 1864 paper Wallace recognized different types of variation in the butterflies of the Papilionidae: (1) simple variability, (2) polymorphism, (3) local forms, (4) coexisting varieties, (5) races/subspecies, and (6) true species.[88] Wallace did not know how to explain the origin and persistence of adaptive polymorphisms, and this is still a puzzling question for evolutionary biologists. However, mimetic polymorphisms in *Papilio* and *Heliconius* species have been a test case for theories of density and frequency-dependent selection. In one recent experiment, researchers created "dummy" artificial prey that closely resemble *Heliconius numata*, and estimated selection (survival rates) of the prey.[89] They found that the most abundant mimetic morphs survived best in their own location, and that predators stamped out any deviation from the color pattern. So predators in different localities selected for multiple signals across space. Although Wallace didn't suggest variable predation as a cause for polymorphisms, he did discuss the role of differences in predators' habits in preying on different species.[90]

Mimicry in Snakes

The cases of mimicry that Wallace described as "perhaps the most remarkable cases yet known" are seen in certain snake species. In his 1867 study Wallace made the observation that the bright stripes of several nonvenomous snakes are quite similar to the gaudy coloration observed in the highly venomous coral snakes found in the same region. He wrote, "[T]he size and form as well as the coloration, are so much alike, that none but a naturalist would distinguish the harmless from the poisonous species" (see figure 5.2). He went on to suggest that this similarity with coral snakes provided protective advantages to the harmless mimics, and in later work explicitly pointed to this adaptation as an example of Batesian mimicry.[91]

This idea was questioned by some, because of the lack of overlap in the geographic ranges of the coral snake species and their putative mimics. In 2016, however, a landmark paper reconstructed the evolution of mimicry in the coral snake, using phenotypic, geographic, and phylogenetic data. The authors demonstrated that the red and black banding seen in mimics evolves in nonvenomous species only when coral snakes in the same geographical region had previously evolved distinctive markings.[92] This paper provided firm evidence for Wallace's Batesian mimicry hypothesis.

Figure 5.2 A coral snake (*A*) and its mimic (*B*). From Wallace (1879i), p. 293.

The mimicry in snakes described by Wallace was not limited to conspicuous aposematic colors, but also extended to inconspicuous coloration and behavioral adaptation. One example is the egg-eating snake *Dasypeltis scabra* in South Africa. This extraordinary species "has neither fangs nor teeth, yet it is very like the Berg adder (*Clothos [Bitis] atropos*), and when alarmed renders itself still more like by flattening out its head and darting forward with a hiss as if to strike a foe."[93] The hypothesized mimicry seen in *Dasypeltis scabra* was also the subject of much criticism in the years following Wallace's publication. The main criticism was that mimics were thought to often occur in localities where there was no model present, or the particular habitats of the models and mimics were quite discrete. A secondary concern was that the case of mimicry involving *Dasypeltis scabra* appeared to conflict with the traditional view of aposematism that conspicuous coloration was a prerequisite for predator avoidance.[94] It was only in 1961 that a thorough review of the localities and color of museum specimens of both *Dasypeltis scabra* and its putative model—small vipers in the genus *Causus*—was made. Doubts over Wallace's suggestion were laid to rest.[95] The paper made the important observation that while egg-eating snakes and their model did occupy discrete habitat types (with the model being found in scrub habitat, and the mimic being found in islands of vegetation within the scrub habitat), the protection provided to the mimic would depend on the range of the predator. If the predator range was large and covered both types of habitat, then the camouflage would be effective because the predator would encounter snakes of the dangerous *Causus* genus, as well as the harmless mimic.

Despite decades of debate and controversy, "perhaps the most remarkable cases yet known" of mimicry have provided critical insights into the evolution and function of mimicry in the wild.

Aggressive Mimicry

Wallace also devised and popularized the notion that mimicry could be used aggressively.[96] This is a type of mimicry used for the purposes of *enhancing* predation. In this section we focus on three examples that Wallace discussed that have led to many recent scientific discoveries.

Orchid Mantids and Floral Mimicry

A form of mimicry that Wallace described as being "one of the most curious" concerned the orchid mantid *Hymenopus coronatus*.[97] Wallace's interest in this species began with a story told to him by Sir Charles Dilke. Dilke recounted how he came upon a pink mantis resembling a flower that would lie in wait for insects, particularly butterflies, using its own body as bait for prey. Wallace later described this phenomenon in *Darwinism*. By that point Wallace had obtained drawings of a similar species from James Wood-Mason, curator of the Indian Museum at Calcutta.[98] These drawings allowed Wallace to suggest that the mimicry was so complete that the different parts of the mantid were all mimicking various parts of the orchid.[99] As one example, he described "the large and oval abdomen looking like the labellum of an orchid." The drawings from Wood-Mason also came with an anecdotal account of a botanist momentarily mistaking the mantid for a flower.[100] Wallace described the mantid as a "living trap, baited in the most alluring manner to catch the unwary flower-haunting insects" and referred to this type of mimicry as "alluring coloration."[101]

The idea that mantids attract pollinators as prey by floral mimicry seems intuitively obvious, but Wallace's ideas went untested for 125 years. In 2014 experimental evidence of pollinator deception by *Hymenopus coronatus* was finally documented.[102] The complexity of deception employed by the orchid mantis was even more remarkable than Wallace had imagined. Comparisons were made of the femoral lobes of the mantis to the shape of petals of plants within the home range that overlapped with the orchid mantis itself; additionally, and with the aid of psychophysiological vision models of bees, birds, and flies, the perceived color of a range of flowers and the perceived color of orchid mantids were compared.[103] If, as suggested by Wallace, the mechanism of orchid mantis deception was mimicry of a flower species, one would expect that the color or shape of the mantis would be similar to a particular species of flower. It isn't! Instead, the color and shape of the mantis show similarities to a range of flower species rather than to one particular species. The mantis thus seems to have evolved to exploit the senses of their prey with a super-flowerlike stimulus.

Cuckoo Mimicry

Instead of building their own nests, female cuckoos lay their eggs in the nests of other birds.[104] Just like caterpillars that mimic twigs, the cuckoo's eggs mimic the color and pattern of their hosts' eggs.[105] To sneak their eggs into the nests of their hosts, cuckoos have evolved plumage that resembles the hosts' predator, sparrowhawks. When cuckoo chicks hatch, they throw out the other eggs in the nest, and impersonate the begging call of their foster parent's entire brood of chicks;[106] the parents then feed the cuckoo chick as if it's their own. This extraordinary set of behaviors had Wallace exclaim of deceitful animals like cuckoos, "They appear like actors or masqueraders dressed up and painted for amusement, or like swindlers endeavouring to pass themselves off for well-known and respectable members of society."[107]

Wallace suggested that while cuckoos lay their eggs in a variety of species' nests, the female cuckoo may select a nest with eggs most closely resembling her own.[108] The reason for this, he posed, was not rejection by the host (though he did suggest that this probably happens occasionally), but instead that it would not disrupt the camouflage of the whole clutch.[109] This explanation for egg mimicry has been overtaken by new evidence on the co-evolutionary arms race between cuckoos and hosts. Selection favors hosts that are more likely to recognize and reject foreign eggs, which in turn selects for cuckoos whose eggs more closely match the eggs of its host.[110] This endless cycle of improvement means that the best-matching cuckoo eggs are found in the nests of the most picky host species.[111]

Wallace believed that cuckoos were "defenseless creatures" seeking protection from predators, but we now know they are far from defenseless in the way they dupe their hosts. In some parts of the world cuckoos mimic hawks, earning them the name "hawk-cuckoo," while in others the cuckoos look like dangerous birds such as drongo-shrikes.[112] Furthermore, hawk-like plumage appears more often in parasitic cuckoos than in nonparasitic ones.[113] So their disguise is much more sinister than it initially appears. It has been suggested that the hawk-like appearance might aid parasitic laying by frightening or luring hosts away from the nest,[114] or by inducing mobbing to help the cuckoo locate host nests. And, in fact, we now know that host neighbors alert other hosts to local cuckoo activity.[115]

According to Wallace, the largest number of cases of mimicry in any bird family comes from the cuckoos.[116] He may well have been correct!

Ant Mimicry

In 1867, Wallace described a genus of small spiders in the tropics that "feed on ants, and they are exactly like ants themselves, which no doubt gives them more opportunity of seizing their prey."[117] Today, over two thousand ant mimics have been described. This phenomenon, known as myrmecomorphy, is found in over two hundred genera. These groups include true bugs, mantids, flies, wasps, katydids, stick insects, beetles, and thirteen families of spiders. While there is still some suggestion that ant mimicry may aid hunting, it is now accepted that the primary reason for ant mimicry among spiders is likely to be predation avoidance. Ants, despite their prevalence, have relatively few obligate predators. This is attributed to their capacity to severely punish a predator with their powerful jaws or paralyzing venom, or to their ability to quickly amass conspecifics when alarmed. They can also be fortified with spines, nodules, chemical sprays, and heavily sclerotized (hardened) segments. Consequently, ant defenses severely reduce their attractiveness as a food item and afford an ant mimic the obvious adaptive benefit of protection from predators.

Deimatic Displays

Perhaps the most startling use of color in animal defense is the deimatic display. These remarkable displays can instill a harmless prey species with "terrifying powers" to startle its predators, and so avoid predation.[118] In the words of Wallace, "A considerable number of quite defenceless insects obtain protection from some of their enemies by having acquired a resemblance to dangerous animals, or by some threatening or unusual appearance."[119] The display results from modification of shape, of habits, of color, or of all combined. Wallace suggested that the extent of this type of defense could range from a simple posturing or tail-raising behavior, right through to elaborate displays like the caterpillar of the royal persimmon moth (*Citheronia regalis*). This impressive animal is a large green caterpillar, with a crown of orange-red tubercles; if disturbed, it erects this and shakes it from side to side. Wallace noted that this defense was so alarming the local people believed the harmless creature to be as dangerous as a rattlesnake. He suggested that this extraordinary behavior might have been born of necessity as the caterpillar grew too large for camouflage, and developed a habit of head shaking to startle predators. In time the effect of this behavior was further aided by the development of its colorful crown.

Since Wallace's original description of "prey terrifying," such displays have been referred to by various names ranging from "startle display" to "dymantic display."[120] The current and most popular term is "deimatic display," from the Greek word meaning "I frighten." This term is derived from "deimatic reaction," coined in 1970 to describe the defensive displays of the praying mantis *Stagmatoptera biocellata*, and later applied to a characteristic intimidation posture that appears to cause predators to pause or withdraw.[121] The more encompassing usage applies to behaviors ranging from musk oxen forming a protective ring around young, to salamanders performing the so-called "unken reflex."

Despite the dramatic nature of this type of behavior, the mechanism of deimatic displays is still poorly understood. The topic has, like many of Wallace's notes, courted much controversy in recent years—not so much because of what he said, but because of how modern researchers view the trait. Different schools of thought suggest the exclusion or inclusion of different behaviors in deimatic displays. The conflict rests largely on the difficulty of understanding how the predator perceives the display: Is the predator terrified by mistaking the sudden appearance of spots for the eyes of a predator? Is the predator reluctant to approach a displaying animal because of the novel change of appearance? Or is the predator "afraid" of the prey's advertisement of the possibility of retaliation? Wallace didn't pose a specific mechanism himself, and clearly there is still much to discover.[122]

Conclusion

In reading Wallace's letters, articles, and manuscripts, it's hard to find an evolutionary topic that he *didn't* discuss and that hasn't been explored sometime over the past 140 years. Wallace's coverage of antipredator defenses is outstanding. He championed coloration as a subject of huge significance to the theory of evolution: "the more we examine it [color] the more convinced we shall become that it must serve some purpose in nature, and that besides charming us by its diversity and beauty, it must be well worthy of our attentive study, and have many secrets to unfold to us."[123] He was correct. The field of adaptive coloration is a thriving research area, with much work still to do—How do predators perceive deimatic displays? Which predators are fooled by ant mimicry? What maintains aposematic and mimetic polymorphism? And why are zebras striped?

Acknowledgments

HMR thanks the Department of Zoology, University of Cambridge, Churchill College, Cambridge; and the Institute of Zoology at the Zoological Society of London, for support while writing this chapter. ED is supported by a studentship at the Department of Biology, University of York, and was supported by the Association for the Study of Animal Behaviour grant awarded to HMR while writing the chapter. We thank Tim Caro and two anonymous referees for constructive comments.

Notes

1. Rosen (2007).
2. Berry (2013).
3. Pain (2013).
4. Wallace (1865a, 1867e, 1889a).
5. Caro (2017).
6. Wallace (1867b).
7. Wallace (1877b).
8. Caro (2017).
9. Caro, Hill, et al. (2008), Caro, Merilaita, et al. (2008).
10. Wallace (1879h).
11. Wallace (1867c).
12. Stevens (2007).
13. Wallace (1867e).
14. Ibid., pp. 12–13.
15. Tate et al. (2016).
16. Wallace (1889a), p. 199.
17. Darwin letter to Wallace, 23 February 1867 (DCP5415); Wallace letter to Darwin, 24 February 1867 (DCP5416); Wallace letter to Darwin, 1 May 1868 (DCP6153); Darwin letter to Wallace, 5 July 1870 (DCP7218).
18. Endler (1984).
19. Wallace letter to Frederick Bates, 2 March 1858 (WCP367).
20. Merilaita et al. (2001).
21. Wallace (1889a), p. 190.
22. Wallace (1858).
23. Ruxton et al. (2004), Stevens and Merilaita (2011).
24. Stevens and Merilaita (2011), Stevens et al. (2014).
25. Schultz (2001).
26. Rowland (2009).
27. Cott (1940).
28. Rowland (2009).
29. Allen et al. (2012).
30. Kelley and Merilaita (2015).
31. Stuart-Fox and Moussalli (2009).
32. Wallace (1877c), p. 389.
33. Wallace (1891a).

34. Barber (1874).
35. Wallace (1891a).
36. Stevens et al. (2014).
37. Stevens et al. (2013), Stevens et al. (2014). For another study, see Greene (1989).
38. Wallace (1877c), p. 389.
39. Wallace (1889a), p. 195.
40. Poulton (1890).
41. Noor et al. (2008).
42. Wallace (1889a).
43. Wallace (1867e).
44. Wallace (1900b).
45. Wallace (1867e), p. 10.
46. De Ruiter (1952).
47. Suzuki and Sakurai (2015).
48. Skelhorn (2015).
49. Caro et al. (2014), Caro (2016).
50. Wallace (1889a), pp. 217–218n.
51. This discussion is thus also related to Wallace's theory of "recognition marks." For an overview see Wallace (1910c), chapter 9.
52. Caro and Stankowich (2015).
53. Caro et al. (2014).
54. Melin et al. (2016).
55. Caro and Stankowich (2015).
56. Behrens (2009).
57. See, for example, Stevens et al. (2011).
58. Hogan et al. (2016).
59. Wallace (1877b).
60. Wallace (1867b).
61. Darwin (1871).
62. Wallace (1889a), p. 232.
63. Poulton (1890).
64. Caldwell and Rubinoff (1983).
65. Poulton (1890).
66. Ruxton et al. (2004).
67. Summers and Clough (2001), Darst et al. (2006).
68. Brodie (1993), Wüster et al. (2004).
69. Dumbacher et al. (1992), Dumbacher and Fleischer (2001).
70. Hauglund et al. (2006), Kauppinen and Mappes (2003).
71. Lyytinen et al. (1999), Nishida (2002).
72. De Cock and Matthysen (2003).
73. Lev-Yadun (2003a, 2003b).

74. Gittleman and Harvey (1980), Roper and Wistow (1986), Roper and Redston (1987).

75. Alatalo and Mappes (1996), Lindström (1999), Mappes and Alatalo (1997).

76. Wallace (1877c), p. 390.

77. Wallace (1889a), p. 232.

78. Puurtinen and Kaitala (2006).

79. Sherratt and Beatty (2003), Jansson and Enquist (2003).

80. Wüster et al. (2004).

81. Wallace letter to Darwin, December? 1860 (DCP2627).

82. Bates (1863).

83. Wallace (1865a).

84. Wallace (1867e).

85. Mallet (1999).

86. See Wallace (1882a, 1883b).

87. Turner et al. (1984).

88. Wallace (1865a).

89. Chouteau et al. (2016).

90. Wallace (1897b).

91. Wallace (1877c).

92. Rabosky et al. (2016).

93. Wallace (1889a), p. 262.

94. Cott (1932).

95. Gans (1961).

96. Wallace (1867e).

97. Wallace (1877c), p. 390.

98. Wood-Mason (1878).

99. Wallace (1877c).

100. Wood-Mason (1878).

101. Wallace (1889a), p. 212.

102. O'Hanlon et al. (2014).

103. Ibid.

104. Davies (2000).

105. Brooke and Davies (1988).

106. Davies et al. (1998).

107. Wallace (1867e), p. 40.

108. Wallace (1889a).

109. Ibid.

110. Davies and Brooke (1989).

111. Stoddard and Stevens (2010).

112. Wallace (1891a).

113. Welbergen and Davies (2008).

114. Wyllie (1981).

115. Thorogood and Davies (2012).
116. Wallace (1879i), p. 294.
117. Wallace (1867e), p. 30.
118. Wallace (1889a).
119. Ibid., p. 209.
120. Edmunds (1974).
121. Maldonado (1970), Edmunds (1974).
122. Skelhorn et al. (2016), Umbers and Mappes (2016).
123. Wallace (1877c), p. 385).

‹ 6 ›

The Many Influences Shaping Wallace's Views on Human Evolution

SHERRIE LYONS

> But now for your Man paper . . . The great leading idea is quite new to me . . . would you like at some future time to have my . . . notes . . . they are at present in a state of chaos.
>
> —DARWIN TO WALLACE, 28 May 1864

> I hope you have not murdered too completely your own and my child.
>
> —DARWIN TO WALLACE, 27 March 1869

Any theory of evolution that does not include human evolution must be considered an incomplete theory, and any theory of human evolution that does not account for the development of a moral sense must also be considered incomplete. For Wallace the moral development of humans was the subject that most fundamentally underlay his thinking about human origins.[1] Wallace played a pivotal role not only in Darwin's making public his general theory of evolution, but also in showing that humans were no exception to his theory. Darwin had been convinced for many years that organisms change through time, discussing his ideas with only a handful of colleagues who were encouraging him to publish as he began working on his "big book," tentatively to be called *Natural Selection*. In 1858 the much younger Wallace, then living in the Malay Archipelago, sent Darwin a brief letter and his paper "On the Tendency of Varieties to Depart Indefinitely from the Original Type," in which he outlined virtually the identical theory of natural selection.[2] Devastated, Darwin wrote Charles Lyell, "I never saw a more striking coincidence; if Wallace had my MS sketch written out in 1842, he could not have made a better short abstract!"[3] Lyell, along with Joseph Hooker not wanting Darwin to lose priority, arranged for Wallace's paper and some extracts from Darwin's unpublished writings to be read before the Linnean Society. Wallace remained in the Malay Archipelago for several more years while Darwin frantically worked to finish a shorter version of the big book that was published the following year: *On the Origin of Species*. While this story is well known, what is perhaps not fully appreciated is that Wallace had been corresponding for a couple of years with Darwin, on a variety of topics. But when Wallace had asked about human evolution, Darwin replied, "You ask whether I shall discuss 'Man';—I think I shall avoid the whole subject, as so surrounded with prejudices, though I fully admit that it is the highest & most interesting problem for the naturalist."[4]

For Wallace, however, the question of human origins was always critical to his thinking and not something that could be avoided. His belief in species change by natural selection was not just the result of his detailed studies of plants and animals, but also a function of his interest in ethnology and human origins. Darwin's and Wallace's views were never identical, and as the years went on, their differences grew. But in 1858 the similarities far outweighed the differences, and we must assume that Wallace, like Darwin, did not think humans were an exception to evolutionary theory. Indeed, Wallace's ideas helped Darwin in his own thinking about human evolution. But much to Darwin's dismay in 1869 Wallace claimed that there were several traits that could not be accounted for by natural selection. The prevailing view among scholars and in more popular writings about Wallace has been that he "changed his mind" (actually, *reversed* himself) on human evolution, and that this change was due to his new involvement with spiritualism.[5] However, such a view is overly simplistic. It does not do justice to the significant issues he raised or to the many factors that contributed to his thinking about human origins. These include his ethnographic studies, his interest in phrenology and mesmerism, and his exposure to the socialist ideas of Robert Owen. Unlike Darwin, Wallace had lived among native peoples for extended periods of time; in fact he was one of the first Europeans of scientific bent to do so. He had a very different view of the "savages" (the term that was commonly used at the time) than Darwin and virtually all Europeans, being remarkably free of the prevailing racist attitudes. This chapter explores how these many influences shaped his views and argues that Wallace did not "change his mind" (i.e., reverse himself).

Rather, in line with the scholarship of Charles H. Smith (and further elaborated on by Martin Fichman), Wallace's later views about human evolution were an extension on his thinking about the role and limitations of natural selection in giving a complete account of the history of life.[6]

The publication of the *Origin* created a storm of controversy, much of which revolved around the question of human origins. However, the dispute predated Darwin, as people had long been fascinated with the humanlike apes. The relationship of these primates to humans and to each other was a source of debate as soon as these creatures became known to Europeans. In 1699 Edward Tyson published a monograph on the "Orang-outang" or "Pygmie" that detailed the remarkable similarities between it and humans. However, the creature he described was neither an orangutan nor a member of the forest peoples in central Africa known as Pygmies. Rather, from Tyson's drawings and descriptions it was clearly a chimpanzee. Although it was more humanlike than any of the humanlike apes, Tyson did not think it was the result of a hybridization between the two species. In his dedication he said that the purpose of his monograph was to demonstrate the state in nature in which the Pygmy was the connecting link between the animal and the rational.[7] Interestingly, the word "orang" means "man" in Malaysian, and an "orang-utan" is a man of the forest. Wallace spent many years in Malaysia, and one of the reasons he went there was to study the orangutan in its native habitat. This was not the chimpanzee of Tyson's drawings, but truly an orangutan. Various other naturalists also had commented on the physical similarities between humans and the apes. The French naturalist Buffon argued that the system of classification was somewhat arbitrary and that Nature's productions were always gradual and marked by "minute shades." Nevertheless, he agreed with Tyson. The ape might physically appear to be intermediate between humans and animals, but "[it] is, in fact, nothing but a real brute, endowed with the external mark of humanity, but deprived of thought, and of every faculty which properly constitutes the human species."[8] In contrast, James Burnett (later known as Lord Monboddo), much to the distress of his eighteenth- and nineteenth-century Scottish and Enlightenment colleagues, claimed that some men were born with tails and that the orangutan was a man. In 1773 in his major treatise *Of the Origin and Progress of Language*, he argued that they deserved to be included in the human family for a variety of reasons. In addition to the obvious similarity in how they looked, he put forth some rather questionable facts. They "live in society, build huts, joined in companies [that] attack elephants, and no doubt carry on other joint undertakings for their sustenance and preservation."[9] Although they lacked speech, they had organs for speech. Monboddo's ideas were not well accepted and were widely criticized. In the second edition he corrected his claim that orangutans looked just like us and walked upright, but he continued to maintain that they were "one of us" because this supported his theory of language. Thus, even before Darwin published his ideas, the humanlike qualities of the apes had captured the imaginations of both naturalists and the general public. Do apes think? Are they moral beings? Can they develop language? However, with the exception of Lamarck, before Darwin no one was suggesting that one species actually transformed into another. Still, how to define the boundary between species was actively debated, and particularly how to define the boundary between the humanlike apes and humans. With the publication of the *Origin*, this discussion intensified. Not only did it lead to debates over the origin of *Homo sapiens*, but it rekindled the eighteenth-century debate over the origin of the different races. Were the different races of a single species, or were they actually different species? With the distinct possibility emerging that humans might truly be related to other primates, the polygenist-monogenist debates were reformulated. Were the present-day races derived from different apelike ancestors or were they just varieties that developed later from one common ancestor? Wallace proposed a brilliant and nuanced answer to the question.

Wallace's Two-Stage Theory of Human Evolution

In 1864 Wallace published a provocative paper, "The Origin of Human Races and the Antiquity of Man Deduced from the Theory of Natural Selection," which provided a subtle and wonderful compromise between the polygenist and monogenist hypotheses. In it he explained why the human body with the exception of the skull looked so similar to the bodies of present-day apes, and at the same time why the skull and mental abilities diverged so widely from these same apes.[10] Wallace proposed that humans, unlike other organisms, had experienced two distinct stages in evolution. In the first stage, human ancestors had evolved like any other creature, according to the principle of natural selection. Those individuals carrying favorable variations in the constant struggle for existence survived, and their form continued to be modified. However, once the human brain had evolved to a certain point and the moral and intellectual capabilities were fairly well developed, natural selection would no longer act on human physical structures, and the body would remain essentially unchanged. Accordingly,

> Humans were able to adapt to harsh and changing environmental conditions by the use of their intellect alone. A change in climate might result in an animal developing a thick coat of fur or a layer of fat, but humans would put on warmer clothes and build a shelter. No change in body structure would be required. A change in the abundance of food might result in an animal changing its diet or bodily weapons such as claws and teeth, or the digestive system might be altered. Humans, however, in an early period had already learned to hunt. Improved trapping and hunting techniques would allow them to survive with no change in body structure. As soon as humans possessed fire many new foods became edible, increasing the available food supply. Once they had developed agriculture and the domestication of various animals, the amount of food was also greatly increased. By intellect alone with no change in the body, humans could survive and adapt, remaining in harmony with the ever-changing environmental conditions. Furthermore, the developed social and sympathetic feelings allowed humans to band together for mutual comfort and protection, and such traits would be preserved and accumulated. Thus, Wallace believed that the more intellectual and moral races would replace the lower ones. Such views also suggested that early humans must have first arisen when the brain size was still fairly small, at a far earlier time than most people thought, and when natural selection would still be acting on the body.[11]

Wallace's two-stage theory of human evolution provided both an explication and a solution to the monogenist-polygenist controversy. Monogenists maintained that the further back in time one went, the more similar the races became. If it were possible to go back to the time to when the first humans emerged as a species, only one race existed. In contrast, the polygenists claimed that the mummies found in ancient Egyptian tombs showed as much difference existed between the Negro and Semitic races then as now—which supported their position. However, Wallace disagreed, finding the Egyptian evidence not very compelling for two reasons. First, he thought that humans had been around far longer than five thousand years and proposed that humans might have existed as far back as the Miocene. At the time, evidence did not exist for such a position; however, the incompleteness of the geological record provided a reasonable explanation for the absence of such data. Second, at a far earlier time racial divergence had already ended because natural selection was no longer acting on the physical structure of these ancestral forms. Once the brain had reached a certain level of development, it protected the body from the action of natural selection. Such an idea supported the monogenist position. Only the mind continued to evolve. This would explain why little structural change would be observed between the races five thousand years ago and now, and would also explain why human bodies were so similar to ape bodies.

This last point also addressed the real stumbling block that underlay the monogenist-polygenist debate, which was not a scientific one about human origins, but a more insidious one concerning the intellectual and moral qualities of the different races. If the polygenists were correct that each race had a separate origin, then in reality they were different species. It would be a small step to claim that some races were superior to others. Such a position in turn could be used to justify slavery, imperialism, and various other racist policies. In arguing for a single origin of all human beings, monogenists could claim that they had scientific proof that the "savage" races were not qualitatively different or inferior to their masters or colonial administrators. Abolitionists and members of other humanitarian reform movements could use the monogenist position to further their own goals. Both Darwin and Wallace were opposed to slavery, and they both knew that the theory of polygenesis had been used as a quasi-scientific argument to justify slavery. But the two men agreed that no matter how diverse the different races seemed to be, they all were derived from the same common stock.

Wallace's ingenious solution to this controversy essentially amounted to a moral monogenesis and a physical polygenesis. He argued that at a time far back in history humans indeed were a single homogeneous race. At that time, our ancestors had a human form, but not yet a well-developed brain. They did not have speech or the sympathetic and moral feelings that today characterize the human species. By the principle of natural selection they developed distinct racial characteristics adapted to their local environments. If one did not consider these ancestors to be fully human until the higher faculties were fairly well developed, then one could assert that there were many distinct races of humans. However, if one thought that our ancestors—whose form was essentially modern, but whose mental faculties were scarcely raised above the brute—still deserved to be considered fully human, then all of humankind shared a common origin. Thus Wallace did not deny that physical and racial differences existed that could have been due to separate origins, but that all depended on how one defined the origin of the human species. The different races all had previously evolved to a certain point in which they all exhibited a shared common humanity. By defining the origin of the human species at that point in time, the different races were simply variants of a single species.

Darwin was very impressed with Wallace's paper, as he also had been contemplating the problem of human evolution. He even offered to give Wallace his own notes on man, claiming that Wallace was far ahead of him in his thinking about the matter. However, Wallace declined the offer. For the next five years the two men continued to correspond, exchanging and developing their views on their joint theory. As mentioned in chapter 2, however, in 1869 Wallace presented a very different explanation for how humans might have evolved to their present state in his review of new editions of Lyell's *Principles of Geology* and *Elements of Geology*, appearing in London's *Quarterly Review*. Wallace now claimed that neither natural selection nor a general theory of biological evolution could "give any account whatever of the origin of sensational or conscious life. . . . [T]he moral and higher intellectual nature of man is as unique a phenomenon as was conscious life."[12] Darwin must have suspected that something was amiss as he wrote Wallace, "I shall be intensely curious to read the *Quarterly*: I hope you have not murdered too completely your own and my child."[13] Unfortunately for Darwin, his fears were justified.

The Limits to Natural Selection

In the *Origin* Darwin had emphasized that natural selection was based on a principle of utility. Natural selection could not preserve a harmful structure; furthermore, it was a principle of present utility and relative perfection only. A structure would not be preserved by natural selection because it was to become valuable for future generations, nor would natural selection accumulate favorable variations to provide a more perfect structure if a less elaborate one could do the job in the ongoing struggle for survival.[14] Nature was a tinkerer, modifying and building on what already existed. Many different factors contributed to Wallace's seemingly new position that natural selection was inadequate to fully explain human evolution, but he grounded his argument based solely on the principle of utility. After reviewing what was known about the prehistoric races and the present-day "savage" races, Wallace concluded that a utilitarian analysis could not explain several traits. In particular, several aspects of humankind's intellectual and moral character seemed inexplicable in terms of natural selection. Natural selection did not anticipate future needs. Why would prehistoric man have needed such a large brain? In addition, three other physical features could not be explained by natural selection: the hand, the loss of hair, and the speech organs. Wallace maintained that none of these traits could be accounted for because they were present in the prehistoric and savage races in a much more highly developed form than was necessary for the struggle for existence. Even Darwin had realized that the loss of hair was probably harmful. Thus it was difficult to explain why such a trait should have been selected for. Wallace surprised everyone by arguing that all these traits *would* be useful—*in the future*—for *civilized* man. While it was true, he thought, that plants and other animal life had developed under strictly utilitarian evolutionary regimes, there was also some higher power that was guiding the action of the "great laws of organic development" in definite directions for special ends.[15] As I have previously written,

> Ironically, Wallace used the same analogy of domestic variation for which he had criticized Darwin in *The Origin* to suggest that a higher intelligence was guiding the process of evolution. Just as humans selected particular variations to produce desirable varieties of fruits, vegetables, and livestock, an "overruling intelligence" had watched over the action of laws of variation and selection, directing the variations, which resulted in "the indefinite advancement of our mental and moral nature." Darwin was horrified; marking his copy of Wallace's review with a triple underlined NO followed by numerous exclamation points.[16]

Critical to understanding Wallace's position is acknowledging that his extensive time spent living with native people resulted in his rejecting the racist thinking of the time, in spite of often referring to them as savages. He realized that at their core they were not different from Victorian gentlemen. They were just as intelligent, expressed the same range of emotions, and had the same moral capabilities. It is worth contrasting Wallace's view of native people with that of Darwin. As HMS *Beagle* cruised the shores of Tierra del Fuego, Darwin observed for the first time "wild men" in their native habitat, and he was truly shocked: "I shall never forget how savage & wild one group was . . . they were absolutely naked & with long streaming hair. Springing from the ground & waving their arms around their heads, they sent forth most hideous yells. Their appearance was so strange, that it was scarcely like that of earthly inhabitants."[17] Darwin could not believe how wide the difference was between the "savage" and civilized man. Did these naked barbarians, their bodies coated with paint, and their often-unintelligible gestures really belong to the same species as Europeans? Darwin thought that the difference was even larger than that between wild and domesticated animals, because humans were capable of greater improvement. The proof of that capability, in fact, had been right aboard the *Beagle*.

Three Fuegians, Jemmy Button, York Minster, and Fuegia Basket, had been living in London during 1830 and 1831, but were returned to their home in early 1833. Darwin had noticed time and again the many little character traits that showed how similar their minds were to Europeans'. In their short time in England they had become quite Anglicized in both manner and dress: Jemmy in his fine London clothes and Fuegia in her English bonnet. After less than fifteen months Jemmy now polished his shoes and made jokes. They were a stark contrast to members of their own tribe. When Darwin compared Jemmy to his cousins, he was amazed to think that Jemmy undoubtedly had behaved the same way as the "miserable savages whom we first met here." Darwin may have been shocked and repulsed, but clearly the Fuegians did not represent some separate species. The difference between Jemmy and his tribe members on the shore suggested to Darwin that if the gap could be bridged between the civilized and savage races—one greater than that between domestic and wild animals—this would provide the foundation for his assertion that no unbridgeable gap existed between animal and humankind. He eventually came to think the present-day primitive races provided a window into the past, exhibiting behavior that was undoubtedly quite similar to that of ancestral primitive races. This would suggest a chain of continuity from an apelike ancestor to primitive human ancestor to present-day humans. However, Wallace had an entirely different view of native people. The so-called higher faculties were already fully developed, in spite of the harsh environment in which these people lived. They did not require a trip to London to learn "civilized" behavior—and that was precisely the problem.

In many ways Wallace was even more of a selectionist than Darwin, as he did not think it was necessary for the natives to have the same traits and abilities that characterized the so-called civilized races. These traits bore no relation to the wants, needs, desires, or welfare of those people in the conditions they were living in, and this would have been equally true for the prehistoric races. Wallace claimed that survival in the conditions they lived in should only produce a brain "a little superior to that of an ape." Why then would they have evolved such capabilities? Wallace now claimed these capabilities existed because they were a provision for humankind's future. He used the same reasoning to explain the structure of the hand, claiming that the higher apes also had hands far more developed than they needed for survival. However, it would be necessary for humans to have such a perfect hand for later developments in the arts and sciences. The "savage" races had the same vocal capabilities as the civilized races, but Wallace claimed they wouldn't need them. However, critical to the advance of civilization was our power of speech. Again, the structures necessary for articulate speech were a provision for future use, not present-day use. But natural selection did not anticipate the future. Wallace thought a variety of other features of external form were also useless, including erect posture, delicate expressive features, beauty and symmetry of form, and smooth naked skin. Nakedness was even harmful to "primitive" and prehistoric races, and thus, once again, natural selection could not account for them. But Wallace suggested that the delicate expressive features and beauty of form were necessary for the refined emotions and aesthetic sense that developed in civilized cultures. Naked skin would stimulate humankind's inventive faculties to devise clothing, which would develop their feelings of modesty and contribute to their moral nature. All of these traits seemed to be preparations for the future.[18]

Wallace insisted this was not a negation of the principle of natural selection, but rather an extension of the general theory of evolution. Just as in artificial selection the process had been supplemented by the conscious selection of humans, the "laws of organic development have been occasionally used for a special end, just as man uses them for his special ends."[19] He continued to base his argument on the grounds of utility alone, maintaining that human beings had a variety of unique traits that simply would not have been useful in the very lowest state of civilization. These included mathematical ability, ability to form abstract ideas, ability to perform complex trains of reasoning, aesthetic qualities, and moral qualities. All "savages" possessed these latent capabilities, but their needs did not require these abilities. Only rarely were they used. They did not need to form or use abstract ideas, and indeed their language contained no words for them. They did not reason on any subject that did not immediately appeal to the senses or foresee beyond the simplest necessities. Wallace did not think that their abilities to design and use weapons were any more ingenious than the tactics of the jaguar while hunting. Other animals seemed to have as much forethought as "savages." Furthermore, "savages" certainly did not need to have the ability to form ideal conceptions of space and time, of eternity and infinity, to have capacity for intense artistic feelings of pleasure in form, color, or composition. The native did not make use of geometry or other abstract ideas concerning form and number. "Natural selection could only have endowed savage man with a brain a little superior to that of an ape, whereas he actually possesses one very little inferior to that of a philosopher."[20] Wallace even doubted whether civilized man made full use of these capabilities. He concluded that they were for the use of civilized man's future, not just for prehistoric man's future.

Although Wallace claimed that primitive humans did not need a big brain to survive, in 1870 he wrote that savages faced great mental challenges of immense complexity and difficulty in their daily life. Referring to Wallace's paper "On Instinct in Man and Animals," Huxley wrote that "the intellectual labour of a 'good hunter or warrior' considerably exceeds that of an ordinary Englishman."[21] Most present-day evolutionary biologists have come to the same conclusion.[22] Thus it is possible to make a utilitarian argument in favor of developing a large brain. The most difficult trait to account for by natural selection, and the one that was consistently cited as separating humans from the rest of the animal world, was conscience, the development of a moral sense. Even if selection were capable of producing the beginnings of human reason and the moral sentiment, intensified social and sympathetic feelings would prevent the beneficial culling of the mentally and morally inferior. Therefore, natural selection would become disengaged. Wallace had recognized this problem with his two-stage theory of human evolution. Although he argued that cooperation could be adaptive, the complex moral behavior of humans could not just be reduced to cooperation. This seemed to be totally inexplicable on the grounds of utility alone. Wallace claimed that while the practice of honesty, benevolence, or truth might be useful to a tribe, the moral sense included more than just that. Rather, it was the feeling of *sanctity* for such concepts as honesty that also was a critical component of morality. This was the essence of why Wallace thought natural selection was inadequate to account for the development of morality. As discussed more fully below, Darwin eventually gave an account of how a moral sense could have evolved that was rooted in helping one's own offspring survive, which then would be expanded to other members of the community. This behavior could be explained on the principle of utility. Even in the *Origin* he had suggested how altruism would be of benefit and would be selected for, in his discussion of social insects such as bees. But Wallace pointed out that tribes made a distinction between actions that were useful and ones that they considered moral. They were often quite different from one another. For example, Wallace did not see how the mystical devotion toward objects of moral worth could have been any aid to survival in the jungle. In fact, all through history untruthfulness had been allowable in love, even laudable in war. Most people didn't seem to value it at all when it came to trade or commerce. A certain amount of untruthfulness was necessary as politeness in both Eastern and Western cultures, and even extreme moralists believed a lie was justifiable to elude an enemy or prevent a crime. At the same time Wallace claimed that untruthfulness was associated with a "mystical sense of wrong" among many tribal peoples as well as the so-called higher classes of civilized people. Citing other observers, Wallace wrote that two "barbarous" hill tribes from Central India, the Kurubars and Santals, were known for their honesty. It was a common saying, "A Kurubar *always* speaks the truth." And "the Santals are the most truthful men I ever met with." He retold an incident in which some Santals had been taken as prisoners. In an uprising they escaped, but some of them returned to the prison and returned money to the guards that had been stolen.[23] Thus it was hard to support the idea of honesty from a totally utilitarian analysis. Since morality was an essential part of human nature, however, Wallace concluded that the feelings humans have about right and wrong came prior to any experience of utility. If that were the case, they could not be the result of natural selection. Rather, a superior intelligence had guided the development of humankind in a definite direction and for a special purpose. For Wallace, the law of natural selection was insufficient alone to produce the "ultimate aim and outcome of all organized existence—intellectual, ever-advancing, spiritual man."[24]

Regardless of what Wallace claimed was the explanation for how certain traits came about, this should not overshadow the significant issues he raised regarding the origin and evolution of humans—issues that are still being debated today. It is what Theodor Roszak called "Wallace's dilemma."[25] For Wallace adaptation by natural selection was essentially a conservative process, moving in a horizontal direction, altering and modifying organisms to their ever-changing local environment. One of the most heated scientific debates in the middle of the nineteenth century concerned the supposed progressive nature of the fossil record. Natural selection explained why species changed over time, but did not explain why the change would necessarily result in organisms of increasing complexity. As we now know, bacteria have been around since the earliest days of life, and likely will be around long after we humans are gone. Many exist in places once thought to be totally inhospitable to life: for example, under conditions of extreme cold or heat. Why would any form more complex arise if evolution were only about mere survival? If we adopt Dawkins's idea of the selfish gene and that evolution is merely about getting one's genes into the next generation, why would sex have evolved?[26] There are tremendous costs to sexual reproduction, not the least of which is that sex means only half of one's genes are carried into the next generation. In addition, one has to find a mate, attract the mate, and compete with rivals. Why grow big antlers or beautiful feathers or engage in brawls?—so much more energy is needed, and it is so much more complicated than just asexually dividing.

Although the idea of common descent had been relatively quickly accepted, throughout Wallace's time natural selection was still quite controversial. Furthermore, many people used the progressive nature of the fossil record as evidence for a revised version of the argument from design. But Wallace had rejected the creationist idea that organisms were adapted to their particular phylogenetic position. He also thought that many human traits were not the result of adaptation to environmental conditions. Nevertheless, he was convinced that organisms were evolving toward higher and higher states of perfection. Darwin may have been horrified that Wallace had indeed murdered their child, but the objections he raised meant Darwin could no longer avoid the topic of human evolution. Even before Wallace had published his views on the insufficiency of natural selection, such issues had been voiced by many others and were not trivial. In 1871 in *The Descent of Man*, Darwin presented a theory of how the moral sense could have evolved.[27] His theory of conscience was not just based on utility, but depended on his claim that a continuum existed between humans and animals. He even suggested the beginnings of religiosity could be found in animals by implying that the underlying emotions of religious belief were incipient in a dog. Were this the case, Victorians would be forced to acknowledge a chain of continuity progressing from animal to apelike ancestor to primitive human ancestor to modern-day "savage" to cultured Englishman. Darwin drew a parallel between the native who imagined that natural objects and agencies were animated by spiritual or living essences and the behavior of his dog—which barked and growled at a parasol blown by the wind. He suggested that the dog must have reasoned that its flight indicated the presence of an invisible agent.[28] Wallace did not find Darwin's explanation for the evolution of religious feelings convincing, in part because of his very different view of native people. Darwin's analysis was extremely racist, not just because of his explanation for what the dog might be imagining when he saw the parasol, but for the further comparisons he made between the "savage's" thought processes and that of a dog. Darwin admitted that the feeling of religious devotion of the savage, a combination of "love, complete submission to an exalted and mysterious superior, a strong sense of dependence, fear, reverence, gratitude, and hope for the future" could only be experienced by someone with a moderately high level of moral and intellectual development. Nevertheless, Darwin thought that it bore a strong resemblance to the simpler emotional response displayed by a dog's worshipful devotion to its master.[29] Darwin was also trying to distance the religious feelings of the "savage" from those of the civilized races in the hopes of preventing objections from pious Victorians as to the origin of humans and their religious beliefs. But regardless of his rhetorical strategy, this was certainly an offensive comparison and one that Wallace would have rejected. Compare this view of the natives to those of Wallace: "The more I see of uncivilized people, the better I think of human nature on the whole, and the essential differences between civilized and savage man seem to disappear."[30] (See figure 6.1.) For Wallace an unbridgeable gap existed between animals and humans when it came to humans' moral and spiritual nature.

Wallace's two-stage theory of human evolution left no doubt that it was the mind and moral faculties that turned the human animal into a human being. He was also quite optimistic about the future of the human race. He ended his 1864 paper with a utopian vision for the future of humankind. As technology continued to advance and spread throughout the world, the mental and moral qualities would continue to develop and eventually a single homogeneous race would exist in which

> perfect freedom of action will be maintained, since the well balanced moral faculties will never permit any one to transgress on the equal freedom of others; restrictive laws will not be wanted, for each man will be guided by the best of laws; a thorough appreciation of the rights, and a perfect sympathy with the feelings, of all about him. . . . [M]ankind will have at length discovered that it was only required of them to develope the capacities of their higher nature, in order to convert this earth, which had so long been the theatre of their unbridled passions, and the scene of unimaginable misery, into as bright a paradise as ever haunted the dreams of seer or poet.[31]

In such a world, slavery and other kinds of repression would be abolished. Eventually, humans would evolve into a new and distinct order of being. In this 1864 paper it is apparent that humans were certainly a part of nature

Figure 6.1 The seal of the Society for Effecting the Abolition of the Slave Trade, created in the late 1700s and in following years featured in a series of popular medallions and broadsides. One presumes Wallace would have approved of its message. Image used by permission of the American Philosophical Society.

subject to the evolutionary process; however, for Wallace the essence of man was his spirit. "If you leave out the spiritual nature of man you are not studying man at all." Such a view suggests that even that early in his intellectual development Wallace may have believed that natural selection provided an incomplete explanation for the presence of the higher intellectual and moral qualities in humankind.

More Than Spiritualism

Wallace's spiritualism certainly played a role in his coming to believe that a higher power was guiding and shaping evolution toward a particular end. His involvement with spiritualism also resulted in his losing credibility with many of his colleagues and undoubtedly contributed to a relative lack of attention by historians of science until fairly recently. But it is worth exploring why Wallace was open to the claims of spiritualism while Darwin, Huxley, and various other members of the Darwinian camp were not. That exploration also reveals many other significant strands that shaped his views on human evolution. Living among native people, particularly on the Amazon, he experienced their religious rituals that were permeated by the presence of spirits, which would certainly have made him more open to the spiritualist phenomena in Victorian drawing rooms. But long before he began his travels, other experiences had contributed to how he viewed human nature, and one of the most important was his interest in phrenology. Although today phrenology is regarded as a classic example of pseudoscience, its history is complicated, and demonstrates that the line between science and marginal science is often not sharp. The founder of phrenological doctrine, Franz Gall, argued that the brain was the organ of the mind and that the mind was localized for function. These two core ideas continue to guide our research in neurobiology today. What is the use of fMRIs to map the brain and its functions if not bringing the basic ideas of Gall into the twenty-first century? As a young man Wallace was exposed to phrenological doctrine by reading George Combe's *Constitution of Man*.[32] He had attended lectures on the topic, purchased a phrenological bust, and had his own head examined on two different occasions. In addition he also became quite interested in mesmerism, attending lectures and demonstrations (as noted in chapter 2), and even succeeded in putting some boys into a trancelike state. That the trance of a mesmerized subject was quite similar to the trance state of the medium might also have primed him to spiritualist phenomena. He became a strong advocate of phreno-mesmerism: "I . . . established to my own satisfaction, the fact that a real effect was produced on the actions and speech of a mesmeric patient by the operator touching the various parts of the head; that the effect corresponded with the natural expression of the emotion due to the phrenological organ situated in that part." He concluded that there "were mysteries connected with the human mind which modern science ignored because it could not explain."[33]

Early on, around 1845, Wallace also read the anonymously penned

Vestiges of the Natural History of Creation (1844). The author was Robert Chambers, a Scottish journalist, publisher, and thoroughgoing phrenologist. Phrenological doctrine permeates *Vestiges*, and the author explicitly acknowledged the importance of Gall in the development of his ideas.[34] Darwin panned the book to his colleagues, but much to his chagrin the author used many of the same kinds of evidence—the fossil record, embryology, classification, and comparative anatomy—as Darwin was to use in developing his own theory of evolution. Even though *Vestiges* is couched in the language of natural theology, the book yet presents a naturalistic and progressive view of the evolution of life. Wallace was quite impressed with *Vestiges*, particularly the sections that dealt with the mental life of animals and humans. Drawing directly from the ideas of Gall, the Vestigestarian argued that animals shared many of the same mental qualities as humans and that their brains were organized in a similar manner. He claimed that Gall had been the only philosopher who had developed a system of mind that was founded on nature. Gall had identified particular parts of the brain, referred to as faculties, that were associated with various capabilities such as imitation and wonder. Chambers, like Gall, believed that these capabilities were far more developed in humans. Nevertheless, he argued that they existed to some degree in the lower animals. In fact there were striking similarities between passages in *Vestiges* and Darwin's *Descent of Man*. Darwin's debt to phrenology was actually enormous.[35] Gall had been accused of being a materialist because he claimed the brain was the organ of the mind. Many people interpreted this to mean that mind was merely an epiphenomenon of the brain. However, Wallace did not agree. He never claimed that brain and mind were identical. In addition, it was precisely those faculties that phrenology maintained were unique to humans, such as benevolence, veneration, conscientiousness, hope, and wonder, that Wallace claimed became highly developed once the evolution of the brain shielded the human body from the action of natural selection. This also suggests that Wallace even in 1864 might have thought that natural selection would not have been powerful enough for those traits to have evolved to their present state.

Wallace's extensive travels also shaped his views on human evolution in ways that are not entirely obvious. His many years wandering in solitude in the jungle meant he did not share in the institutions of middle-class social and intellectual life. But Wallace had never aspired to a middle-class existence. His working-class background, his exposure to the socialist ideals of Robert Owen, and his disenchantment with traditional Christian theology had already made him somewhat critical of life in Britain. His disillusionment only became more pronounced as the years went on. While staying in a small village in South America, he wrote a poem about the natives:

> They are a peaceful race; few serious crimes
> Are known among them; they nor rob nor murder,
> And all the complicated villanies
> Of man called civilized are here unknown.[36]

He became convinced that many unsavory behaviors such as lying and stealing that were attributed to the natives in the Malay Archipelago were actually the results of these people coming into contact with the Europeans:

> Each man scrupulously respects the rights of his fellow, and any infraction of those rights rarely or never takes place. . . . There are none of those wide distinctions, of education and ignorance, wealth and poverty, master and servant; . . . there is not that severe competition and struggle for existence, or for wealth, which the dense population of civilized countries inevitably creates. All incitements to great crimes are thus wanting, and petty ones are repressed, partly by the influence of public opinion, but chiefly by that natural sense of justice and of his neighbour's right which seems to be, in some degree, inherent in every race of man.[37]

Wallace was appalled at the behaviors of the "civilized" races who, driven by greed, were creating a society where corruption and dishonesty were rampant. He could not abide the "survival of the fittest" mentality that dominated life in Britain. For the supposedly better-off, but poor, man in England, life was often worse than for the Malays or Burmese before they had any modern inventions. He became quite pessimistic about the civilized nations and believed that it was the mediocre in regard to both morality and intelligence who were succeeding best in life and multiplying. Wallace also thought that many natives clearly had superior mental and moral qualities. He continually complained about his young assistant Charles Allen, who after twelve months of constant instruction and practice showed no sign of improvement. "I believe he never will improve. Day after day I have to look over everything he does and tell him of the same faults. . . . He never, too, by any chance, puts anything away after him. When done with, everything is thrown on the floor."[38] He finally let Allen go and had no desire to have another assistant that came from Britain. In 1855 he hired a young Malay named Ali who became his most able and trusted servant and was absolutely critical to his success. Though he was initially hired as a handyman and cook, it soon became apparent that Ali was capable of much more. Wallace described him as his most faithful companion in virtually all of his journeys. Together they walked, netted insects, shot birds, and shared their evening meals. Ali helped teach Wallace native languages. He also trained and oversaw many other natives to help with the work. Wallace credited Ali with saving his life, nursing him when he lay delirious with fever on Ternate. As a result of living among these local tribes for years Wallace concluded that colonial rule of them by the Europeans was "an absurdity."[39] His experience with them also meant that in spite of his pessimism about the so-called civilized races, he remained convinced that the human species continued its moral and intellectual advancement.

Conclusion

Near the end of his life Wallace listed in his autobiography what he considered his ten most important ideas. These included the creation of the science of island biogeography, his identification of the causes of glacial epochs and accompanying geographical changes, and his arguing for the permanence of continents and deep seas in contrast to commonly held theories of "vanished continents" such as Atlantis. Natural selection made the list, but Wallace did not necessarily consider it his most important achievement. Since the discovery of it came out of his interest in understanding the distribution of organisms, contrasted with Darwin's desire to find a mechanism that explained how species change *adaptively*, it is important to acknowledge the many other interests that contributed to Wallace's view of human evolution. Wallace firmly believed that humans' place in nature was "strictly due to the action of natural law," but he also wanted to develop an evolutionary model that was much broader than just explaining change within the physical/biological realm. He wanted the boundaries of science to be extended to include phenomena that could not be explained in strictly materialistic terms. He regretted that he had titled an essay "The Scientific Aspect of the Supernatural." As he wrote, "Supernatural" was a misleading term because "all the phenomena, however extraordinary, [are] really 'natural' and involving no alteration whatsoever in the ordinary laws of nature."[40]

Wallace's pointing out what he regarded as the limitations of natural selection for humans highlights a tension that permeates evolutionary theorizing. Many ardent selectionists continue to argue that every trait is a product of natural selection, but often their explanations, particularly when it comes to human evolution, seem little more than wishful adaptive storytelling. Others such as Stephen Gould and Richard Lewontin argue that traits that have been evolved for one purpose often get repurposed for something else.[41] We have numerous examples that this indeed has been the case, such as feathers initially evolved for thermoregulation that eventually were used to support flight, and bones from the jaws of reptiles that eventually evolved into the bones of the middle ear. We now know that an incredible amount of diversity and developmental plasticity is possible by making use of relatively few genes that have been conserved through almost the entire history of life.[42] We also know that once our ancestors had evolved to a certain point, culture took on a life of its own, and many of our current behaviors are neither the product of natural selection nor necessarily adaptive. We all need to eat to survive, but that does not explain the variety of taboos surrounding food or the development of Michelin-star restaurants. Sex is necessary for reproduction, but does natural selection explain the Kama Sutra? Our large brain is ultimately what makes possible all the rich cultural diversity that exists in the world today. Our brain is not a cultural artifact, however; it is a biological organ, and it is the ultimate adaptation that makes us human. Yet in spite of the vast increase in our understanding of evolution since Wallace's time, we still do not have fully adequate explanations for the origin of consciousness and all the other abilities that consciousness has made possible: language, art, music, abstract thinking, mathematical reasoning, and a highly developed moral sense. Evolutionists and cultural anthropologists claim that music and art served to bind communities together, but that still does not explain how or why those capabilities developed initially.

What is the future of our species? Wallace postulated that humans would eventually evolve into pure spiritual beings. For many readers this may seem patently ridiculous. But at the other end of the spectrum, some futurists now claim that *Homo sapiens* will evolve into cyborgs.[43] In spite of the fact that many people seem glued to their cell phones, and the advent of the Apple watch suggests that we are already merging with our technological gadgets, to this writer such a future seems equally unlikely. Rather, more in line with Wallace's view expressed in his 1864 paper, if our species is to survive, the key will be in our developing, for lack of a better phrase, a "higher consciousness." If we don't expand our empathy and concern not only for our fellow human beings but for all life on Earth, we will be doomed to extinction. "Nature Red in Tooth and Claw" has dominated evolutionary theorizing ever since Darwin. This legacy, however, owes more to Herbert Spencer, who described natural selection as "survival of the fittest," and the prevailing laissez-faire capitalism that characterized the thinking of the time. But Darwin always acknowledged the importance of cooperation and made a compelling case for the evolution of a moral sense in *Descent of Man*. Today, research in primatology and neurobiology provides evidence that empathy and kindness are an essential part of our nature. The work of Franz de Waal has shown that chimpanzees exhibit behaviors that can provide a basis for building a moral code and that monkeys have a sense of fairness.[44] Antonio Damasio argues that emotion plays a critical role in rational decision making.[45] V. S. Ramachandran claimed that mirror neurons make imitative learning possible and were the driving force behind what he called "The Great Leap Forward" in human evolution. He suggested that a sophisticated mirror neuron system made possible the emergence of protolanguage that was facilitated by mapping phonemes onto tongue movements, which then made possible a "theory of other minds" and the ability to "adopt another's point of view."[46] He dubbed mirror neurons Dalai Lama or empathy neurons. And even the Dalai Lama himself has gotten into the act. He and other Buddhist monks have been in a dialogue with neurobiologists for about thirty years, mainly under the auspices of the Mind and Life Institute.[47] The Dalai Lama believes that by understanding the true nature of the human psyche it will be possible to find ways of transforming our thoughts, emotions, and their underlying properties. Furthermore, he maintains that at our most fundamental level, human beings are compassionate and that cooperation rather than conflict is at the heart of our basic existence. With the discovery of neuroplasticity, we know it is possible to train the mind to be happy, and that empathy and compassion are the keys to happiness.[48] Wallace would have been pleased that modern research gives some hope that his utopian vision for the future of humankind might be realized:

> no individual . . . will be inferior to the noblest specimens of existing humanity. Each will then work out his own happiness in relation to that of his fellows; . . . the well balanced moral faculties will never permit any one to transgress on the equal freedom of others; restrictive laws will not be wanted, for each man will be guided by the best of laws; a thorough appreciation of the rights and a perfect sympathy with the feelings, of all about him. . . . [M]ankind will have at length discovered that it was only required of them to develop the capacities of their higher nature;

in order to convert this earth, which had so long been the theater of their unbridled passions, and the scene of unimaginable misery, into as bright a paradise as ever haunted the dreams of seer or poet.[49]

Acknowledgments

My thanks to Martin Fichman and Charles Smith, whose writings on Wallace and many conversations have helped enormously in my efforts to understand him.

Notes

1. Space constraints here prevent us from outlining all the anthropology-related matters that Wallace considered. For Wallace's support of the "mouth-gesture" theory of the origin of language, see Wallace (1895b) and Paget (1951); for his studies on the orangutan, see Wallace (1856a, 1856b), van Wyhe and Kjærgaard (2015) and Camerini (1996); for how technology is about "work," not "things" (the "Wallace insight"), see Drucker (2010); on his ethnological classifications, see Wallace (1867d, 1879a, 1879e), Ballard (2008), Goh (2007), and Vetter (1999); on his attention to premodern humans and archeology, see Wallace (1887c, 1887e, 1891b); on the "degeneration hypothesis," see Wallace (1874a, 1876b) and Kottler (1974).

2. Wallace (1858).

3. Darwin letter to Lyell, 18 June 1858, in Darwin (1887), vol. 1, p. 473.

4. Darwin letter to Wallace, 22 December 1857, in Darwin (1887), vol. 1, p. 467.

5. Earlier prevailing theories recognized a reversal of opinion based either on Wallace's adoption of spiritualism (e.g., Kottler 1974, Stepan 1982), or his inability to reconcile natural selection with his views on social/moral evolution (e.g., R. Smith 1972, Schwartz 1984, Oppenheim 1985), or both (Malinchak 1987, Benton 2008, 2013).

6. See Smith (2004a) and chapter 2 in this volume, and Fichman (2004).

7. Tyson (1699).

8. Buffon, quoted in Greene (1959), p. 182.

9. Lord Monboddo, quoted in Cloyd (1972), p. 44.

10. Wallace (1864b).

11. Lyons (2009), pp. 122–123.

12. Wallace (1869d), p. 391.

13. Darwin letter to Wallace, 27 March 1869, quoted in Marchant (1916a), p. 197.

14. Darwin (1859/1976), pp. 227–229.

15. Wallace (1869d), p. 393. See Kottler (1974), p. 152.

16. Lyons (2009), p. 126.

17. Darwin (1839/2001), p. 194.

18. Wallace develops these ideas most fully in his essay "The Limits of Natural Selection as Applied to Man" (1870c).

19. Wallace (1875b), p. 370.

20. Ibid., p. 356.

21. Huxley (1871/1893), p. 176. See Wallace (1875b), pp. 207–208.

22. See Diamond (2005), p. 20. The New Guineans impressed Diamond as being more intelligent and alert than the average European or American.

23. Wallace (1875b), pp. 353–354.

24. Ibid., p. 360.

25. Roszak (2004).

26. Dawkins (1976).

27. Darwin (1871/1981).

28. *Ibid.*, p. 67.

29. Ibid.

30. Wallace (1905b), vol. 1, pp. 342–343.

31. Wallace (1864b), pp. clxix–clxx.

32. Combe (1835).

33. Wallace (1905b), vol. 1, pp. 235–236.

34. [Chambers] (1844/1969), p. 341.

35. For a further explication of this idea, see Lyons (2009), pp. 61–71.

36. Wallace (1853/1870), p. 260.

37. Wallace (1869/1962), p. 456.

38. Wallace letter to Frances (Fanny) Sims, 25 June 1855, quoted in van Wyhe (2013), p. 127.

39. Marchant (1916a), p. 389.

40. Wallace (1905b), vol. 2, p. 280. See also chapter 2 in this volume.

41. Gould and Lewontin (1978).

42. Carroll (2005).

43. See Harari (2015), Jastrow (1978), and Kurzweil (1999).

44. See de Waal (1996, 2009).

45. Damasio (1994).

46. Ramachandran (2000).

47. Mind and Life Institute, https://www.mindandlife.org.

48. See Harrington and Zajonc (2006), Begley (2007), Davidson and Harrington (2002), and Lyons (2007).

49. Wallace (1864b), pp. clxix–clxx.

‹ 7 ›

Wallace as Social Critic, Sociologist, and Societal "Prophet"

MARTIN FICHMAN

In October 1910 Wallace provided a list of six key areas of Victorian society in which change (or a lack thereof) in the previous half century might serve as barometers of positive social evolution: (1) land reform, (2) socialism and poverty, (3) religious toleration, (4) educational reform, (5) prison reform, and (6) militarism and the "lower races." This list affords a snapshot of the sociopolitical issues to which he devoted much of his intellectual and practical energy during the last decades of his life. Wallace felt that Great Britain had witnessed "admirable" progress in the government's land reform measures and moves toward a more equitable state of land ownership and taxation. He viewed developments tending toward an ultimately more just social system and an abolition of poverty to have been equally significant, though (of course) incomplete. Concerning religious toleration, Wallace noted that "[a]gnosticism, and even atheism, no longer render their advocates liable to fine or imprisonment, or . . . even to any disqualification for high public services." Yet, for Wallace, "a vast mass of religious intolerance still exists among us."[1]

Areas that Wallace regarded as still being in a state of "absolute failure" in Britain were education and the prison system. Most critically, Wallace perceived developments in "the vast subjects of militarism and the rights of the lower races" to have been retrograde. He cited the enormous buildup of armaments globally and "the parallel increase of capitalism and the greed for wealth and power" to pose even more significant challenges for "real humanitarian progress" in 1910 than they did in 1860. Despite this rather gloomy balance sheet, Wallace felt that the "great advances" in land reform, social justice, and reduction in the general level of poverty "indicate the opening of a new era in which the better instincts of the people will have opportunities for development." However rosy and vague this prognosis might seem, Wallace never deviated from his conviction that "[e]quality of opportunity from birth to manhood and womanhood for every British subject" was the most appropriate "touchstone" whereby all proposals for social, political, or economic legislation must be tested and measured.[2]

This chapter highlights Wallace's major efforts during roughly the last half century of his life to advance the cause of social justice and greater political and economic equity. It is divided into three sections: (1) Wallace's views on social justice, "social economy," and social evolution; (2) Wallace on eugenics; and (3) Wallace and the antivaccination movement in late Victorian Britain. A brief assessment of Wallace's success as a social prophet will form the chapter's conclusion.

Social Evolution

Evolution and Sociopolitics

Social, political, and economic implications of evolutionary thought on the pressing questions of Victorian industrial society came to occupy a greater portion of Wallace's life and work from the 1860s onward. His impassioned stands on land nationalization, socialism, vaccination, eugenics, and other incendiary topics expanded his reputation as a diverse and innovative thinker. After the presentation of Wallace's and Darwin's famed theory of evolution by natural selection on 1 July 1858, Wallace quickly distinguished himself as a prolific author of books and articles on that subject as well as on biogeography, geology, and climate (especially in the new field of glacial phenomena). Later he became an active and influential social theorist and reformer in a turbulent period in British history.

The relationship between evolutionary theory and social and political ideology is one of the most intensively studied aspects of the cultural context of Victorian science.[3] Thus Wallace's attempt to integrate his biological discoveries with concerns for societal reform is hardly remarkable. What is notable is the degree to which Wallace made this integration a public initiative. Unlike most of his scientific colleagues, who either confided their political opinions to their immediate circle of family and friends or to their diaries and personal journals, Wallace's political and social views were a matter of public record. His outspokenness on key issues such as capitalism, poverty, and the "woman question" made him a fixture in the Victorian cultural wars. Analysis of the fundamental concepts that underlay Wallace's political vision makes it clear why he could not remain silent on the burning controversies of his day: he believed it was the *duty*, not merely a peripheral activity, of the scientist to "go public" on questions of poverty, land nationalization, gender roles, and kindred explosive subjects. His passion to use whatever tools lay at hand—science, education, lecturing, spiritualism—to effect radical and lasting societal change was as much a part of his core vision as were his detailed scientific observations.

Wallace's perspective on the public role of the intellectual strongly parallels that of the influential nineteenth-century English philosopher and social analyst Herbert Spencer. Although the two disagreed on many points of theoretical and practical detail, they were united by a conviction that it was the duty of the scientist to engage directly in social and political controversies. For both Spencer and Wallace, ethics constituted the starting point of any legitimate sociopolitical philosophy.[4] Wallace was wholly sympathetic to Spencer's goal of establishing a new ethics for technological society. Unlike Spencer, however, Wallace insisted that any such new ethical system not be restricted by a framework of narrow scientific naturalism.[5]

Wallace had earlier established to his satisfaction that human culture was not predicated, except in certain obvious aspects, on biologically determined behavior. He thus freed himself from the constraints of a strictly naturalistic explanation for the history of cultural developments from humanity's origins through to the Victorian period. At the same time, however, he believed that his and Darwin's theory of evolution by natural selection *did* provide an important framework for understanding human evolution, especially in its early stages. Wallace could now articulate an emergentist conception of both individual and societal evolution. In particular, he accepted the possibility that human evolution might involve phenomena that were not completely reducible to, or explicable by, mechanistic materialism. Wallace deployed natural selection as a powerful explanatory tool for examining and, hopefully, changing human activity.[6] Wallace's consistency of vision is evident from the publication in 1864 of "The Origin of Human Races and the Antiquity of Man Deduced from the Theory of 'Natural Selection'" to the publication in 1913 of his two last books, *Social Environment and Moral Progress* and *The Revolt of Democracy*.[7]

Social Economy and Social Justice

Wallace's critique of late-nineteenth-century political economy was fundamental to his sociological views. As he declared in 1894, "The science of Political Economy has now guided, and often governed, the civilised world for near a century, but it may be doubted if the world is much the better for its guidance."[8] Wallace could hardly deny that owing to certain dictates of the political economy of Victorian capitalism, Great Britain, the nations of Western Europe generally, and North America were enormously richer than they were in the eighteenth century. He accepted that this increase in national wealth was due, in large measure, to a growing application of the physical sciences and technology to economic development, notably to the production and use of steam power and, more recently, electricity. But Wallace also complained that despite the spread of steam power and labor-saving machinery—the backbone of the Industrial Revolution in the West—"the contrasts of riches and poverty, the gulf between rich and poor, is greater—far greater—than ever before in the world's history."[9]

Wallace's emphasis on the striking inequity in income and wealth that characterized most advanced industrial societies then is perhaps even more apt in our own age of tech billionaires and inflated CEO salaries of the major corporations (the phenomenon of the so-called one percent that typifies many early twenty-first-century economies). Although living standards have improved for most individuals, at least in the developed and developing nations, the "inequality gap" has, arguably, in many respects grown larger in those populations. Wallace's critique thus retains its full force. His solution to this dilemma was to enunciate a vision of "a new science of Social Economy, in the place of the old and altogether insufficient science of Political Economy."[10]

Wallace's central objection to industrial capitalism as an engine of wealth creation was that "it takes no heed who has the wealth."[11] In his view, much of the wealth created by the labor of the majority seemed to accumulate for the benefit of a comparative few in the very top fraction of society. For Wallace, political economy ignored altogether "the social or moral results of this wealth accumulation." Its very success as an economic paradigm, its intense focus on fiscal characteristics, seemed to preclude any attention to questions of right or wrong in social and political arrangements. Wallace sought to reintegrate what he saw as the basic concept of social *justice* into industrial society. To do otherwise, he believed, would be to submit to the dictates of a powerful industrial capitalism allied to unregulated technoscientific advance. This failure to promote social justice, Wallace predicted (in *The Wonderful Century*), would permit "the struggle for existence [to] become more fierce than ever before; and [allow] year by year an ever-increasing proportion of our people [to] sink into paupers' graves."[12] Even correcting for Wallace's rhetorical flourishes, his opinion here remains cogent.

Although a proponent of a broad and nondoctrinaire socialism as the path to social justice, Wallace felt that continental varieties of socialism could not be a guide for Great Britain in the late Victorian period.[13] Instead, he proposed an alternative sociopolitical system he termed Social Economy. He viewed it as a transitional national model that, "while securing many of the beneficial results of Socialism, will preserve all the advantages of individual self-dependence and healthy rivalry, and will so educate and develop social feelings, that if any advance in the direction of Socialism is then desired, it will no longer be impracticable." Wallace left for later the details of this future system of "organized cooperation." But his various articles on contemporary social issues and several books (notably *The Wonderful Century* [1898], *Studies Scientific and Social* [1900], and *Social Environment and Moral Progress* [1913]) suggest, in one way or another, that a "true MORAL and SOCIAL ECONOMY [would] have *human labour* saved, wasted *time* utilised, *health* improved, and *happiness* increased."[14] Wallace placed his hopes for a prosperous social order—one in which monotony, disease, and discontent would be the exception rather than the rule—on a population largely composed of "the independent self-employing worker." He knew that this was a tall order given the entrenched power of industrial capitalism, but this goal remained firm over the last decades of his long life.

Wallace read widely in the political, economic, and sociological (though not yet called by that name) literature of his day. He was particularly influenced by the work of figures such as Herbert Spencer, John Stuart Mill, Henry George, Benjamin Kidd, Edward Bellamy, and the Hungarian economist and journalist Theodor Hertzka. Hertzka's 1890 work *Freeland, a Social Anticipation* depicts a society based on free use of the nation's land and (approximate) equality in the distribution of the national wealth. Wallace felt that *Freeland*, like his favorite utopian work, Bellamy's *Looking Backward*, employed nothing "which is not true to nature, and quite possible, even if rather improbable."[15] Wallace also had the highest admiration for the Russian novelist and reformer Leo Tolstoy, whose spiritual and reformist ideas Wallace found congenial. He declared that Tolstoy was a "great thinker and moralist who is doing more than any other living writer to expose the evils of all government by force." Wallace said he owed his espousal of a "voluntary socialism"—and his opposition to any attempt to establish a "compulsory socialism"—"mainly to Tolstoy."[16] One sentence from Tolstoy's *Anna Karenina* evokes the same antipathy to the reigning political science as Wallace's own: "[Levin's intended theoretical treatise on the Russian economy and people] would not merely revolutionize political economy, but completely destroy it as a discipline, and lay the foundation of new discipline, dealing with the relation of people to the land."[17] Wallace's own extensive periodical writings on these subjects provide intriguing insights into his conception of a more just society. His concept of social justice was linked inextricably to his efforts to refine an understanding of social economy.

Wallace's ideas on social economy first appeared in his writings on nationalization of the land in the early 1880s and became more overt in the last two decades of his life.[18] Many of the ideas that are described in these later writings, however, made their first appearance in a lengthy speech Wallace delivered at a conference organized by a wealthy businessman in 1885 in Prince's Hall, Piccadilly. Entitled "How to Cause Wealth to Be More Equally Distributed," Wallace's talk provided an outline of the principles of a *social economy* that would establish the equity he believed was ignored or destroyed by the nineteenth-century devotion to a solely *political economy*. (Later that year, Wallace expanded on this distinction in the book *Bad Times*.) Wallace noted that in the thirty-year period between 1849 and 1878, the productive wealth of England had doubled while the population had increased by only about 40 percent. He argued that an immense problem had yet to be confronted, namely, to "determine what practicable and just changes in our social economy will tend, naturally, to bring about a more equable division of wealth between capitalists and labourers, or perhaps more accurately, between the actual producers of wealth and the rest of the community." One egregious example Wallace highlighted was that the vast majority of semiskilled or unskilled laborers, even when fully employed, barely earned sufficient to provide themselves a decent life. Using the language of the political economists, Wallace pointedly remarked that wages ever tend to the minimum necessary to support bare existence; "hence the poverty and pauperism of labourers." Wallace invoked Henry George to declare that if the "great political wrong" clearly and specifically supporting the current social inequities was unmasked and fearlessly put right, then no longer would the "actual producers of wealth in the wealthiest country in the world . . . continue to live without enjoying a fair and adequate share of the wealth which they create."[19]

For Wallace, the major impediment to ameliorating the lives of workers is a misapplied Mathusianism. This argues that any permanent rise in wages will only be nullified by population growth that is the inevitable result of the poor multiplying more rapidly *because* they are better off. Wallace, however, believed this prediction is both superficial and unfounded "since the universal testimony of all inquirers is, that by improving the status of the labourer, and raising his standard of comfort, you increase his morality and delay the period of marriage." Wallace argued, accordingly, that the rate of population growth would be actually reduced in a society where wealth is more equitably distributed (in line with what later came to be known as the "demographic transition"). Wallace reminded his audience that the persistence of widespread poverty in the midst of increasing societal wealth and productivity was at root *both* a political and social injustice. He declared that this appalling inequity so "permeates our whole civilisation, and so contaminates it that all the powers of nature which during the past century science has enlisted in our service" have only served to increase, rather than diminish, human suffering.[20]

Wallace decided that a political act was necessary as a first step toward solving Britain's grossly unequal distribution of social wealth. He had in mind an act of Parliament that would enable those large numbers of urban workers unable to secure a living wage to return to homesteads in their native rural districts. With protected land tenure, laborers could produce their own food rather than depending on daily industrial wages that often pushed them into urban poverty. Moreover, such an act would necessarily raise the standard of wages for the remaining urban workers (as well as lower their rents) by sharply reducing the competition for factory jobs. Wallace believed that if government could partially reverse the trend of Britain's population migration from rural to urban centers, a more salubrious and economically efficacious distribution of agriculture and industry would result. As an additional benefit, such an act would "increase the food-supply of the country, especially in dairy-produce, poultry, fruit, and vegetables," and thus reduce Britain's increasing reliance on food imported from abroad. This theme of reversing the trend of rural depopulation was an important one for Wallace, and figured, to some degree, in many of his writings on land nationalization.[21]

Still, it would be a mistake to think of Wallace as a naïve "rural nostalgic." He was always alert to the specific advantages of town and city living, such as concerts, theaters, clubs, and well-stocked libraries, and wanted these in *any* envisioned community. As David Stack points out, rather than advocating a mere return to the land, Wallace wanted to "foster a thoroughly 'modern' ecological connection between man, nature, and the land." Stack further suggests that if Wallace's praise of the virtues of self-sufficient agricultural towns is "rephrased" in terms of "locally-sourced produce," "fair trade," and "anti-globalization," his alternatives to the large urban agglomerations of contemporary societies would appear decidedly apt—if not crucial—for city dwellers of the early twenty-first century.[22]

Significantly, Wallace felt that spiritualism, like socialism, could be an important adjunct to working toward a social economy. In an address to the International Congress of Spiritualists in June 1898, he integrated spiritualism into the broader context of his social and political worldview. Having openly declared himself a socialist in 1890,[23] Wallace now wanted to show the public that the ethical teachings of spiritualism were fully compatible with his socialist political philosophy, as well as other nonsocialist causes such as land nationalization. He argued that one of the most potent teachings of spiritualism was that all men and women should strive to work for improved social conditions for the mass of humanity in this life, not only for the benefit of the afterlife. Wallace fully endorsed spiritualism's emphasis that the "higher law" demanded all people be guaranteed equality of opportunity and social justice in their lives. Thus, spiritualists' commitment to social reform and social duty was an excellent example for truly civilized social and political action—even if one did not accept all of the aspects of full-blown spiritualism.[24] As he told his fellow spiritualists two years later, "If you will continually keep *this duty* before you, asking yourselves *how* you can best further this great Cause, your spirit-guides will, I feel sure, impress you how you should act so that the New Century may witness the birth, and perhaps even the maturity, of a truly moral and spiritual civilisation."[25]

Social Evolution

The precise nature of the interaction between spiritualism and the theory of natural selection in Wallace's views on human evolution has been the subject of (often heated) scholarly dispute. Basically, Wallace thought that man's moral and ethical attributes had stagnated—or improved only marginally on occasion—since the time when those qualities had first appeared. He believed that this "higher nature of man arose at some far distant epoch, and though it has developed in various directions, does not yet seem to have elevated the whole race much above its earliest condition, at that time when, by the influx of some portion of the spirit of the Deity, man became a 'living soul.'"[26] Any discussion of Wallace's basic approach to social evolution must, therefore, take these features of his evolutionary theory into account. We now turn to the development of Wallace's later ideas concerning social evolution; his understanding of the respective roles played by natural selection and sexual selection in the specific case of human evolution, touched upon here, will be more fully discussed in the section on eugenics.

Wallace never set down a detailed concept of social evolution in any one text. In a review of Benjamin Kidd's widely read and controversial *Social Evolution* (1894), however, Wallace provided insight into his own mature ideas concerning that complex topic. He regarded Kidd as having provided a novel theory of social progress that was "thoroughly scientific in its methods," based as it was on the most recent findings in evolution by natural selection. Equally crucial for Wallace was Kidd's demonstration of the central role in social evolution played by advancing democracy. Although Kidd was ambivalent as to social evolution's future, Wallace felt that his theory supported the "avowedly socialistic legislation for which the times appear to be ripening."[27]

But the most original and "ingenious" feature—indeed the "keynote"—of Kidd's work, for Wallace, was the declaration that "religion is not, as the scientific urge, a mere system of superstition and error, a clog on the wheels of progress, or the enemy of science and enlightenment." On the contrary, both thinkers were in agreement that religion had been one of the most important agencies in social development. Wallace endorsed Kidd's view that the ethical teachings of religion were closely bound up with "that portion of our nature to which all recent social advance is due, and which will inevitably decide the course of our future progress." Wallace was quick to point out that Kidd's concept of religion was (as was Wallace's own) totally removed from any "dogmatic religion." Instead, both men's definitions include only "those great ethical principles which have always formed part of religious teaching, and whose influence is in great part due to it."[28] Wallace was also pleased with Kidd's emphasis that no progress in human history has been achieved without "some form of selection." Kidd's book was ambiguous as to the precise mode of selection that brought about this progress. Wallace, in contrast, espoused a form of sexual selection—specifically, female choice—among humans as instrumental in whatever progress had been, and might in the future be, achieved. It will be recalled that Wallace's invocation of sexual selection (in this specific evolutionary domain only) was intended, among other purposes, to obviate any argument against socialism on the grounds that it removes the individual struggle for existence.[29]

Wallace embraced Kidd's central thesis that societal evolution is due primarily to the continuous action of religious beliefs in history. In particular, Wallace thought that Kidd's book was a clear, if rather prolix, demonstration that societies and civilizations have prevailed in the struggle for existence in proportion as they have been efficiently organized—and that this "organisation has always rested on some form of religious sanction." In the early development of European civilization, for example, the doctrines of the divine right of kings and of subjection to popes and bishops had served to weld together tribes and peoples, checking the supremacy of brute force. As Wallace interpreted Kidd, such doctrines were efficient in subordinating "the many to the few, which was essential to the production and accumulation of wealth, to the growth of the arts, and to the firm establishment of that national unity which is the most important factor in the growth of civilisation." Although this was the primary role of religious belief in the early development of European civilization, Wallace endorsed Kidd's view that religious influence during the last few centuries operated differently, but equally importantly. As nations became more advanced in education and the arts, and a considerable middle class arose whose interests were opposed to those of the warrior caste and to constant war and bloodshed, the ethical side of all religious teaching began to have more influence. More specifically, ideas of justice and mercy, and of the inherent rights of man independent of class or caste, acquired for the first time some real effect throughout all ranks of society. Wallace read Kidd as seeing this process of social evolution as culminating during the nineteenth century in the abolition of slavery and of many class and religious privileges, in the spread of general education, and in the grant of almost universal (male) suffrage. As Kidd expressed it, social evolution is primarily "the story of the political and social enfranchisement of the masses of the people hitherto universally excluded from participation in the rivalry of existence on terms of equality." Most importantly, Wallace felt Kidd argued convincingly that recent social evolution was due to an ever-growing fund of altruism, a development of humanitarian feelings, and a deepening sense of justice—in short, the ethical precepts of modern religion.

Wallace perceived one crucial problem with Kidd's overall theory. It is worth quoting Wallace's reservation in full here since his clarification reflects a critical element in the development of his own theory of evolution by natural selection:

> [Kidd's] fundamental doctrine is that all human progress is due to selection in the struggle for existence, whether that struggle acts most severely upon individuals or upon communities. But it is not shown *how* the rude struggles of the two thousand years terminating in the sixteenth century could have had any tendency to increase and develop these altruistic and ethical sentiments. . . . The natural possessors of such sentiments were usually buried in religious houses, and, as a rule, left no descendants. All selection seems rather to have tended to the extermination of the possessors of humane and altruistic sentiments, not to their continuous preservation and increase. Yet nothing is more certain than that they *do* now prevail to an extent never before known, and if they have not been developed by selection they must have been inherent in the race, developed perhaps at some earlier period, and have lain dormant till a more peaceful and more intellectual epoch called for their manifestation.[30]

This is, of course, Wallace's striking emendation of natural selection theory, namely, the role of "surplusage of power." What he means here is, more simply, the notion that certain traits appeared at a time when they were beyond the actual needs of humankind at that stage in their evolution *but* would become useful at a later stage. Wallace gives as an example what he regarded as the large brain of prehistoric humans and savages.[31] Wallace's use of concepts such as planning, prevision, and higher intelligence in his evolutionary theory brings up the question of teleology—or purpose—in the philosophy of biology.[32] This question is beyond the scope of the present discussion, but suffice it to say that there has been extensive controversy on teleology and evolutionary biology from Wallace's, and Darwin's, day to the present. As biological knowledge increases both in scope and complexity, Wallace's teleological views maintain their distinctive relevance to fundamental issues in contemporary evolutionary thought.[33]

This positive critique of Kidd's book foreshadowed Wallace's own model of (human) social evolution: (1) sexual selection by female choice, (2) a gradual replacement of the existing competitive and antagonistic social system by one based on "co-ordination for the equal good of all,"[34] (3) movement toward a society based on *equality of opportunity*, both politically and socially, and (4) the triumph of an ethical movement in which "the very highest" of human qualities and attributes find their truest expression.[35] Wallace's adoption of sexual selection as an essential complement to natural selection among humans will be discussed more fully in the section on eugenics. His belief in the principle of political and social equality of opportunity for all members of society was tied fundamentally to his insistence that what he termed true individualism was the essential preliminary of real societal advance. This in turn was a basis for Wallace's hope that a social order would emerge whose members would all work, voluntarily but firmly, toward a cooperative goal. He did give some clues as to how these benevolent—and moral—goals could be realized through the evolutionary process.

In a crucial essay on societal evolution, published in his *Studies Scientific and Social* (1900), Wallace focused in on what he saw as a viable and effective potential transition from the pernicious Victorian social order to a more just society. Although Wallace felt that this society of the future would be "some form of socialism," he understood that most citizens in late-nineteenth-century Britain disliked "the very idea of socialism, because they think it can only be founded by compulsion." To disabuse them of this notion, Wallace offered an alternative: a voluntary, transitional "organization of labour for the good of all." He cited the post office and the railways as contemporary models for organizing individuals and institutions for the equal benefit of the whole community. If this model could somehow incorporate a sense of voluntary competition, under strictly equal conditions, then Wallace believed Victorian society, at all levels, could actually "develop all the forces and all the best qualities of humanity, in order to prepare us for that voluntary organization which will be adopted when we are ready for it, but which cannot be profitably forced on before we are thus prepared." Late-nineteenth-century society, according to Wallace, denied most people the opportunity for the full development of their individual capacities. He saw a misdirected energy of individuals forced to compete against each other in order to produce an accumulation of wealth largely siphoned off by those few in power. Worse, this unprecedented accumulated wealth was often used for purposes either hurtful to many, if not most, people (such as unsafe working conditions and environmental pollution), or not directed toward any positive social good (such as the increasing gap in incomes, or misguided national military adventurism). In sum, there "can be no *true* individualism, no *fair* competition, without equality of opportunity for all. This alone is social justice, and by this alone can the best that is in each nation be developed and utilized for the benefit of all its citizens."[36]

Wallace supported Spencer's famous designation of the "law of social justice [as] deduced from the law of the survival of the fittest," that is, that individuals best adapted to the conditions of their existence shall prosper most, and those least adapted will prosper least. However, since Wallace abhorred even the suggestion of destruction of those deemed less fit, he was left only with a hope that that an appropriate education would somehow encourage people either to delay marriage until a much later age than was then the norm, or refrain from reproduction altogether. At the same time, those who were more fit (however redefined) would prosper and leave more offspring. This state of affairs, it appeared to Wallace, was clearly not the case in Victorian Britain. He believed, however, that there was a way out of this dilemma. As an example, Wallace indicated that such a desirable, beneficent competitive process actually operated in accordance with the rules governing, ideally, various sports events or tournaments. There the event organizers strive for equality of conditions for the competitors. Participants are, usually, all of roughly equal age and in good health, and have undergone nearly identical training. Further, it is universally recognized that the skill or endurance of each competitor can be ascertained only under equal or fair initial conditions. However, and this is a profound qualification that Wallace brings forward, this sports model—to which Wallace (ideally) likens Spencer's law of social justice—was prevented from operating by those very forces that promote capitalist industrial advance. In the "real battle of life, failure in which often means continuous hardship, want, or premature death, with the loss to friends and to the community of all those higher qualities or talents which were undeveloped through want of leisure or opportunity, we make no attempt whatever to give fair play to all alike." Wallace maintained that the fact that fewer inventors and original thinkers have sprung from the "ranks of peasants and mechanics" than from the wealthier segments of the population indicates that Victorian society actually obstructed the law of social justice. In order for this law to operate appropriately, he declared, what is needed "in order to profit by all the skill, and talent, and genius that may exist in our whole population, is that all should have the education and the opportunities for developing whatever abilities they may possess, which are now accessible only to the higher and the wealthier classes." Wallace believed that only in this way could the fundamental principle of organic evolution actually function in "the development of the social organism."

The full application of this principle of equality of opportunity would mean that all people should have the best education they are capable of receiving, that they might achieve the fullest potential of their particular intellectual and physical qualities. Drawing on his own less than orthodox academic history (and his own not altogether happy experiences as a marker and grader for many years in various educational settings), Wallace asserted that this did not "mean that all shall have the *same* education . . . but that all shall be so trained as to develop fully all that is best in them. It must be an adaptive education, modified in accordance with the peculiar mental and physical nature of the pupils, not a rigid routine applied to all alike, as is too often the case now." In words that sound eerily prescient in today's era of massive student debt, Wallace noted that "equality of opportunity [both ethically and practically] requires that all shall have an endowment to support them during the transition period between education and profitable employment."

Wallace further pointed out, provocatively, that the unequal distribution of wealth and inheritance in any society is a detriment to further social evolution. He believed that inherited wealth is, often, both a major impediment to full societal well-being and injurious or unhelpful to its recipients, enabling "them to live permanently in idleness and luxury." In Wallace's vision of a just society, "when all were well educated and well trained and were all given an equal start in life, and when every one knew that however great an amount of wealth he might accumulate he would not be allowed [by legal prohibition against inheritance] to give or bequeath it to others in order that they might be free to live lives of idleness or pleasure, the mad race for wealth and luxury would be greatly diminished in intensity." He also imagined that such citizens would be content with a modest retirement income. And, most significantly, as work of every kind would have to be done by persons who were as "well educated and as refined as their employers," while only a small minority could, or would care to, possibly become employers, the greatest incentive would exist toward the voluntary association of workers for their common good. Wallace concluded his prospectus for a transition to a future, more just, society with the insistence that all legislation be based, henceforth, on the "adoption of the principle of Equality of Opportunity as our guide." This principle, Wallace would continue to argue in the last fifteen years of his life, "is therefore well fitted to become the watchword of the social reformers of the Twentieth Century."[37] From the vantage point of the early twenty-first century, it is clearly questionable whether Wallace's transitional period in social evolution has yet occurred. Yet his brief but powerful essay on "true individualism" and social advance still retains its force for our own pressing, and often polarizing, discussions about social justice in a technological age.

Wallace realized that his was a utopian view of social evolution. But his years of living in the Amazon and the Malay Archipelago among nonindustrialized cultures had convinced him that his stated goals were worthy and viable.[38] There is, to be sure, a hopeful streak that seems to have been part of Wallace's emotional and intellectual makeup. Thus, when confronting the question of precisely *how* to initiate the transition to a just society, he relied—as he had since first encountering them—on Bellamy's utopian (and widely read) novels, particularly *Equality*.

In an essay published in 1905 in the London socialist newspaper the *Clarion*, Wallace points to chapter 38 of *Equality*—"The Transition Period"—as a guide to progressive, but gradual, electoral change. A progressively leaning Parliament could begin by passing measures municipalizing all of the local public services: water and gas supply, all forms of electrical supply, liquor sales, and milk supply. Wallace indicated, also, that such general services as railroads and canals, and the working of coal and other minerals—as well as land itself—could be nationalized. He followed Bellamy in emphasizing that the adoption by increasing numbers of officials and citizens of the "great ethical principle . . . that all inequality of inheritance is unjust" was a prerequisite for such positive political change. Wallace clarified: the consequent establishment of absolute "equality of opportunity" would not necessarily lead to socialism, "but rather to a perfect *individualism* under equal and fair conditions, and this would almost certainly, as I have urged elsewhere, bring about universal *co-operation*, which might or might not lead on to Socialism."[39] Wallace's emphasis on cooperation at both the individual and group levels, irrespective of a particular political ideology, is suggestive of certain contemporary critiques of unregulated technological advance. These critiques, increasingly germane in our age of technological overindulgence, mirror Wallace's sustained effort to articulate a public strategy that would balance techno-scientific goals and demands against human social and moral, as well as environmental, well-being. Wallace would not find it at all odd to question the ever-increasing financial and corporate demands today for accelerating technological progress. Indeed, he might likely accord a higher priority to such fundamental collective goals as enhancing social justice, equalizing educational opportunities, reducing income inequality and unemployment, and more effectively maintaining human health (mentally as well as physically).[40]

Wallace on Eugenics

However innovative he was in other domains, Wallace's views on marital and sexual matters were rather conservative. He felt that, in general, heterosexual permanent marriage was "the true human relation."[41] The various eugenics schemes put forward in the latter half of the nineteenth century, unsurprisingly, elicited a frank but complex response from him. While deploring any system of involuntary or coercive negative eugenics, that is, the removal from the reproductive pool of "unfit" individuals, Wallace did propose his own particular version of voluntary positive eugenics, that is, the encouragement of those individuals with (what he believed were) superior traits to produce offspring. As was so often the case with Wallace, he borrowed elements from existing concepts and movements—in this case, hereditarianism, egalitarianism, socialism, and libertarian antistatism—to forge his own views on "human breeding." Though Wallace objected, quite vehemently at times,[42] to being labeled an advocate of eugenics, his mature views on human improvement render his position on the late Victorian eugenics/antieugenics spectrum complicated.[43] Two essays from the early 1890s constitute Wallace's most explicit statement of his views relating to eugenics and are the primary focus of this section.

"Human Selection" (1890) is Wallace's clearest rendering of what he believed to be the appropriate relationship between evolutionary biology and political ideology. He opens "Human Selection" with the following words: "In one of my latest conversations with Darwin he expressed himself very gloomily on the future of humanity, on the ground that in our modern civilisation natural selection had no play, and the fittest did not survive. Those who succeed in the race for wealth are by no means the best or the most intelligent, and it is notorious that our population is more largely renewed in each generation from the lower than from the middle and upper classes."[44] Wallace's politics were often antithetical to Darwin's, but he agreed that there was an undoubted check to progress in social evolution. Wallace dismissed as possible solutions to this evolutionary dilemma any proposals based solely on beneficial environmental influences such as education or hygiene. Though these could produce improvements in any given generation, Wallace held that they could not of themselves lead to a sustained improvement of humanity. Implicit in such proposals was the Lamarckian concept "that whatever improvement was effected in individuals was transmitted to their progeny, and that it would be thus possible to effect a continuous advance in physical, moral, and intellectual qualities without any selection of the better or elimination of the inferior types."[45] The inheritance of acquired characteristics was still accepted by many evolutionists. Under the rubric of neo-Lamarckism, it underlay certain biologically oriented reformist speculations in the 1880s and 1890s.[46] Wallace rejected all versions of Lamarckism, old and new. In an interview in 1894, Wallace indicated that he also rejected, for similar reasons, "the theory of Pangenesis, advanced by Darwin, and supported by his great reputation."[47] Wallace maintained that the researches of Francis Galton and August Weismann had demolished all theories of the inheritance of acquired traits. According to Weismann, the hereditary material (germ cells in the ovaries and testes that produce egg and sperm) cannot be modified by changes undergone by the remaining body cells (comprising the somatoplasm).[48] Wallace accepted Weismann's controversial but ultimately influential hypothesis and concluded that there remained "some form of selection as the only possible means of improving the race."[49]

In 1894, Wallace clarified what he meant by saying education alone cannot produce permanently (i.e., heritable) "good qualities." Rather, education, or the environment more generally, "simply develops the inherent faculties of a child, it does not impart those faculties. Good environment will enable such noble qualities as the child may possess to develop advantageously, so also will education. In like manner, bad environment will have a tendency to bring to light and strengthen the baser elements of the child's nature. But the influence of environment or of education upon the parent is not transmitted to the offspring."[50] For Wallace, it is only when the results of appropriate education are internalized in the human species, so to speak, by "enlightened mate choice," that permanent social and ethical moral progress becomes possible. In the same interview, Wallace explicitly—and caustically—ties his biological views to his political ones:

> If it is thought that this non-inheritance of the results of education and training is prejudicial to human progress, we must remember it also prevents the continuous degradation of humanity by the inheritance of those vicious practices and degrading habits which the deplorable conditions of our modern social system undoubtedly foster in the bulk of mankind. Throughout trade and commerce, lying and deceit abound to such an extent that it has come to be considered essential to success. It is surely a blessing if this kind of thing does not produce inherited deterioration

in the next generation. We have little to lose in not having the effects of our present social system transmitted. Education has been so bad for two thousand years that we should be a degraded race altogether, if acquired character were inherited.[51]

Wallace detested what he termed "artificial selection," under which he included such schemes as Galton's eugenics as well as systems of "enforced" or involuntary eugenics. Among Galton's proposals was "a system of marks for family merit." Those individuals who rated well in health, intellect, and morals would be encouraged, by state subsidies, to marry early and raise large families. Wallace argued that while such "positive eugenics" might increase slightly the number of excellent human specimens, it would be socially ineffective and evolutionarily insignificant. Positive eugenics would leave the bulk of the population unaffected and fail to "diminish the rate at which the lower types tend to supplant . . . the higher."[52] Given the limited knowledge of human inheritance, Wallace declared, artificial selection was not only scientifically dubious, but also culturally pernicious. Eugenics, by perpetuating class distinctions, would postpone social reform and afford quasi-scientific excuses for keeping people "in the positions Nature intended them to occupy." Negative eugenics, or the prevention or discouragement of procreation by those deemed unfit, seemed to Wallace "a mere excuse for establishing a medical tyranny. And we have enough of this kind of tyranny already. . . . [T]he world does not want the eugenist to set it straight. . . . Eugenics is simply the meddlesome interference of an arrogant scientific priestcraft."[53]

Wallace had initially welcomed Galton's *Hereditary Genius* (1869) as providing important data in support of evolutionary biology. His later antipathy to Galton's eugenics stemmed from what he regarded as its cultural authoritarianism. Wallace also distanced himself from Galton because of the latter's general lack of empathy, if not rudeness, to those he regarded as not being of his social standing. Galton's views on gender were a further irritant to Wallace. Galton never regarded women as intellectual equals. He was also insulting to ambitious, but socially less privileged, men. Finally, his descriptions of his own African explorations are replete with crude and arrogant comments about indigenous peoples he likened to baboons, pigs, and dogs. While Galton was hardly alone among Victorian explorers in expressing such Eurocentric attitudes, he was among the most extreme.[54] All such opinions were anathema to Wallace. His habitual empathy, as manifested in his views on women and his deep appreciation of the cultures of indigenous peoples, separated him from Galton both on scientific and personal levels.

To Wallace, neo-Lamarckism, eugenics, and the capitalist apologias extracted from evolutionary theory were not merely biologically dubious. They also proceeded from fundamentally objectionable social premises. Wallace maintained that all were predicated on class distinctions and economic inequities. Their advocates ignored or failed to confront the central fact that Victorian culture frustrated, rather than facilitated, genuine evolutionary advance. Wallace thought that the key to permanent human betterment lay in the operation of a benevolent biological selection. Socialism, he believed, would provide the sufficient—and necessary[55]—conditions for that selective force:

> when we have cleansed the Augean stable of our existing social organization, and have made such arrangements that *all* shall contribute their share of either physical or mental labour, and that all workers shall reap the *full* reward of their work, the future of the race will be ensured by those laws of human development that have led to the slow but continuous advance in the higher qualities of human nature. When men and women are alike free to follow their best impulses; when idleness and vicious or useless luxury on the one hand, oppressive labour and starvation on the other, are alike unknown; when all receive the best and most thorough education that the state of civilisation and knowledge at the time will admit; when the standard of public opinion is set by the wisest and the best, and that standard is systematically inculcated on the young; then we shall find that a system of selection will come spontaneously into action which will steadily tend to eliminate the lower and more degraded types of man, and thus continuously raise the average standard of the race.[56]

Socialism, by removing disparities of wealth and rank, would eliminate the economic and political prejudices that, Wallace claimed, dominated the selection of reproductive partners in Victorian society. In their place, mate choice would focus on those higher moral and intellectual traits often neglected (or rendered subservient) in competitive capitalist society.

That the selective process Wallace envisioned as the key to further human evolution is a form of female sexual selection is, at first sight, surprising. One of the major theoretical disagreements between Wallace and Darwin had stemmed precisely from Wallace's refusal to accord any scientific status to female choice as an agent of evolution. By 1890, however, Wallace had come to regard female choice *among humans* as a necessary agent of any additional societal evolution.

Two years later, in an essay on possible societal advance, he elaborated on his thesis that female sexual selection was the key mechanism for effecting any positive, and permanent, change in the human condition. In "Human Progress: Past and Future" (1892), Wallace argued that sexual selection under socialism afforded a potent means through which to permanently ameliorate human society. The advance in material civilization in historical times was undoubted. But Wallace questioned whether there had been a corresponding advance in human mental and moral nature. He granted that "during the whole course of human history the struggle of tribe with tribe and race with race has inevitably caused the destruction of the weaker and the lower, leaving the stronger and higher, whether physically or mentally stronger, to survive." He doubted whether such a process did, or ought to, operate under the conditions of modern civilization. Wallace fulminated that the system of inherited wealth—"which gives to the weak and vicious an undue advantage both in the certainty of subsistence without labor, and in the greater opportunity for early marriage and leaving a numerous offspring"—had unfortunate consequences for human evolution.[57] He also dismissed as "unmitigated humbug" any talk about hereditary class distinctions being rooted in Nature.[58] And though the preservation of the weak or malformed could be construed as interference with the course of nature,[59] the cultivation of sympathetic feelings "has improved us morally by the continuous development of the characteristic and crowning grace of our human, as distinguished from our animal, nature."[60] The fact that some who in infancy were weak or physically deformed later exhibited superior mental qualities thus also afforded an ethical sanction for civilization's protection of the weak.

In the same essay Wallace reiterates his concern that capitalist social systems were retarding evolution's "general advance." He stresses that the widespread modern trust in education and environmental reform as the main engines of human betterment is misplaced. Under socialism, the mistaken belief in the hereditary transmission "of the effects of training, of habits, and of general surroundings" would be replaced by the powerful action of sexual selection. Wallace wryly adds that Weismann's argument against the inheritance of acquired traits, whether physical or cultural, is cause for relief, not despair: the debauched practices of the wealthy and the sordid habits of the oppressed workers in Victorian society could not produce any further degradation of humanity. Wallace was, however, cognizant of what were commonly perceived to be the pessimistic cultural consequences of Weismannian biology. He sought to allay the fears of those with Lamarckian leanings, that is, colleagues such as Lester Ward and Joseph LeConte. They believed that Weismann's germ-plasm theory of heredity doomed to failure all proposals for human betterment "except by methods which are revolting to our higher nature." Far from negating the influence of education and of beautiful and healthful surroundings, Wallace asserted, a socialist society would treasure them. Sexual selection, when informed by an ethos of freedom and human dignity guaranteed by economic equality, necessarily entailed "that education has the greatest value for the improvement of mankind." Moreover, Wallace declared that for the first time in human history "selection of the fittest may be ensured by more powerful and effective agencies than the destruction of the weak and the helpless." Hence would arise a continuous improvement of the race, far more certain and more beneficial than could be brought about by the compulsory or inhuman means so often advocated by those he regarded as neo-Malthusians.[61]

In the 1898 article "Darwinism in Sociology" Wallace continued his attack on all those eugenics schemes that involved any form of state or group coercion. Ever the astute biologist as well as social reformer, Wallace repeated his conviction that although earlier phases of human evolution operated through forces that were similar to those operating on the entire biological world, at that point in evolutionary history when human evolution became "guided" by higher intelligences in certain key aspects, arguments based on parallels to evolutionary forces at work in the animal or vegetable kingdoms lost their value. Instead,

> in social and civilised man the mental and moral nature rules over the physical; and as I have just shown, a new and higher kind of selection will come into action as soon as he learns how to subordinate the latter to the former, and how to so organise his social state as to satisfy the economic requirements of all. . . . The method of bringing about social excellence here suggested, works with nature, not against her. It depends upon the natural play of the higher qualities of human nature, and will therefore be both self-acting and efficient; while any forcible intervention of authority will be as certain to produce evil as it must be powerless for good.[62]

Wallace deemed this principle of sexual selection under socialism to be "by far the most important of the new ideas I have given to the world."[63] The principle was not, however, a new idea of Wallace's. He borrowed it almost verbatim from Bellamy. But Wallace's particular deployment of sexual selection under socialism is highly significant and was crucial for the development of his vision of social evolution. Malthus had long ago provided Wallace with one stimulus to bring various lines of thought together. Bellamy and Henry George now gave Wallace an equally powerful stimulus. He was able to refashion a number of conceptual and ideological themes that had existed in sometimes uneasy proximity in his earlier life and career. Wallace had long embraced certain socialist ideas, dating from his youthful attendance at Owenite lectures in the working-class halls of science and mechanics' institutes. The ending of his 1864 essay "The Origin of Human Races" had echoed that Owenite vision.[64] But Wallace "modified" the ending in an 1870 version, which emphasized spiritualist rather than socialist themes.[65] He never abandoned his belief in the guidance of spiritual intelligences as agents in human evolution. Nor did he waver in his insistence that spiritualist claims could be verified empirically and thus constituted a body of demonstrable scientific knowledge.[66] Nonetheless, Wallace clearly appreciated the polemical advantages of the unambiguously *naturalistic* mechanism of sexual selection within the broader framework of his evolutionary teleology. Bellamy and George legitimated Wallace's ongoing effort to forge a synthesis of social progressionism with biological progressionism.

Wallace commented further on his views on human sexual selection in an 1893 interview titled "Woman and Natural Selection."[67] When asked what he thought the effect of the notable changes in women's education, career opportunities, and family roles in the latter half of the nineteenth century might have on human progress, he replied without hesitation that "the effect will be entirely beneficial to the race. Women at the present time, in all civilised countries, are showing a determination to secure their personal, social, and political freedom. The great part which they are destined to play in the future of humanity has begun to force itself upon their attention." Wallace maintained that when both men and women were free to follow their best impulses, and when both sexes received the best and most thorough education then available, the resulting society would be ready for a notable advance: "[W]e shall find that a system of human selection will come spontaneously into action which will bring about a reformed humanity." This form of human selection would, Wallace continued, bring "about a continuous advance in the average status of the race." Mainly through the agency of (hopefully enlightened) female choice in marriage, Wallace predicted the "man of degraded taste or feeble intellect" would have little chance of finding a wife and his undesirable traits die out with him. Conversely, "men of spotless character and reputation, [and] the most beautiful in body and mind" would most readily secure wives and transmit their inheritable traits to future generations. The result over several generations would be to bring the average person "up to the level of those who are now the more advanced of the race."

Wallace realized that for this form of enlightened female choice to work to the most desirable collective end, women must be enabled to become independently financially secure and thus freer to choose to marry or not. Since a number of women might well choose careers over marriage (this was a century or so before paid maternity leave and other twentieth-century developments affecting childbearing!), the smaller remaining number of females who did desire to marry and have families would, in effect, do so and choose the best men. Since improvements in working conditions would also result in fewer male deaths due to industrial accidents, the pool of men seeking wives would also gradually increase. Wallace hoped a steady state would be achieved in which the number of "men desiring wives will be in excess of women wanting husbands, resulting in a society in which women would guarantee 'the improvement of the race.'" Perhaps speaking more freely in a comfortable interview in his home, Wallace again invoked evolutionary biology in the cause of social advance. He declared that the emancipation and equality of women would be a crucial step forward in societal evolution and would be consonant with biological science. Giving women "the power of selection in marriage" would, in fact, "cleanse society of the unfit." As Wallace explained, "This method of improvement by the gradual elimination of the worst is the most direct method, for it is of much greater importance to get rid of the lowest types of humanity than to raise the highest a little higher." He reminded his interviewer that the method by which the animal and vegetable worlds have been improved and developed has been through weeding out. Wallace said that he regarded survival of the fittest as really the extinction of the unfit. He claimed that "[n]atural selection in the world of nature is achieving this on an enormous scale, because owing to the rapid increase of most organisms a large proportion of the unfit are destroyed."[68]

Wallace was never shy about expressing his opinions on any subject, but he was especially candid in this interview. His hopes for the efficacy of (enlightened) female choice among humans, as an adjunct to natural selection, is clear evidence of Wallace's espousal of a certain form of eugenics, a concept he otherwise often railed against. Admittedly, the system here advocated is far less harsh than the more notorious eugenics schemes usually advocated in the late Victorian era—and practiced to ghastly effect by the Nazis—but Wallace's views render his attitude toward eugenics more complex than is usually thought.[69] In any case, Wallace was now positioned to reject any proposals, including involuntary eugenics schemes, for social and/or biological amelioration that accepted Victorian competitive capitalism as their premise. He declared that all such schemes "attempt to deal at once, and by direct legislative enactment, with the most important and most vital of all human relations, regardless of the fact that our present phase of social development is not only extremely imperfect, but vicious and rotten at the core." Wallace painted as obscene the contrast between newspaper accounts of the lives of wealthy Victorians—"with their endless round of pleasure and luxury, their almost inconceivable wastefulness and extravagance"—and the simultaneous reportage of the deplorable conditions of millions of workers, including children. In 1890 he cited the detailed and "absolutely incontestable evidence"[70] presented in the reports on the House of Lords' commission on sweating describing workplace environments.[71] Wallace also pointed to the even more awful circumstances of individuals unable to find any kind of work at all. He wondered how any thinking person could "admit for a moment that, in a society so constituted that these overwhelming contrasts of luxury and privation are looked upon as necessities, and are treated by the Legislature as matters with which it has practically nothing to do, there is the smallest probability that we can deal successfully with such tremendous social problems?"[72] The applicability of Wallace's observation to contemporary issues is clear.

Interestingly, T. H. Huxley—no foe of capitalist imperialism, and who regarded Wallace's socialism as anathema—expressed similar concerns about the harsh consequences for many individuals should any Victorian eugenics proposals be implemented. Huxley argued that eugenic intervention would destroy the bonds of social sympathy.[73] Of course, this was precisely what most eugenics advocates considered the virtue of their schemes: scientific experts would manage societal evolution. Wallace's opposition to involuntary eugenics schemes was not shared by all socialists. Karl Pearson, like certain Fabians, saw eugenics as compatible with an "elitist socialism"—a planned socialism by middle-class experts and administrators.[74] The key to these Fabian schemes for reform of society was the recruitment of socialist intellectuals to serve as experts and managers of the new order. Wallace was wary of these models of centrally managed socialist societies. He sought, instead, to accommodate his long-standing insistence on the sanctity of the individual with his newer allegiance to the socialist and collectivist vision for a more just society. Wallace's espousal of a form of "voluntary eugenics"—female choice in human sexual selection—was his attempt to resolve this tension.

Wallace and Antivaccinationism

Wallace was one of several important Victorian scientists who exposed certain critical shortcomings in nineteenth- and early-twentieth-century arguments in favor of vaccination and compulsory vaccination laws. Wallace's critique of vaccination science, featuring statistical arguments, aimed to demonstrate the inadequate basis for many widely held views on the effectiveness of vaccination and evidence for subsequent immunity. Indeed, both pro- and antivaccination arguments would prove inconclusive in the light of (then) contemporary standards of evidence. During this period both sides invested much effort in collating and analyzing statistical data sets that either supported or refuted the claims of vaccination's effectiveness. While each side presented "controlled" case studies to support their assertions, without an unambiguous test to measure or demonstrate vaccination's effectiveness, Wallace and the antivaccinationists continued to mount credible statistical critiques of late Victorian vaccination science. Six years before his death in 1913, the antivaccinationists successfully secured a conscience clause that effectively dismantled the compulsory vaccination laws. Wallace's ideas were significant in spurring efforts to secure more conclusive evidentiary standards for vaccination in the twentieth and early twenty-first centuries.

In the early 1880s, Wallace launched himself into the center of a politicized and polarized debate over unpopular compulsory vaccination laws that forced all English infants to be vaccinated before the "tender" age of three months. In 1883, he wrote his first public letter denouncing this practice, to the International Anti-vaccination Congress held in Berne, Switzerland. This public denunciation of England's increasingly strict compulsory Vaccination Acts (1853, 1867, 1871) was succeeded by a series of booklets, pamphlets, book chapters, and articles detailing Wallace's years of intensive research into the question of vaccination's effectiveness, and his original statistical work on the issue.[75] Because he was admired by Charles Darwin, Sir Charles Lyell, Joseph Dalton Hooker, and the American philosopher of science Charles Sanders Peirce as having one of the keenest minds of the Victorian age, Wallace's public conversion to the antivaccination camp was a coup d'état for the various English antivaccination leagues. It gave them a new scientific foothold in the public debates over the utility of vaccination.

Yet Wallace's cogent and influential defense of antivaccinationism has been relegated to the historical sidelines, as has the pervasive and powerful antivaccination movement itself. Recent work by historians Jennifer Keelan and Nadja Durbach,[76] however—along with a growing body of articles by authors from various disciplines, including immunology—has begun to alter the situation dramatically. It now appears that the magnitude and significance of the vaccination debates (1870–1914) helped redefine the boundaries of medical expertise and state control over its citizens' bodies.[77] Meanwhile, historians of science in recent years have begun to free themselves from the caricature of Wallace as a brilliant scientist who unfortunately "lapsed" into nonscientific or questionable crusades.[78] In fact, Wallace's objection to mandatory vaccination programs was shared by a considerable portion of the population in both Europe and North America. His name and formidable power of argumentation became an important tool for the antivaccination movement.

It was precisely Wallace's scientific experience combined with his social and cultural activism that sharpened his alertness to the flaws in the medical arguments supporting vaccination and the compulsory vaccination acts. Assessing Wallace's antivaccination activities as part of a parallel scientific discourse is an important step toward understanding the significance of Victorian debates over vaccination. Provaccinationists argued that vaccination was effective, smallpox was ubiquitous, and that the risk of catching smallpox and dying from the disease greatly outweighed the rare complications from vaccination itself. The few reported deaths from vaccination, they argued, were necessary to protect the interests of the general public. Antivaccinationists produced a compelling risk calculus of their own: smallpox was neither ubiquitous nor infection inevitable, vaccination did not provide sufficient protection and was as risky as smallpox itself, and there were safer and more reasonable alternatives to a state-enforced compulsory medical intervention.

Antivaccination science had credibility in the public realm, and to be an antivaccinationist was not seen as being anti-science. Science and its specialized disciplines were in the process of becoming professions, but this process was still in its early stages in the Victorian period.[79] For Wallace's time, then, a posed dichotomy between science and nonscience is both less useful and less historically accurate than the terms good science/poor (faulty) science. The logic for compulsory vaccination crossed the permeable boundaries between scientific reasoning and political reasoning. Since vaccination policy targeted populations, not individuals, it embedded the implementation and enforcement of compulsory vaccination into the social and political machinery of the state. Thus, for Wallace, any attempt to separate the scientific reasoning for vaccination from the sociopolitical reasoning would have no practical meaning. Wallace's investigations into vaccination reflected his holistic approach to the natural and social world, or what has been described as Wallace's evolutionary cosmology. He opposed attempts to apply arbitrarily reductionist solutions to complex phenomena—including the presumed control of a dynamic, evolving disease with a single intervention like vaccination.

In *The Wonderful Century: Its Successes and Failures* (1898), Wallace provides an idiosyncratic but penetrating ranking of the nineteenth century's great breakthroughs, as well as its notable defects. Alongside many glorious symbols of progress, Wallace lists what he regards as the nineteenth century's most egregious failures: the neglect of phrenology, the opposition to hypnotism and psychical research, militarism ("the curse of civilization"), the plunder of the earth (environmental destruction), and mandatory vaccination programs. The last topic constitutes the single longest chapter in the book (110 pages): "Vaccination a Delusion: Its Penal Enforcement a Crime."[80] Wallace's antivaccinationism is notable for two major reasons. First, Wallace (in writings starting from the early 1880s) developed a convincing critique of some of the most frequently deployed theoretical and, particularly, statistical arguments of the provaccination movement.[81] Second, he embedded his scientific critique of vaccination within the broader framework of an ethical condemnation of certain aspects of the emerging medical establishment—especially its increasing ties to state-sanctioned monopolistic and interventionist politics and practices. As he declared on the last day of his testimony before the 1890 Royal Commission on Vaccination, "Liberty is in my mind a far greater and more important thing than science."[82]

As regards vaccination, Wallace rejected the idea that medical expertise, especially clinical data tabulated by practitioners in the field, should be unquestioned. Instead, he held that recommendations made by medical authorities must be measured by a more objective audience, and carefully weighed against personal liberty. He believed that surrendering political decisions on the health of the nation to vaccinators was a grave problem exacerbated by the developing specialization in medicine (and corresponding arguments that only full-time vaccinators could adequately judge the effects of vaccination). In a lengthy discussion of why some provaccination statistics were useless, and to justify the superiority of his own conclusions, Wallace argued that vaccinators could not objectively assess the effects of vaccination because they had a financial stake in supporting the practice.[83] These same arguments have resurfaced in the debates surrounding vaccination at the start of the twenty-first century.

Vaccination Science and Statistics

Many late Victorian "men of science" held that smallpox vaccination, if performed competently with good lymph, provided immunity against smallpox. How much protection it granted, and for how long, however, was openly debated. Wallace was not the only high-profile scientific figure to adopt antivaccinationism. Charles Creighton (1847–1927), the leading Victorian epidemiologist, and Edgar March Crookshank (1858–1928), a noted pathologist, both mounted several troubling lines of evidence suggesting that the early-nineteenth-century experiments on vaccination and immunity were inconclusive in the light of the natural history of smallpox, late-nineteenth-century pathology, new bacteriological taxonomy, and contemporary standards of evidence. They criticized the assertion that Edward Jenner and his contemporaries had successfully performed rigorous "controlled" experiments proving that an infection with vaccine lymph protects against smallpox. In turn, Creighton and Crookshank provided an alternative interpretive framework for the existing epidemiology and vaccine statistics.[84]

It is not surprising, then, that antivaccinationists had a receptive audience for their claims that vaccination was an expensive distraction from more proven approaches to disease control. These included rigorous sanitation and quarantine of infectious cases, improved personal health through better nutrition and exercise, and a cleaner environment. The statistics produced to support these holistic medical ideas about smallpox were widely disseminated in the form of brochures, pamphlets, and scientific articles. Wallace's contributions to a more accurate use of statistics in the contentious disputes over vaccination science were innovative and influential in shaping the course of late Victorian debates, and have had a lasting impact.[85] Statistics themselves, however, could not resolve the vaccination controversy; cultural arguments contextualizing statistics such as those

Wallace provided were also highly significant. The antivaccination logic that framed antivaccination statistics had an intuitive appeal for those like Wallace who were interested in broad social and political reform. Antivaccinationism attracted followers with interests clustered around several key reform movements, notably those of a philosophical and sociopolitical character. Five of the most common of these allegiances were to (1) social/socialist reformism, (2) spiritualism, (3) Swedenborgianism, (4) vegetarianism, and (5) antivivisectionism. As Durbach has shown, these movements, in addition to theoretical and philosophical kinship, provided antivaccinationists with potent models for institutional organization (leagues, organized debates, mass meetings, pressure-group tactics). By the 1860s and, increasingly, in the last three decades of the nineteenth century, the antivaccination movement operated quite effectively at both national and local levels, with membership that was significant in numbers and distributed widely throughout England.[86] These broader ideological allegiances provided an important context for Wallace's statistical research. This cluster of affinities made him open to arguments that vaccine injuries were underreported, that vaccination science was supported by a particular interventionist medical tradition that had a long history of making patients worse rather than better, that universal vaccination was a simplistic approach to the complex problem of infectious disease, and that compulsory vaccination placed an unjust burden on the poor and working class (by the heavy fines often imposed for noncompliance). Collectively, these affinities made him suspicious of high-handed government medical experts and any polemically motivated narrow "scientific consensus," such as the rejection of spiritualism.

Wallace, Social Reformism, and the Antivaccination Movement

In the early 1870s a pandemic of smallpox swept across Europe and the United Kingdom. The epidemic aggravated existing tensions over compulsory vaccination and further polarized popular opinion about vaccination along class lines. When enforcing vaccination, local magistrates targeted the working class because it was both politically and economically awkward to enforce the law among the gentry (their social superiors). This reinforced the notion that hostility to vaccination was largely a working-class phenomenon, and played on traditional cultural stereotypes that portrayed the working class as the locus of disease. After the 1853 act was passed making vaccination compulsory, the legislation was expanded: first in 1867 to allow officials to repeatedly fine recalcitrant parents, then in 1871, following the great epidemic, whereupon every Board of Guardians across the country was required to hire vaccination officers to enforce compulsion. Fines for refusing to vaccinate a single child often surpassed £30, in those days an astronomical sum for a laborer. In the cases of working-class resisters, fines were paid by auctioning off the family's possessions or sending the father to jail—events that evoked sympathy for the antivaccinationists.

Additionally, resistance to vaccination became a galvanizing issue for a variety of so-called identity politics movements in the late Victorian period (as well as today). Many people came to interpret compulsory vaccination as a violation of their bodies: a form of political tyranny. This struggle of workers to gain literal control over their own bodies also neatly intersected with reform movements of the late nineteenth century that emphasized the need to nurture self-discipline, temperance, and moral reform among the lower classes. Prominent antivaccination activists were often simultaneously involved in universal suffrage, early animal-rights activism (antivivisectionism), and holistic food reforms such as vegetarianism and the whole-food movement (whole-grain breads versus white flour). They also frequently were supporters of restricting the consumption (among the working class) of expensive luxury items associated with modern urban life: tobacco, coffee, and tea.[87] Although Wallace never gave up meat entirely, he deemed vegetarianism to be, ultimately, the best diet for humans.[88] Wallace's attitudes here are clearly relevant to our own society.[89]

With respect to antivivisection, Wallace believed that the capacity for pain and suffering of humans was far more intense than that of all other animals, with the possible exception of the larger mammals and birds. He was, nonetheless, opposed to vivisection, owing to his ethical convictions. Wallace insisted that

> [t]he moral argument against vivisection remains, whether the animals suffer as much as we do or only half as much. The bad effect on the operator and on the students and spectators remains; the undoubted fact that the practice tends to produce a callousness and a passion for experiment, which leads to unauthorised experiments in hospitals on unprotected patients, remains; . . . and, finally, the iniquity of its use to demonstrate already-established facts to physiological students in hundreds of colleges and schools all over the world, remains. I myself am thankful to be able to believe that even the highest animals below ourselves do not feel so

acutely as we do; but that fact does not in any way remove my fundamental disgust at vivisection as being brutalising and immoral.[90]

Antivaccinationism and the Limits to State Intervention

Some laissez-faire economists and free-trade proponents, English economist Alfred Milnes (1849–1921) among them, objected to compulsory government-supervised vaccination on the grounds that it undermined the promotion of a strong, independent citizenry. These economists argued that all citizens had an absolute right (1) to choose medical treatment and (2) in turn to determine, in good conscience, the best route to good health for themselves and their children.[91] But the situation was complicated, extending to public health considerations. Most antivaccinationists, including Wallace, held conventional views on government's role as an agent supporting large public works projects—especially those promoting a generally salubrious living environment. State intervention was seen as the most effective way to ensure a clean water supply, functional sewers, and a minimization of overcrowding in urban housing. Antivaccinationists subscribed to widely held medical theories that contagions were opportunistic, most afflicting the bodies of the young, malnourished, and unclean, and spreading quickly through densely packed tenement housing. Wallace's own evolving position with respect to both liberalism and socialism provides a striking case study of the crucial importance of political ideology and activity within the antivaccination movement.

Wallace's commitment to obliterating socioeconomic and class inequities in Britain dated from his youth.[92] This commitment became more overt in the turbulent period 1870–1900, when Wallace became a leader in the fight for land nationalization as well as an outspoken advocate for socialism.[93] In these efforts, Wallace sought policies that would mitigate the widespread pauperism, vice, disease, and crime of large portions of the English laboring classes, "which strike foreigners with the greatest astonishment." The fact that many landholders were also magistrates, Wallace argued, furthered their power to coerce their tenants into conformity with their own political and religious opinions.[94] Wallace's views on land reform thus paralleled his attitude toward vaccination policies, and both became meshed with his socialism. A fundamental component of Wallace's socialism was the sanctity with which he clothed the concepts of individualism and personal "home privacy." Such concerns led Wallace to oppose mandatory vaccination schemes vigorously and to regard their penal enforcement as nothing short of a "crime" committed by the state against its citizens, notably its poorer members. He specifically underlined the phrase "liberty is as dear as equality or fraternity" in his annotated copy of Bellamy's *Looking Backward*.[95] The opponents of compulsory vaccination were particularly skillful in deploying the concept of the "rights of citizens" to agitate not only for political and legal equality, but also against infringement on the "sanctity of the home."[96]

Spiritualism, Swedenborgianism, and Antivaccinationism

To round out the nexus of convictions held by many individual antivaccinationists, mention must be made of their spiritualist and Swedenborgian beliefs. During the Victorian period—as, indeed, during most eras, including ours—medical and religious beliefs and practices were often intertwined. As nineteenth-century antivaccinationism is frequently characterized as an alternative or dissenting medical movement, it is scarcely surprising that religious nonconformity often motivated its members. Two of the most potent avenues of dissent in the late Victorian period were the controversial doctrines of the eighteenth-century Swedish scientist and theologian Emanuel Swedenborg (1688–1772), and spiritualism. Wallace shared an interest in these with many antivaccinationists.

Wallace's spiritualist convictions have been one of the most intensively studied aspects of his life, particularly after the late 1860s.[97] Wallace credited the prominent antivaccinationist William Tebb, also a spiritualist, as providing a major stimulus to the writing and publication of several of Wallace's tracts defending antivaccination in the 1880s.[98] A number of Swedenborgians, for example William White and James Garth Wilkinson, were drawn to the antivaccination movement, in part, because the procedure and aftereffects of vaccination seemed particularly odious—and avoidable—intrusions into one's body. Within the framework of Swedenborg's theories of the relationship between physical and spiritual health, damaging the body could also endanger the soul. White, a Swedenborgian bookseller, is significant since he, with Tebb (a wealthy merchant) and the pharmaceutical chemist William Young, founded the socially diverse London Society for the Abolition of Compulsory Vaccination in February 1880.[99] Wallace probably first learned of Swedenborg from Robert Dale Owen's *Footfalls on the Boundaries of Another World* (1861). Dale Owen's book, and subsequent writings, elaborated a tamer version of the doctrines of Swedenborg.

Dale Owen integrated essentials of Swedenborgianism with spiritualism and political reformism, especially Owenite socialism. Wallace's early exposure to Swedenborg subsequently influenced certain of his own views on the mind-body connection. Wallace, like his contemporaries the American pragmatists William James and Charles Sanders Peirce, refashioned Swedenborgianism into a comprehensive philosophy of nature more consonant with the findings of nineteenth-century science, notably evolutionary theory.[100]

Wallace's interest in the metaphysical mind-body connection clearly influenced his understanding of the complex nature of health and disease, including whether any single materialistic medical procedure could be effective in combating smallpox. Wallace's concept of smallpox epidemics was conventional: namely, that the spread of the disease was facilitated by overcrowded, unclean, and insalubrious environments. While anyone was susceptible to the disease, it disproportionately struck the poor, the young, the sick, and the aged. The complex relationship between environment and the reproduction of the smallpox contagion in its host, apparent to Wallace, made any simple medical intervention to eradicate smallpox seem naïve. The disputed theoretical claims surrounding vaccine-induced immunity, especially taxonomical issues whereby an infection with one disease gave protection against another disease (e.g., cowpox against smallpox), made Wallace suspect that smallpox vaccine-based immunity was highly unlikely. Any vaccine effect was a product of the wishful thinking of the medical profession. There were far more obvious explanations for susceptibility to the disease and, at the same time, for barriers or protections from smallpox. The vast social and cultural changes that brought about the Victorian sanitary movement, added to improved wages and nutrition for the poor and working classes, made it seem highly improbable that the dramatic decline in smallpox from the mid-eighteenth century to the late nineteenth century had been brought about solely, or even principally, by a medical intervention, vaccination. Medical historian Anne Hardy has argued convincingly that the decline of smallpox and other contagious diseases was caused by a multiplicity of factors affecting nineteenth-century disease ecology (including sanitary reforms and other public, and individual, health measures), though she also acknowledges a role for vaccination.[101]

Swedenborgianism itself did not lead to antivaccinationism, any more than to antivivisectionism or socialism. Still, the constellation of ideologies described above allows us to profile the Victorian antivaccinationist, and to locate skepticism about the procedure in the fertile ground of these sociopolitical and cultural reform movements. The latter intrinsically shaped how individuals like Wallace approached the contentious issue of causality between vaccination and epidemic diseases like smallpox. The antivaccinationists' worldview made certain solutions to the smallpox problem appear more logical than others. This made them ask fundamentally different questions from their provaccination rivals, and in turn led them to instrumentally distinct statistical analyses. While each side presented "controlled" case studies to support their assertions, without an unambiguous test to measure or demonstrate vaccination's effectiveness, the antivaccinationists continued to mount credible statistical critiques of vaccination science. Clearly, the impact of antivaccinationism at the close of the nineteenth century demonstrates the problem with casting it as antiscience: pro- and antivaccinationists alike were influencing the early development of this particular branch of medical science by aiding and defining what data were or were not germane to it.

Wallace never wavered in his belief that smallpox vaccination was useless and likely dangerous. Six years before his death, the antivaccinationists successfully secured a conscience clause that effectively dismantled the compulsory vaccination laws. Statistical arguments against vaccination technologies continued to dog efforts to expand the vaccination program in England. In 1904, the eminent statistician Karl Pearson published a critique of the typhoid vaccine that made advocacy for vaccination appear old-fashioned, paternalistic, and ultimately unscientific.[102] To be sure, advances in immunology and serology in the interwar years resolved many of the issues that plagued Victorian vaccine science, statistics, and epidemiology. However, these advances came too late to convert the antivaccinationists of Wallace's generation. The public face of vaccination was transformed, not by statistics, but by the gradual implementation of successful immunotherapies such as diphtheria and tetanus antitoxins.[103]

The advances in twentieth-century immunotherapies and the rise of bacteriological medicine and serology have, however, tended to obscure—or minimize—the way statistics were used to establish a causal relationship between the implementation of vaccination and the decline of smallpox in the nineteenth century. Until recently, medical historians generally have downplayed or dismissed the scientific arguments made by figures like Wallace, Creighton, and Crookshank, for the reasons discussed here. The impressive success of twentieth-century vaccination, therefore, cast a long shadow over its more shaky theoretical and empirical beginnings. Both sides in the debate over vaccination's efficacy argued that accurate statistics would serve as an objective arbitrator in the discussion. It is helpful to remember that credible statistical analyses, such as those of Wallace and his antivaccination colleagues, did exist, and threw doubt on certain of the aspirational claims regarding the ultimately beneficial effects of vaccination. Statistics served both masters, and the antivaccinationists used them as a powerful weapon in nearly all their print material and broadsheets. Through today's mass media, including the pervasive internet, both pro- and antivaccination forces compete for the public's attention, presenting new, sometimes inconclusive, and increasingly complex medical information and experimental data on the effectiveness of novel vaccines.[104]

Wallace's critique of Victorian provaccination statistics focused on measurement bias, errors in determining categories of data, and how these categories (immune versus not immune) corresponded to measurable interventions (vaccinated versus unvaccinated). In addition, antivaccinationists targeted the underlying theoretical assumptions made by provaccinationists, thereby destabilizing their reading and construction of the data. Significantly, late-nineteenth- and early-twentieth-century antivaccinationists questioned the simplicity of the concept of reducing the complex measurement of immunity to the binary of "vaccinated versus unvaccinated." This and Wallace's pointed remark that physicians were hardly objective gatekeepers of vaccination science still resonate today in antivaccination writings.[105] The possible connection of some physicians to pharmaceutical companies and their inducements adds fuel to this particular issue.

Wallace's critique of vaccination science and epidemiology—a critique of the aims and goals of vaccination as *the* method to combat contagious disease—has been reborn in today's biopolitics of universal vaccination. Opposition to vaccination, often based on arguments structurally similar to those of Wallace and his contemporaries, has resurfaced since the 1980s. Present-day antivaccinationists emphasize, among other complications, vaccine toxicity and side effects, the troublesome link between vaccine advocacy and pursuit of financial profits by the massive and politically potent pharmaceutical industry, and the age shift in infectious disease occurrence. This last refers to the fact that vaccine-induced immunity tends to diminish over time, as opposed to natural infection, which tends to induce lifelong immunity. Thus there results a larger population of nonimmune adults who, as research shows, often have severe effects from diseases that in children are usually mild.[106] If the term "vaccine hesitancy" is used in place of outright "vaccine refusal," then the persistence of Wallace-like doubts about the benefits and safety of particular (especially newly introduced) vaccines becomes clearer still. Even people who today are vaccinated can hesitate to embrace certain aspects of vaccination policy and vaccination consequences.[107] Such resistance has generated public debate over the safety and necessity of routine childhood vaccination, and has had a significant impact on public health policy.[108] It is, in fact, often still technically difficult to provide a clear demonstration of the effectiveness of vaccination using statistics derived from the poorly controlled setting of human populations. This paucity of evidence, one way or the other, has also recently fueled bitter controversies concerning autism. Since diagnoses of autism (and related disorders such as Asperger's syndrome) have been on the rise in recent years and the diagnosis of autism is usually made in children of roughly the same age at which most vaccines are given, it is not surprising that hypotheses of possible associations of autism with vaccination of children for other diseases have surfaced. That the verdict is still out only makes the plethora of information and advice available via the mass media (especially the internet) confusing.[109]

To be sure, modern provaccinationists have the advantage of having a more developed science of immunology than did their Victorian predecessors—as well as laboratory models of immunity that are more rigorous and accessible. However, all public scientific demonstrations rely on interpretive frameworks that are assailable when the underlying theory is not intuitive to broad sectors of the population. The popular understanding of the *cause* of infectious disease has been dominated by military metaphors whereby hostile pathogens invade vulnerable hosts. Similarly, immunity has been cast in terms of self versus nonself. Highly successful programs to eradicate the "enemy," such as smallpox and polio, captured the public's imagination and led to optimistic predictions of the end of infectious diseases altogether. However, since the last few decades of the twentieth century, this optimism has been tempered by a resurgence of vaccine-resistant diseases in developing countries, novel or emerging infectious agents such as Ebola and HIV, and highly adaptable germs (antibiotic-resistant tuberculosis and influenza). The provaccinationist mantra has, as a consequence, been somewhat chastened.[110] Parental anxiety concerning the number and efficacy of recommended/mandated vaccinations has increased to the point where one-third or more of American parents (it is estimated) now opt out totally, or delay at least one vaccine for their children. In many countries, citizens find themselves caught in the middle of the still passionate debate between antivaccinators and vaccine proponents. Concerns regarding (1) the safety of vaccines, especially with what are termed "vaccine adverse events" such as joint pain, or some neurological disorders manifesting occasionally after vaccination of children and some adults with MMR vaccine (for immunization against measles, mumps, and rubella); (2) the complex vaccine schedules for certain vaccines; and (3) the idea that some vaccines may be seen as medical "enhancement" rather than explicitly required medical "therapy" for some diseases, are all fed by legitimate scientific data. Additionally, there are the more dramatic (if sometimes quite exaggerated) claims made by the media and, now, abundantly available on the internet. Physicians, child-care workers, school educators, and health officials at various levels can no longer avoid or ignore scientific information or cultural, religious, and philosophical viewpoints that may differ from their own training and experience.[111] The decision whether to vaccinate is a legitimate but complex and challenging issue that should, ideally, be approached with a sense of what is best for the individual or family in their particular setting at a particular time.

As the public necessarily becomes more familiar with the protean nature of many pathogens and their co-evolution with the environment, Wallace's evolutionary cosmology—a Swedenborgian ecological perspective on the "world of life"—appears increasingly germane to twenty-first-century medical concerns. Diseases with simple external causes and cures are understood to be the exception to the rule, and multifactorial paradigms now dominate the culture and science of medicine.[112] Cancer, autoimmune diseases, and ever-evolving germs have taken center stage in the theater of disease. They invoke metaphors drawn from environmentalism and molecular biology, where the boundary between humans and their environment is more permeable. Health and disease are now understood as emergent properties of complex and constantly evolving immunological and regulatory processes. As a consequence, the underlying rationale of universal vaccination (generating specific immunity to a specific disease contagion) may seem simplistic to doctors, biologists, and public health authorities, as well as to the general public. The twenty-first-century resurgences in pertussis infections (whooping cough), for instance, are attributed to a combination of waning immunity from the existing pertussis vaccination, and bacterial mutations that elude it.[113] The balancing of risks and rewards of each specific vaccine available for each disease targeted today is a complex task. While opposition to *all* vaccination is, of course, a less tenable option now than it was in the late Victorian era, Wallace's insights into medical decision making by individuals and government are as pertinent at the start of the twenty-first century as they were in the late nineteenth century.

Conclusion: Wallace as "Social Prophet"

Wallace's attention to the social and political details of Victorian culture became increasingly explicit during the last three decades of his life. He had always been alert to societal and philosophical issues, but this interest assumed a more critical and urgent nature as the outlines of late-nineteenth-century industrial society took shape. Wallace delighted in many of the scientific and technological advances of his age. Indeed, he had taken a leading role in the transformation of the biological sciences with Darwin and other contemporary theorists. But he was deeply troubled by what he perceived as the continuing—if not growing—inequities and injustices of his world. Rather than sharing the optimism that was characteristic of many (though not all) of the leading thinkers and pundits of his time, Wallace was alarmed by certain of the environmental and cultural impacts of science and technology. Moreover, he perceived that many of these impacts, if unchecked or unregulated, could exacerbate, instead of lessen or abolish, existing societal problems—as the boosters of unlimited techno-scientific change declared. Where many influential late Victorian commentators saw mainly "progress" in industrial growth, Wallace observed increasing urban squalor, environmental degradation and pollution, and growing economic inequality. Further, he was convinced that these problems were fundamentally social and moral at root. His visions of the present, and future, were primarily social and ethical, not technological. Wallace's later writings increasingly took on an anticipatory, "prophetic," and critical tone. How does he measure up, then, as a social prophet?

In certain of his pronouncements on industrial society, Wallace appears remarkably astute: (1) Growing economic and social inequality is stark in many of the so-called "advanced" nations and emerging societies today. (2) Conversely, gender inequality seems to have declined markedly, and Wallace's "woman of the future" looks a good deal like the women who have assumed more significant political, economic, and educational roles in contemporary societies. (3) Environmental pollution and degradation have continued apace since the late nineteenth century. (4) Science's military function has magnified, and the "vampire of war" has even sharper fangs.[114] (5) The role of religion and spirituality as a force in social evolution remains unmistakable. (6) Vaccination continues to be a contested, or at least debated, medical practice, while the ties between biomedical science and big business have intensified. With respect to Wallace's posited solutions to these problems—excepting, obviously, point 2—the verdict is still out. His universal remedy, a voluntary socialism whereby all live and work for the common good, seems noble but has yet to find viable expression. However, Wallace's critical insights into the problems and potentials for technological societies remain as cogent at the start of the twenty-first century as they were when he first articulated his concerns.

Acknowledgments

I want to thank Charles Smith, James Costa, and Kathleen Lowrey for their comments on earlier versions of this essay.

Notes

1. Wallace (1910a).

2. Ibid.

3. Bowler (1993), Lightman (1997), Hale (2014).

4. Spencer had, in fact, given Wallace a copy of his *Data of Ethics* (1879).

5. Wallace's annotated copy of *Data of Ethics* is at Edinburgh University, Special Collections. It is here that epistemological concerns intersect most clearly with Wallace's emerging evolutionary teleology.

6. Wallace acknowledged, however, that the process was subject to other influences, especially in its latest stages.

7. Wallace (1864b, 1913a 1913c).

8. Wallace (1894f), p. 177.

9. Ibid., p. 178.

10. Ibid., p. 179.

11. Ibid., p. 182.

12. Wallace (1898d), p. 379.

13. Blaazer (1992), p. 98.

14. Wallace (1894f), p. 191.

15. Wallace (1892b), p. 24.

16. Wallace (1901b).

17. Tolstoy (1878/2014), part 3, chapter 30.

18. Notable instances include Wallace (1885c, 1893a, 1894d, 1894f, 1898d, 1900a, 1905a, 1909c, 1913a, 1913c).

19. Wallace (1885a), pp. 369–372.

20. Ibid., pp. 389–391.

21. For example, Wallace (1882b, 1885c, 1900d, 1905b, 1911, 1913c).

22. Stack (2008), pp. 302–303.

23. Wallace declared himself a socialist most memorably in his essay "Human

Selection" (1890d), though he had also announced the news ten months earlier in a letter that appeared in *Land and Labor* (1889c).

24. Wallace (1898b).

25. Wallace (1900e).

26. Wallace (1913a), p. 102.

27. Wallace (1894c), p. 549.

28. Ibid.

29. Wallace (1890d).

30. Wallace (1894c), p. 550.

31. Wallace (1870b), pp. 338–344.

32. Interestingly, Wallace's idea of surplusage of power—or the role of some higher intelligence operating in the evolution of human intellectual and moral faculties—was influential among certain "theologically motivated" biologists and philosophers in Russia during the period 1880–1930; see Levit and Polatavko (2013), p. 296.

33. For a fine introduction to these issues, see Ayala and Arp (2010), pp. 337–390.

34. Wallace (1913a), p. 174.

35. Wallace (1894c).

36. Wallace (1900a), vol. 2, pp. 510–513.

37. Ibid., pp. 514–520.

38. Wallace (1857a), p. 204.

39. Wallace (1905a).

40. See, for example, Agar (2015), pp. 106–108, 166–195. On the practicality of the cooperative (or collective) interest/ideal—and why it is not "like the ideal of group hugs as a solution to conflict in the Middle East (a nice but dangerously naïve idea)"—Agar argues that it is not necessary that every individual or every nation conform with it for the ideal to be efficacious. Rather, those who defend the ideal should have some realistic expectation that they can bring human behavior in greater accord with it. Agar cites nuclear power, particularly nuclear weaponry, as an example of how competition at both individual and societal levels can bring about enormous technological success in a relatively brief time (during and shortly after World War II). In contrast, the problems of nuclear wastes and hazardous nuclear power plants can best be approached in a spirit of cooperation to achieve a collective solution that is in the interest of all citizens of all nations. According to Agar, if the cooperative model of technological progress means accepting a slower pace of technological change than has been the norm in the late nineteenth, twentieth, and (especially) early twenty-first centuries, such a cultural shift might avoid some of the bleaker scenarios of global technological catastrophe that have been predicted for our own civilization (pp. 98–99).

41. Wallace (1890d), p. 331.

42. Wallace (1912).

43. Paul (2008), pp. 263, 273–276.

44. Wallace (1890d), p. 325.

45. Ibid., p. 325.

46. Bowler (1993).
47. Wallace (1894a), p. 82.
48. Haig (2007), pp. 415–417.
49. Wallace (1890d), p. 326.
50. Wallace (1894a), p. 84.
51. Ibid., p. 85.
52. Wallace (1890d), p. 328.
53. Wallace (1912), p. 663.
54. Fancher (1998), pp. 108–109, 112.
55. Wallace (1905b), vol. 2, p. 266.
56. Wallace (1890d), pp. 330–331.
57. Wallace (1892a), pp. 147–148.
58. Wallace (1917).
59. Wallace (1892a), pp. 148, 157.
60. Wallace (1890d), p. 337.
61. Ibid., pp. 332–337; Wallace (1892a), pp. 156–159.
62. Wallace (1898c), p. 59. It should be noted, however, that Wallace is only condemning all eugenics schemes that involve coercion, that is, "involuntary eugenics." His attitude to a more benevolent "voluntary eugenics" is more complex and nuanced, as is clear from our discussion of Wallace's espousal of a specific form of human selection via female choice.
63. Wallace (1905b), vol. 2, p. 389.
64. Wallace (1864b), pp. clxix–clxx.
65. Wallace (1870b), pp. 329–331. The new version had the new title "The Development of Human Races under the Law of Natural Selection."
66. Oppenheim (1985), pp. 316, 320.
67. Wallace (1893i).
68. For Wallace on the "elimination of the unfit" see Smith (2012a).
69. Wallace (1893i). Wallace repeated these views a few years later in another article (1898c). See also Paul (2008).
70. Wallace (1890d), p. 330.
71. A sweatshop is a factory or workshop, especially in the clothing industry, where up to one hundred manual workers were employed at very low wages for long hours and under crowded, unsanitary, and usually dangerous conditions. Between 1850 and 1900, sweatshops attracted the rural poor to rapidly growing cities, and attracted immigrants to places such as London's and New York City's garment districts. It is to these workplaces that Wallace refers; see House of Lords (1888–1889). The problem persists to this day, notably in the overseas workshops (in impoverished nations like Cambodia, Haiti, and Bangladesh) producing clothing for companies in developed countries; see *International New York Times* (2013). For a powerful global perspective on exploitation of workers in the vast, contemporary textile industry, see Beckert (2015), chapter 13, "The Return of the Global South."

72. Wallace (1890d), p. 330.

73. Paradis (1989), pp. 47–48.

74. MacKenzie (1981), pp. 75–79. The Fabian Society had been formed, in part, by disillusioned middle-class Liberals. They rejected Marx's theory of revolution and replaced that with an ideology of a "managed" (top-down) evolutionary socialism.

75. Wallace (1898a).

76. Keelan (2004), Durbach (2005).

77. Keelan (2004); Durbach (2005); Colgrove (2005), pp. 167–191.

78. For example, Fichman (2004), Smith (2008b), and Stack (2008).

79. Barton (2003), pp. 73–119.

80. Wallace (1898d), pp. 143–149, 213–324.

81. Fichman and Keelan (2007).

82. Wallace (1890a), p. 127.

83. Wallace (1898d), p. 222.

84. Crookshank (1891), p. 128.

85. Fichman and Keelan (2007).

86. Durbach (2005), pp. 38, 41–47, 122–23, 143–149.

87. Keelan (2004), pp. 155, 168, 170–171; Durbach (2005), pp. 41–46, 122–123.

88. Wallace (1909d), p. 282.

89. It has been suggested that health-based reform movements can be understood as "hygienic ideologies," that is, idea systems that focus on aspects of personal hygiene (broadly construed) as a necessary, and moral, foundation for human progress. Such hygienic ideologies invite acceptance by incorporating both certain "universal" facts about humans and nature, and the values and anxieties peculiar to distinct eras. See Engs (2000), pp. 4–5. From this perspective, the complex of Victorian movements that linked Wallace to certain of his contemporaries has clear parallels to present-day reform packages that combine vegetarianism, animal-rights activism, antismoking and antivaccination crusades, and environmental awareness in various permutations.

90. Wallace (1910c), p. 381. It is worth noting that Darwin was both provaccinationist and not opposed to vivisection.

91. Milnes (1897).

92. Jones (2002), pp. 73–96.

93. Gaffney (1997), pp. 609–615; Fichman (2004), pp. 211–213. See also chapter 8.

94. Wallace (1882b), pp. 100, 129–135, 176–179.

95. Fichman (2004), pp. 250–252.

96. Fichman (2004), pp. 328–330; Durbach (2005), pp. 36 ff., 69, 87–89.

97. Barrow (1986), pp. 186–188; Owen (1990), pp. 131–132; Smith (2008b). Both Barrow and Owen discuss the link between Victorian spiritualism and antivaccinationism.

98. Wallace (1905b), vol. 2, pp. 351–352.

99. Durbach (2005), pp. 39–41, 45–46.

100. Fichman (2004), pp. 160–161.

101. Hardy (1993).

102. Hardy (2000).

103. The development of sulpha drugs and antibiotics played a similar role in making the "magic bullet" ideology of universal vaccination to combat infectious disease more intuitive to subsequent generations of both scientists and the public.

104. Chatterjee (2013), pp. 11–12.

105. Yaqub et al. (2014), p. 6.

106. Link (2005), especially chapter 4.

107. Yaqub et al. (2014), p. 11.

108. Leask and Kinnersley (2015), pp. 180–182.

109. Chatterjee (2013), pp. 181–183, 202–203.

110. Link (2005), pp. 124–126.

111. Largent (2012), chapter 6.

112. Fichman and Keelan (2007), p. 605.

113. Mooi et al. (2014), pp. 685–694.

114. For example, nuclear proliferation, chemical and biological weapons, and drone aircraft.

‹ 8 ›

Land and Economics

DAVID COLLARD

Wallace was an evolutionary biologist, not a political economist, and does not appear in standard histories of economics.[1] He was definitely an outsider, especially since economics had begun to think of itself as a new "science." However, the so-called marginalist revolution, triggered by the work of William Stanley Jevons and others in the 1870s and consolidated by Alfred Marshall in the 1890s, had not yet displaced the more conventional "classical" economics of Adam Smith, David Ricardo and John Stuart Mill. Between the new and the old there was a window of opportunity for the enthusiastic amateur.[2] But Wallace, a brilliant Victorian polymath, was far too innovative to be just another amateur. Moreover, he was no friend of political economy: "The science of Political Economy has now guided, and often governed, the civilised world for near a century, but it may be doubted if the world is much better for its guidance . . . so narrow [is] its scope, so powerless for good, so utterly divorced from all considerations of morality, of justice, even of broad and enlightened expediency."[3]

The feeling was somewhat reciprocated. Thus Marshall, though respectful to a natural scientist of Wallace's eminence, was determined that he should have nothing to do with an 1886 proposal for a new *Economic Journal*. Of a potential sponsor he wrote, "[H]e might want more room for people like Wallace—to say nothing of Hyndman—than either Keynes or I should think right."[4] Wallace and his friend Henry Hyndman, founder of the Social Democratic Federation, not to mention Henry George, were, then, obviously persona non grata to the new economic establishment. By the 1890s Wallace had indeed firmly nailed his colors to the socialist mast.[5] And it could be argued that he had long been a budding socialist: he had, in his youth, read widely on the subject, attended radical lectures in the Tottenham Court Road, and been prominent in the debates of the Working Men's Institute in Neath. This implicit socialism must have stayed with him during his golden years as an evolutionary biologist, but though his early 1880s program for land nationalization was socialist, he remained a reforming radical more in the image of John Stuart Mill than an outright socialist. His socialism became explicit, it seems, as a result of reading the works of Edward Bellamy.[6] Although Wallace retained his strong belief in a long-term cooperative socialism, he came to accept the usefulness of reforms such as those of the great Liberal government in the early twentieth century. Wallace would never have been happy with a dirigiste version of socialism, since it would have offended his strong attachment to individual liberty, an attachment that also drew him to the antivaccination campaign.

Wallace's fulminations against political economy might, in a lesser man, be taken merely as the outpourings of a pamphleteer. But he was, in fact, deeply serious about the need for political economy to be replaced by a new, broader, "social economy," and in spite of his letter Marshall did take Wallace seriously (though he disagreed with him) on land nationalization. Unfortunately Wallace never systematized his diverse views on political economy; still, although he remained an outsider, his positions on this and other economic issues are well worth investigating.[7]

Wallace on Well-Being

Wallace's notion of economic welfare was surprisingly similar to that of modern nonmainstream economics. Traditionally, economists have tended to use such measures as "gross domestic product per capita" as indicators of economic welfare. But many economists and other social scientists have more recently sought a richer notion of welfare, often embodied in the term "well-being" and sometimes measured as subjective "happiness." Even the World Bank has for several years published a human development index that takes into account measures of health and literacy. Measurements of well-being, it is now widely recognized, should account for levels of access to clean water, sanitation, fresh air, decent housing, and economic security. Importantly they should also recognize "agency," that is to say the ability of people to control their own lives.[8] Finally, it is now acknowledged that widening inequality between labor and capital, always a central concern for Wallace, remains to be addressed.[9]

Well-being, in the modern sense, is precisely the term Wallace himself used. "The final and absolute test of good government," he wrote, "is the well-being and contentment of the people."[10] He had a broad appreciation of the elements that should constitute a good life, being able to compare the lives of "natives" with the lives of workers in his own industrialized society. Wallace, with his realistic and thoroughly based knowledge of "primitive" peoples, had few illusions about them. But he had been much taken by the unencumbered beauty of some of the tribes on the Rio Negro, at one point poetically rejecting the values of civilization:

> I'd be an Indian here, and live content
> To fish, and hunt, and paddle my canoe,
> And see my children grow, like young wild fawns,
> In health of body and in peace of mind,
> Rich without wealth, and happy without gold![11]

This somewhat Arcadian view of well-being was retained, as it were, in the back of Wallace's mind: a powerful image indeed. Something of the same vision was shared by other Victorian "romantics." William Morris, for example, advocated the revival of handicrafts, was for a time a member of the Social Democratic Federation, and organized the Socialist League. John Ruskin, champion of the Pre-Raphaelite movement and a socialist of sorts, offered a moral critique of capitalism. These "reactionary" responses to capitalism were often satirized, as was Wallace's advocacy of "three acres and a cow" for all.[12]

Wallace's own Arcadian vision was set out in his book *Land Nationalisation* and is worth quoting in full.

> Surround the poorest cottage with a spacious vegetable garden, with fruit and shade trees, with room for keeping pigs and poultry, and for storing the house-refuse and manure at some distance from the dwelling, and give the occupier a permanent tenure at a low quit-rent, and the result is absolutely invariable. . . . Under such conditions the poorest cot would soon be improved and made into a comfortable dwelling; the surplus fruit, vegetables, eggs, bacon, and other produce would benefit all the dwellers in the neighbouring towns.[13]

Such imaginings of well-being contrasted with the actual conditions in British cities, whose harsh realities had already been described by Engels.[14] "The modern slave," wrote Wallace, "is forced to live in huge cities far removed from the health-giving influences and spontaneous gifts of nature."[15] In a well-known passage at the end of *The Malay Archipelago* he claimed that "although we have progressed vastly beyond the savage state in intellectual achievements, we have not advanced equally in morals."[16] He was appalled by the pinched, half-starved proletariat of supposedly "civilized" cities: we had created a state of "human misery and crime" in which people were worse off than "the savage in the midst of his tribe." There was too much poverty in the midst of plenty.

What could be done to improve well-being? Wallace, being Wallace, was not content merely to comment. His always fertile mind and tireless energy drove him to look for practical solutions and, what is more, to campaign for them. One issue dominated all others for Wallace: the great question of land reform. The first section of this chapter on economics is therefore devoted to land nationalization, campaigning and arguing for which took up a good deal of Wallace's energy from the early 1880s onward. The second part treats of a number of miscellaneous issues that attracted Wallace's attention. All of them, however, were also concerned with well-being, and all illustrate the originality that Wallace brought to bear on whatever topic concerned him. All were live issues in the fourth quarter of the nineteenth century, and all are still relevant now.

Land Nationalization

Land nationalization was, for Wallace, by far the most important economic and social question to be addressed: he devoted a huge amount of effort to it, both intellectually and as a campaigner, so it is appropriate that it features prominently in this chapter. I would claim that Wallace's treatment of land reform shows him at his reforming best. It illustrates his ability to examine the facts, to provide a technical analysis, to follow through with a proposed reform, and then to campaign for it.

The Background

The rent of land had long been recognized as a special case by classical economists. Unlike labor and capital, rental income, it was acknowledged, is not generated by individual effort but by the general progress of society.

"Rent," argued Ricardo, "is that portion of the produce of the earth, which is paid to the landlord for the use of the original and indestructible powers of the soil."[17] Importantly, "a tax on rent would affect rent only; it would fall wholly on landlords, and could not be shifted to any class of consumers." A tax on rents would therefore be almost ideal: its incidence would be clear, it would create no disincentive effects, and it would be borne by the least popular class of society.[18] As we shall see, the case for taxing rent was later made very powerfully by Henry George. Although the British land tax was not a tax on pure economic rent, it was broadly approved of by leading writers such as Smith, Ricardo, and McCulloch on the ground that it probably did little harm. Mill went further than most of his contemporaries by giving his approval to a betterment tax on capital gains in land and a tax on inheritance. Somewhat surprisingly, classical economists were even prepared to manipulate the price of land to encourage economic development. The promoter Edward Gibbon Wakefield persuaded Mill to endorse his scheme to charge a high price, even for "free" land in Australia, so as to encourage large concentrations of population.[19]

Rents and capital gains from land, it seemed, could be taxed and the price of land manipulated. But no one of consequence advocated state ownership. For one thing it was believed that the state would be a very poor manager of land: this was one of the "heavyweight" arguments against nationalization. The other was its cost. Still, a state might want to take over private land for public purposes, and its right to do this was rarely challenged if proper procedures were followed and full compensation paid to those with legal title. Such interventions in a free market for land were accepted, but universal public ownership would be too expensive to contemplate even if it were desirable.

The main treatment of landownership available to Wallace was the impressive section on "peasant proprietors" in Mill's *Principles of Political Economy*.[20] Peasant proprietorship was poorly understood in England but very familiar throughout mainland Europe. In a similar manner to that later adopted by Wallace, Mill quoted Sismondi and other writers, most of whom had positive stories to tell of peasant proprietorship. If it worked well, it offered an attractive Arcadian vision of sturdy independence and self-reliance. But were such systems inefficient? Mill's sources generally found no evidence of this. Wallace, following Mill, held to the view that small-scale cultivation was efficient. Archaic though it now seems, the nineteenth-century debate over peasant proprietorship provides a useful backdrop to the evolution of Wallace's views.

The question became political mainly because of the Irish problem. What was to be done about the dire state of the population, leading to poverty, dispossession, and mass emigration? The Irish leader Charles Stewart Parnell was a champion of peasant proprietorships in place of tenants at will who had no security of tenure and went uncompensated for any improvements they were able to make. Parnell was president of the Irish Land League, a sister body to Mill's Land Tenure Reform Association. His demands were captured in the slogan "the three Fs": fixity of tenure, fair rents, and freedom to sell. This last meant that although proprietors rented, they were, in effect, operating in a competitive market, and it was intended to ensure fair compensation for improvements. As it happened, at least two of the three Fs, known as the "Ulster custom," were already being observed in some areas. The Ulster custom was legalized in the areas where it operated under the Irish Land Act of 1870 and was extended generally under the Irish Land Act of 1881. This was a great time for "leagues" and the like. Mill's own society was the Land Tenure Reform Association, founded in 1871. Many of its objectives were to be retained by Wallace: its most prominent proposals were (clause 4) to tax the unearned increment in land values: "the Society do not propose to disturb the land-owners in their past acquisitions. But they assert the right of the State to all such accessions of income in the future,"[21] and (clause 5) to encourage agricultural cooperatives. Like Parnell, the league stopped short of advocating land nationalization. This left some members, who wanted to go whole hog, dissatisfied. However, they could find a home in yet another organization, the Land and Labor League. Apart from this dissatisfied minority, most were content with a reform program of security of tenure, fair rents, and compensation for improvements.

The reasons that these mainstream reform movements steered clear of outright nationalization were obvious. Nationalization commanded only minority support, and embracing it would only harm more modest objectives with a reasonable hope of being attained. Without compensation it would be unfair and confiscatory. With compensation it would be impossibly expensive. Furthermore, the state was notoriously bad at managing the land, and state interference in the running of farms could only do harm. All in all, the prospects for anyone advocating nationalization were distinctly unpromising.

Wallace's Proposals on Land Nationalization

Wallace had a special relationship with the land. His early years in the Neath valley working for his older brother William were later to prove valuable as a solid basis for the alleged practicality of his land-nationalization program, though he was not at that time immediately struck by the injustice of enclosures.[22] Later, he had experienced the great tropical forests of the Amazon and the Malay Archipelago. The vast, generous sweeps of land must have seemed to him to belong to mankind rather than to rent-grabbing landlords. In 1892, responding to a report in the *Geographical Journal* "that it was at first difficult to instill into natives the idea of individual ownership of land," he commented, "[P]oor, benighted natives!"[23] His somewhat surprising diatribes against civilization in *The Malay Archipelago* showed that Wallace was "ripe" for reform. Mill was not slow to seize the opportunity and wrote to Wallace in July 1870 to urge him to attend a meeting of his Land Tenure Reform Association. He agreed to add a proposal of Wallace's to the association's program: "the important point which you suggested in your letter to me, viz., the right of the State to take possession (with a view to their preservation) of all natural objects or artificial constructions which are of historical or artistic interest. If you will propose this I will support it, and I think there will be no difficulty in getting it put into the programme."[24]

Wallace joined Mill's committee and became an active member. He was at first persuaded by the arguments of Mill and others that nationalization would be expensive and bureaucratic. But on the Irish question, he was not convinced that peasant proprietorship was the way forward. If there were a free market in land, current inequality would, he argued, be restored within a generation.

Mill died in 1873. By the end of the 1870s Wallace was well established as a biologist and was busy producing scientific papers as well as, unexpectedly, defending spiritualism. Convinced of the case for land nationalization, he could merely have sent off one of his many letters to the editor. But this would not do. A major educational program was called for. So Wallace helped start the new Land Nationalisation Society (LNS), becoming its first president in 1881 and delivering an important series of presidential addresses over the next fifteen years or so. There being a pressing need for a "handbook" for reformers, Wallace wrote it! It was to have been published by Macmillan, but Wallace's account of the Sutherland family's brutal clearances proved too much and it was, instead, put out by Trübner.[25] The admirable nineteenth-century habit of long, descriptive book titles is especially illuminating in this case. The full title is *Land Nationalisation; Its Necessity and Its Aims. Being a Comparison of the System of Landlord and Tenant with That of Occupying Ownership in Their Influence on the Well-Being of the People*. The book was dedicated to "The Working Men of England." The full dedication includes quotes from various eminent writers (James Anthony Froude, John Stuart Mill, Henry George, Francis W. Newman, Herbert Spencer, and William E. Gladstone), all to the effect that it was morally right for the state to acquire land. For moral justification Wallace relied heavily on Herbert Spencer's *Social Statics*:[26] "The present writer had his attention forcibly drawn to this great question about eighteen years ago, by the perusal of Herbert Spencer's demonstration . . . of the immorality and impolicy of private property in land."[27]

Wallace's perusal of Spencer's book must have taken place during the gap between his two great expeditions. Spencer's principal argument was a reductio ad absurdum. Suppose all land to be privately owned. Those without land would have no place to go without being trespassers: there would be no place for them on the planet. This rather abstract piece of logic was buttressed by Wallace's examples of dispossession in Ireland and the Scottish clearances. It probably resonated better with his audience, one feels, than did Spencer's philosophical argument. Even so, Wallace seemed to continue to feel the need for Spencer's support even after Spencer had withdrawn his advocacy of land nationalization.[28]

Wallace's scheme for land nationalization was rather ingenious in seeking to meet the usual objections head-on. I set out the usual criticisms and Wallace's responses:

- The state would be incompetent and probably corrupt at managing the land. Under Wallace's scheme, the land would be owned by the state and rented out to tenants. The state would have no role in the actual management of farms. Wallace, it should be noted, was not hostile to markets.
- It couldn't be afforded. Under Wallace' scheme the quit-rents paid by tenants to the state would finance the payment of annuities to landowners. There would be no cost to the state apart from administration.
- It would be confiscatory. To some extent this was true, as the annuities paid to landowners were terminable rather than perpetual annuities. They would run for the lifetimes of landlords and their immediate successors only. Over a couple of generations the land would revert to the

state. Wallace was in any case strongly opposed to undated securities, as they give their owners economic power over future generations.

- Land maps were in a very unsatisfactory state. A new map should be drawn up, Wallace suggested. Drawing on his own experience as a young land surveyor, he estimated that this would be fairly straightforward.
- Small-scale proprietors could not raise capital. Wallace suggested that local authorities or co-ops could provide loans. It was essential to Wallace's vision that the large tenants did not simply swallow up the small.
- Land reform would not help the urban proletariat. Indeed it would, Wallace argued, because a successful move back to the land would both increase the output of food and create a scarcity of urban labor.

Though in old age Wallace was prepared to make some compromises on these, he continued to campaign for the main principle:

> This all-embracing system of land-robbery, for which nothing is too great and nothing too small; which has absorbed meadow and forest, moor and mountain, which has appropriated most of our rivers and lakes and the fish that live in them; which often claims the very seashore and rocky coasts of our island home, fencing them off from the wayfarer who seeks the solace of their health-giving air and wild beauty, while making the peasant pay for his seaweed manure and the fisherman for his bait of shell-fish; which has desolated whole counties to replace men by sheep or cattle, and has destroyed fields and cottages to make a wilderness for deer and grouse; which has stolen the commons and filched the roadside wastes; which has driven the laboring poor into the cities, and has thus been the primary and chief cause of the lifelong misery, disease and early death of thousands who might have lived lives of honest toil and comparative well-being had they been permitted free access to land in their native villages;—it is the advocates and beneficiaries of this inhuman system who, when a partial restitution of their unholy gains is proposed, are the loudest in their cries of "robbery!"[29]

Readers' Guide to Land Nationalization (first edition 1882)

Chapter 1. Wallace acknowledges the vast increase of productive power in the nineteenth century: steam power, railways, iron and steel, etc. However, he challenges the view that "the accumulation of *capital* by individuals, necessarily advances the well-being of the whole community" (p. 11). On the contrary any surplus is gobbled up by rents, and there are huge increases in inequality, exacerbated by the immoral practice of inheritance, which enables the rich to perpetuate their dubious claims on future generations. Meanwhile urban workers are condemned to poor wages and miserable living conditions. The answer, says Wallace, lies in land reform.

Chapter 2. Here Wallace contrasts modern landlordism with a (somewhat dewy-eyed) view of the feudal system in which owners had obligations, such as supplying soldiers, in exchange for the right to charge rents. Land had become mere property, and Englishmen were guilty of trespass whenever they ventured onto private land. Wallace offers a general description of the eviction of tenants in Ireland and depopulation of the Highlands of Scotland.

Chapter 3. Wallace reproduces harrowing tales of evictions in Ireland after the famine. The behavior of land agents was sometimes particularly brutal, but Wallace insists that the evils perpetrated were due not to evil individuals but to an evil system.

Chapter 4. In this chapter Wallace offers similar horrific tales of Highland clearances. He seems to think that a mere recital of the facts will cause decent people to rise up and demand change. Again Wallace draws upon his own experience: "A considerable acquaintance with savage life in both hemispheres enables the present writer to assert that the people we term *un*civilised rarely tolerate such a state of things" (p. 90). And again, the "factors" or agents are merely the instruments of oppression. The system is in itself fundamentally wrong.

Chapter 5. This chapter is mainly about the precarious lives of the short-term tenant or tenant-at-will. He pays high house rents, has little or no garden and no access to common land: "the commons and roadside wastes from which he formerly obtained fuel for winter, with food and litter for a cow, a donkey, geese or poultry, have almost all been enclosed" (p. 112). So laborers are deprived of a free choice of "house and home," which are essential to well-being. Into this chapter Wallace also inserts some of his other pet topics such as his advocacy of "the right to roam," his defense of the protection of ancient monuments, and his condemnation of the export of coal and other minerals.

Chapter 6. Here Wallace defends "occupying ownership" against "landlordism." Drawing (as already explained) on work by Mill and Sismondi, he describes the advantages of occupying ownership and the evil results of landlordism across Europe: "wherever we find the land cultivated by its owners or permanent occupiers, *there* we find industry, economy, great productiveness, content and comfort" (p. 155). It might be argued that the economies of large-scale agriculture were beyond the reach of owner-occupiers.[30] This, Wallace claimed, was an irrelevance, as large scale was a question of capital, not of ownership, and much "capital" could be provided by the labor of the tenant himself. In this chapter Wallace is, of course, keeping his powder dry, since "ownership" under his system was not to be ownership as normally understood but instead a long-term secure rental from the state.

Chapter 7. Henry George's *Progress and Poverty* is introduced at this point. George claimed that most of the benefits of progress had been siphoned off by rents. Landlords, then, were the common enemy of both labor and capital. Wallace and George were in hearty agreement on this point, and there was little to suggest the gulf that was later to open up between them.

Chapter 8. Here is the essence of the book. Partly it is a summary of the previous chapters. But it also contains the core of Wallace's actual proposal. All land would be nationalized. This was, of course, further than most other reformers, including Mill and George, were prepared to go. Importantly Wallace distinguishes between the "quit-rent" paid for the actual land itself and the "tenant right" paid for farm buildings, improvements, equipment, etc. Upon nationalization, tenants would pay a quit-rent to the state instead of to the landlord. They would then *virtually* be owner-occupiers. Although owning the land, the state would play no further part in running it: there would be no state management "with its inevitable evils of patronage, waste, and favoritism" (p. 193). Landlords would be compensated not at full market value (a frequent stumbling block) but by means of terminable annuities equal in value to expected quit-rents, but only for their own lifetimes or the lifetimes of living heirs. This would be morally acceptable since the unborn cannot possibly have expectations of proprietary rights. At this stage, notes Benton, Wallace was "meticulous in his respect for the rights of the owners."[31] Over time (a couple of generations or more), however, the land would effectively have been nationalized with little cost to the state and little interference in the process of farming. The tenant right was to be rather more problematic for Wallace. Most tenants would be unable to buy out their tenant right, so Wallace suggested loans from local authorities, cooperative banks, or the state.

Campaigning for Land Nationalization

Wallace agreed with those critics who claimed that outright purchase would be prohibitively expensive. This was one of the reasons why a potential ally, Henry Fawcett, had come to reject nationalization.[32] Wallace professed himself amazed that an economist as sympathetic to the laboring poor as Fawcett should reject land nationalization. But he accepted the view that

> the idea of confiscating all the landed property of the kingdom, and reducing all who derive their income wholly or mainly from land to a state of destitution, is not only grossly unjust, but would be utterly subversive of the very end for which nationalisation is proposed—the well-being of the whole community. . . . We next come to the supposed alternative, to buy out the landlords with hard cash, or state securities, at the full market price of their lands; and here, too, I have always accepted Professor Fawcett's proof that the thing is financially impossible.[33]

It was for this very reason that Wallace sought a method of buy-out that would be nonconfiscatory and fair. An article in the *Contemporary Review*[34] showed that he had not yet worked out the "terminable annuity"[35] device or the phasing of his scheme. That this was work in progress may be seen from a series of letters to the *Mark Lane Express* in which Wallace envisaged a ten-year initial survey period after which all land would be nationalized.[36] Quit-rents, as terminable annuities, would be paid not only to the present owners but to their immediate heirs over perhaps three generations. Wallace was treading a fine line. The longer these terminable annuities ran, the more they became like perpetual annuities: a hundred-year terminable annuity at 5 percent would be worth as much as 99 percent of its perpetual value. On the other hand, the shorter they ran, the more they looked like direct confiscation! On the whole, Wallace stuck to his guns here, but he did indicate a willingness to consider other arrangements.[37] Wallace chaired the first meeting of the LNS, an event attended by Henry George. He elaborated his scheme in successive presidential addresses to the LNS but did not substantially depart from his initial proposals. Already in 1882 he was "fire-fighting" allegations that the LNS was opposed to technical progress: of course not, he responded, it was simply that the society did not wish the benefits of progress to be expropriated by landlords. On the great question of whether land nationalization was strictly necessary when a simple land tax would suffice, Wallace and George had already moved apart,[38] and a further drift was to follow. After his American trip Wallace came to see that a tax on rent plus a free market in labor would lead to speculation. In his presidential address of 1884–1885, he consolidated his position against critics by recalling his youthful credentials as a surveyor and valuer.[39] There had been some land reform but not nearly enough. Like most reformers he had to clarify his position on occasion. Wallace had to remind his members that he was thoroughly in favor of free markets for ordinary commodities: "We want no such laws to interfere between buyers and sellers of wheat or cotton, or beef or boots, or of any other commodity whatever."[40]

Wallace stressed that the LNS role was educative: always the optimist, the tide of opinion was, he felt, running in its favor. What would be the position of private residences? This was a little more delicate. On death, properties would automatically fall to the state. Otherwise homeowners would become leaseholders with security of tenure. Wallace rather neglected, one feels, the questionable role of the state as perpetual landlord and seemed to back away from nationalizing owner-occupied (as opposed to tenanted) houses. As for agrarian land, Wallace continued to push his somewhat Arcadian proposal whereby the urban proletariat could "acquire" plots of their choice wherever they chose. By 1891 he felt obliged to defend the *priority* he had given to land reform. Was it not necessary, asked some of his socialist critics, to nationalize capital at the same time? No, said Wallace, because land and labor, unlike capital (which could always be rebuilt), were the truly fundamental factors of production. Like Mill, Wallace had noticed how rapidly an energetic and technically sophisticated population could rebuild its capital stock.[41] Deal with land first, and other forms of nationalization would follow. Railway nationalization, for instance, could be dealt with by issuing a version of the terminable annuity. The raising of capital by small tenants continued to be a problem. Wallace's proposals on agricultural capital had been strongly attacked by Alfred Marshall[42]—so much so that by 1893 Wallace was even suggesting the offer of loans by lot or on a first-come, first-served basis.[43] By the time he came to deliver his presidential address in December 1893, one can see traces of doubt and disappointment at the slow progress of the LNS campaign.

However, he seized on the success of the allotment movement as a kind of nationalization. Unlike nationalization, that movement, by then over a hundred years old, was widely supported by progressive thinkers both in Britain and on the Continent. Wallace was heartened by the Small Holdings and Allotment Act of 1908 which enabled parish councils to acquire land and initiate small tenancies. He came to recognize that it would be necessary "for many years to come" for full market compensation.[44] Though still a keen land nationalizer, Wallace was turning his attention to other matters, such as security of homeowners through rents tribunals and parish councils, and the problems of unemployment and socialism. Reforms of this sort were, of course, no substitute for the real thing, nationalization, but were useful improvements to the lives of workingmen.

Marshall's Critique of Wallace on Land Nationalization

Alfred Marshall, perhaps the most eminent of neoclassical economists, came to criticize Wallace as a consequence of the latter's championship of Henry George.[45] Indeed, Martin Fichman has argued that Wallace's and George's campaigns were perceived as parts of one and the same political program.[46] Wallace wrote to Darwin that he considered George's book *Progress and Poverty*[47] to be "the most startling novel and original book of the last twenty years, and if I mistake not will in the future rank as making an advance in political and social science equal to that made by Adam Smith a century ago."[48]

Darwin was reluctant to be drawn into discussion of a book that, though popular, was not of the first rank. George's work was cleverly constructed, leading the reader to expect him to advocate land nationalization. "We have examined all the remedies [for poverty among plenty], short of the abolition of private property in land. . . . There is but one way to remove an evil—and that is, to remove its cause. . . . We must make land common property."[49] But then it turns out that confiscation is not necessary. All that is required is a tax on land to replace all other taxes. Although Wallace welcomed George's critique of landowners, he was of course disappointed by his failure to embrace land nationalization. Further, Wallace came to see that George was mainly interested in landlords versus capitalists rather than landlords versus workers.

By the early 1880s it was clear that George had a strong popular following not just in the United States but also in Britain.[50] Marshall, normally reluctant to indulge in controversy, had strongly opposed George at a public meeting in Oxford in 1883. Anxious to demolish George's case, Marshall gave three public lectures at Bristol in 1883. In the course of these lectures Marshall explicitly singled out "Mr. Wallace's plan" for an apparently devastating critique.[51] Marshall claimed, correctly, that Wallace's plan involved an element of confiscation, since it replaced perpetual quit-rents with lifetime annuities. He was correct in asserting that Wallace's scheme would involve an element of confiscation, but Wallace would not have regarded this as a valid objection as it was, after all, based on the immorality of landlordism. A problem remained for Wallace, however. As time went on he became increasingly tolerant of inheritance, being prepared to allow inheritance by those with reasonable expectations, grandchildren as well as children. This made the process less confiscatory but, of course, delayed eventual public ownership. Marshall objected even more to the other elements of Wallace's scheme: "his plan contains two other proposals potent for good and evil. The first is that all the buildings and other improvements must be bought by the tenant; and the second is that anyone may, subject to some slight conditions, select five acres for his own occupation out of anybody else's farm. I think this second proposal would lead to blackmailing and other kinds of oppression, and I shall not consider it further."[52]

Wallace's second proposal would allow practically anyone to apply to rent up to five acres of land having reasonable access to the highway. This could have devastating consequences for the viability of some existing farms. Marshall concentrated his fire on the first proposal, which, he argued, would favor rich farmers over poor even if municipal loans were available.

> I respect [Mr. Wallace's plan] as an earnest attempt to do good. But not to speak of its violence, I think it would increase the disadvantages of the poor man; impoverish all the farming class, hinder the intelligent labourer from becoming a small farmer, and the small farmer from becoming a large one. It would cause wealth and accident rather than ability to determine the importance of the post which a man held in the farming world; it might throw England out of her place as the pioneer of agricultural improvement; while the hired laborer would probably find that the farmer pinched for capital became nearly as hard a master as the peasant proprietor himself.[53]

Wallace seems to have taken some of Marshall's points on board. In one of his LNS presidential addresses (1893) he recognized that not everyone was capable of being a successful smallholder: he would need "a certain amount of knowledge, of experience, of prudence, and of industry. . . . [W]e do undoubtedly require some process of selection of the best men for this great experiment."[54] He also came to deny that very large initial capital expenditures would be necessary: improvements, fencing, etc., would be produced gradually by the efforts of the smallholder himself.

Third, it might not be necessary to raise one's own capital or even to borrow capital if it could be provided on a rental basis. Fourth, quit-rents would be set by local tribunals at a level that achieved "the maximum of well-being for the cultivators," rather than maximum rents and small-holders who would have complete security of tenure so long as they paid their quit-rents. Wallace, in responding to Marshall, clearly felt that he had some patching up to do.

An interesting exchange of letters arose from Marshall's Bristol critique of Henry George. George Stigler comments, in his introduction to them, "The Wallace-Marshall letters are concerned only with the historical facts on the trend of the living standards of the working classes. They display Marshall's confident handling of the historical materials and remind us of his one-time intention to write a general economic history. Wallace would perhaps accept today the verdict of natural selection that his land reform proposals were ill-suited to their environment."[55]

There were two points of dispute in these letters. The first was the notoriously tricky question of whether real wages had been rising over the last century, particularly between the 1830s and the 1880s. George, Wallace, and most socialist writers had argued that living standards had not significantly improved over the century and that pauperism was increasing. This was, of course, an essential plank in the reformists' program. The exchange makes it clear that Wallace's definition of living standards, or what he called well-being, was rather wider than the conventional economic one. His criticisms of Robert Giffen, the statistician, are interesting:[56] that he had minimized the importance of rent increases and security of tenure, that the decreased cost of clothing was illusory as quality had declined (e.g., corduroy and boots used to last two or three times as long), that distance to work and therefore travel time and costs had increased, and so on. Wallace was only partly concerned with statistics of living standards: he was much more concerned with the quality of life; brown bread rather than white, availability of firewood and meat, leisure time, security of tenure, etc. For evidence, Wallace drew upon his own early experience of the wages of carpenters when his brother John had been apprenticed to a master builder. Marshall's first letter suggests that Wallace had stimulated him to include some of these other factors.

The second point of dispute concerned the relative earnings of labor and capital. Wallace had taken on board George's view that labor and capital were the common enemies of landlords and that real wages and the real rate of return on capital tended to move together. Marshall's position, as stated in the reported lecture, was that they must move in opposite directions. It was, perhaps, unwise of Wallace to lock horns with the increasingly eminent Marshall. But as it happens, both were correct. In the process of economic development overall factor productivity rises, so real wages and real rates of return would both be increasing. On the other hand, other things being equal (in Marshall's much-used phrase), an increase in labor relative to capital would drive down the *relative* return to labor. Hidden in this rather arid discussion, however, Wallace made an important point about international comparisons of the real wage and well-being. Neither Marshall nor Wallace seems to have moved his position much in consequence of these exchanges.

Wallace's Proposals in Context

Modern economists have tended to treat land nationalization as a rather archaic topic: indeed, Mark Blaug in an influential book on the history of economics states that "the topic has no contemporary interest and may be skipped."[57] But while some of the issues that Wallace and his LNS grappled with are more or less resolved, others are not. It is convenient to distinguish between "old" (principally Europe) and "new" countries. In old countries landownership is now characteristically well defined and protected by the law. There is more or less a "free" market in land. Nationalization or public ownership of private land is, generally speaking, off the agenda, and there is little taste for a revival of "public ownership" campaigns on Wallace's model.

In view of Wallace's highly critical comments on the Highland clearances in Scotland, however, it is worth noticing that the issue of land reform is still alive in that country. A very recent report for the Scottish government puts "public good" at the center of policy.[58] Compared with Wallace's position, there is one clear-cut major difference: it contains no proposal for general land nationalization. But the report does seek to meet many of Wallace's objectives, since the guiding principle is the well-being of the public good rather than the sanctity of private property. This is not a trivial point, as feudal tenure was not finally abolished until the twentieth century. The right of landowners to enjoy their property is acknowledged but subject to the overriding interest of the public good: as Wallace had argued, landownership must be in the interest of the many, including the urban population. Thus rights of way and public access are to be strengthened, and tax privileges are to be removed for shooting and deer hunting: communities are to be given the power to buy neglected land, tenants are to have the priority right to purchase, and so on. Many such rights are now enjoyed in developed economies; indeed, the European Human Rights Convention emphasizes the "public interest" in any land reform. A further Wallacean point is that land must be managed sustainably. Finally, there is one embarrassing coincidence between the recent report and Wallace's policy: a complete land register should be compiled in ten years, precisely the period Wallace thought would be required!

To review the many issues surrounding land reforms in "new" countries is beyond the scope of this chapter. However, it is worth illustrating a few Wallacean concerns.

Compensation. When real revolutions occur, all land falls under the ownership of the new regime, which may or may not respect existing legal title. There is usually no compensation even in the form of Wallace's terminable annuities. Short of total revolution, the situation can be quite complicated. Consider the case of Zimbabwe, formerly a British colony. In order to finance the redistribution of land to a "land-hungry" peasantry, the British government agreed in 1980 to underwrite part of the costs. But this agreement was canceled by the Blair government, leading to a "fast track" land reform program in which land was taken violently from white farmers without compensation. In this case a colonial power had, for a while, been at hand to ease the compensation problem.

State farming. Wallace would not have been surprised by the failures of collective farms in the Soviet Union or Mao Zedong's China. Under the Great Leap Forward all of Wallace's fears about direct state control were realized. In post-revolutionary Russia, peasants were forced to work on collectives to increase the surplus of food available to the urban populations. To some degree the policy worked, but the large-scale collectives became inefficient and the peasants, now agricultural workers rather than peasant proprietors, did not thrive. The Chinese and Russian cases both illustrate Wallace's point that the state should not become involved in the actual running of agricultural units. If public ownership of land could permit Arcadia, it could also produce Hell.

Clearances. The treatment of indigenous peoples such as Native Americans, Australian Aboriginals, and New Zealand Maoris recalls Wallace's outrage at the notorious "clearances" in Scotland and Ireland. Generally their lands, to which they had no legal title, were simply seized by colonists. Where "reparation" has been attempted, indigenous peoples have often ended up with inferior land.

Credit. One of the many problems in reforming agriculture in Brazil and many other countries is lack of credit for agricultural capital: one of Wallace's recommended ways out of this difficulty was the formation of agricultural co-ops, an increasingly promising way forward. Co-ops have also been used successfully in parts of South India and Bangladesh as a way of achieving some of the economies of large scale—bulk purchase of seed and fertilizer, effective marketing, exchange of information, and so on.

Sustainability. Wallace's strictures on excessive mining or logging, formulated with reference to Britain, apply a fortiori on a world scale.

That issues of land reform continue to present deep moral and practical problems on a global scale would have been no surprise to Wallace. So although (to return to Blaug's point) nationalization as a topic may be "skipped," the issues surrounding it may not.

Other Forays into Economics

The fecundity of Wallace's mind led him to some rather unexpected places. Though diverse, these forays were coherent in that he was always concerned with the well-being of the ordinary working person, and always seeking practical remedies. Apart from his little book *Bad Times*, these ephemera largely took the form of letters to the press and relatively short articles.[59]

I have selected the following topics from Wallace's writings: some are covered more briefly than others.

- *Paper money.* Wallace was opposed to inflation as it reduced living standards. But he was also opposed to gold and was a pioneering advocate of a stable paper currency. His proposal on this was remarkably "modern."
- *Depressions.* The increase in general prosperity had been accompanied by increases in poverty, unemployment, and inequality. His analysis had similarities to the later "under-consumption" theories of the sort popularized by John A. Hobson, but his most striking contribution was to suggest inequality as a cause of persistent unemployment.
- *Reciprocity in trade.* Wallace was in favor of free trade (with important exceptions) but only if it was pursued by all. He suggested a mechanism for achieving this.
- *Capital markets.* Wallace favored "small" capital markets and was critical of limited liability and joint-stock companies, as they encouraged speculation.
- *Taxation.* Wallace suggested that a smooth tax function, that is, one

without "steps," would be preferable and suggested a form that this might take for income taxes and death duties.

- *Immigration*. This should be limited if it reduced the wages of workers, or where poverty and unemployment were present.
- *Speculation*. Wallace was strongly against the development of anonymous financial markets.
- *The environment*. Wallace deplored the ill effects of industrialization (now known as "externalities"), both nationally and globally.

Not all of these were areas in which Wallace had much expertise. But they illustrate that whenever the great man brought his mind to bear on a subject the outcome would invariably be interesting and, sometimes, useful.

Wallace's Scheme for Paper Money

Late in the 1890s, Wallace expressed his implacable hostility to gold:

> it must never be forgotten that the whole enormous amount of human labor expended in the search for and the production of gold; the ships which carry out the thousands of explorers, diggers, and speculators; the tools, implements and machinery they use; their houses, food, and clothing, as well as the countless gallons of liquor of various qualities which they consume, are all, so far as the well-being of the community is concerned, absolutely wasted. Gold is not wealth; it is neither a necessary nor a luxury of life, in the true sense of the word.[60]

In the same spirit Keynes was later to remark that the gold standard was a "barbarous relic."[61] Whether or not a gold standard was to be followed, it was widely agreed that some statistical basis for comparing price levels (and hence living standards) was needed. Wallace was by no means the first to attempt a calculation of price-index numbers. Eighteenth- and even seventeenth-century predecessors may be found: Joseph Lowe had constructed an index of wage rates as early as 1823.[62] Wallace's own contribution did not come until the end of the century, and by then the level of technical discussion among statisticians was already quite sophisticated. It is perhaps surprising therefore to find Wallace's contribution being singled out for praise by Irving Fisher, possibly the greatest of twentieth-century monetary theorists: "In 1898, another well-known Englishman, *Alfred Russel Wallace*, the naturalist, advocated the use of an index number in order to ar-

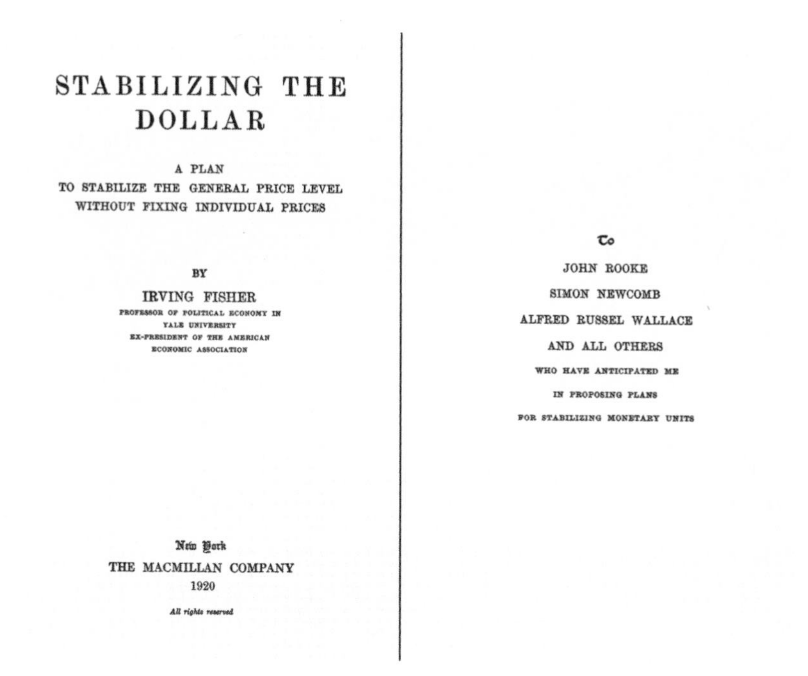

STABILIZING THE DOLLAR

A PLAN
TO STABILIZE THE GENERAL PRICE LEVEL
WITHOUT FIXING INDIVIDUAL PRICES

BY

IRVING FISHER
PROFESSOR OF POLITICAL ECONOMY IN
YALE UNIVERSITY
EX-PRESIDENT OF THE AMERICAN
ECONOMIC ASSOCIATION

New York
THE MACMILLAN COMPANY
1920

All rights reserved

To
JOHN ROOKE
SIMON NEWCOMB
ALFRED RUSSEL WALLACE
AND ALL OTHERS
WHO HAVE ANTICIPATED ME
IN PROPOSING PLANS
FOR STABILIZING MONETARY UNITS

Figure 8.1 Title and dedication pages to Irving Fisher's *Stabilizing the Dollar* (1920).

rive at a stable money, which, as he explained, could very well be in the form of an inconvertible paper currency. In the opinion of this man of science, the index, representative of the necessaries of life, should include: 'Food, clothing, houses, fuel and literature.'"[63]

Fisher went even further. He had already dedicated his book *Stabilizing the Dollar* to "John Rook, Simon Newcomb, Alfred Russel Wallace and all others who have anticipated me in proposing plans for stabilizing monetary units" (figure 8.1).[64]

Fisher's reference to "plans for stabilizing monetary units" explains his emphasis on Wallace. For Wallace had not merely concocted an index; he had proposed a complete approach to the administration of a paper currency. Fisher's "compensated dollar" plan was controversial and, as Patinkin has pointed out, inconsistent with his own quantity theory of money.[65] Patinkin notes that Wallace "as an outsider to economics" was not attached to gold and that his plan was actually more in the spirit of the quantity theory than Fisher's own plan! Here, Patinkin writes, "was a true anticipator of the Chicago School of the 1930s."

Fisher's citing of Wallace, then, was based on something other than his

Table 1. Proportions of standard products consumed and their value on the average years: 1890–1896

Product	Amount	Value
Bread	10,000 pounds	£50
Meat	4,000 pounds	£200
Sugar	1,500 pounds	£10
Tea	500 pounds	£40
Timber	1,000 cubic feet	£100
Coal	200 tons	£200
Etc., etc., etc.	. . .	£600

Wallace's illustrative commodity composition. After Wallace (1898h).

intelligent, though not particularly original, price index. Taken in its entirety, his contribution was impressive. Wallace's monetary "regime" was made up of the following: (1) paper money, (2) a zero-inflation target, (3) a "scientific" price index, (4) an independent monetary authority, and (5) a monetary "rule." Wallace's target was zero inflation, since undue inflation always reduced the living standards of the poor. His base-weighted price index was averaged over about seven years. The relative "weight" attached to each price would be determined by its expenditure relative to total expenditure. Thus the price of bread would get a weight of 50/600 or one-twelfth (see table 1).

Wallace was careful not to claim too much for his index, which he realized was based on his own empirical observations rather than a scientific price survey: "These proportions and prices are put down at a mere guess, but when obtained as accurately as possible for the whole of the 50 or more commodities chosen, we shall have, as a result, that these quantities of these commodities have, on the average of the last seven (or 10 or 20) years' cost a certain gross sum."[66]

Wallace, as was his habit, was not content to theorize but sought to put his construction into practice. His monetary regime would require an independent body: "To do this, a Minister, or Commissioner of Currency, with a sufficient staff of clerks, will be appointed, whose duty it will be to have regular returns made of the market prices of the standard commodities week by week, and to have the averages calculated." This independent body would implement the following simple monetary rule: "If during any month or quarter these averages are seen to fall continuously, that is, everything becomes cheaper, he will advise the Treasury to issue more notes which they will bring into circulation (by using them to pay salaries and current expenses) till the fall is checked and the true average reached. When, on the other hand, the standard goods show a rise in price, it indicates that there is a slight surplus of the currency, which is to be checked by cancelling old notes, as they come back to the Treasury." Wallace also took the trouble to consider the form of paper money: "Here, then we should have a most useful and portable currency—which could be issued for any amounts in very thin or tough cards about the size of railway tickets, and of different colors for the different denominations—and which would be a stable measure of value as well as a convenient instrument of exchange." This monetary regime would effectively replace the gold standard: "the great point is, that by regulating the amount of notes issued in the way above described, this money would become a real *measure of value*, which gold can never be as long as its production is a matter of private speculation, and its cost, and consequent value in exchange, liable to indefinite variation." So although Wallace devoted nothing like the time and energy to monetary reform he had to land nationalization, he made a pioneering contribution to monetary reform.

Assessment. By the middle of the nineteenth century paper money had acquired a poor reputation. To some degree this was due to John Law, who had advocated the creation of paper money (in theory backed by land).[67] Law gave insufficient weight to the inflationary effects of his policies, and his experiments in France and in Louisiana and Mississippi ended in disaster. A fear that the issue of inconvertible paper would lead to inflation was confirmed, for orthodox British economists, by the experience of the Napoleonic Wars. After various crises (at the rate of approximately one per decade) the Bank Charter Act (1844) was passed to limit overissue. Unfortunately the act did not quite solve the problem since when there were trade crises, it could be suspended by Parliament to avoid choking off the supply of money. Otherwise there was no mechanism for ensuring that the supply of money (determined largely by gold mining) would meet the needs of trade. Was gold suitable as the basis for money supply? Wallace clearly thought not.

Monetary authorities in the twentieth century have broadly followed the path Wallace advocated: that is, a retreat from convertibility. Notionally, governments continued to issue paper money based in theory on "convertibility" but in practice on "confidence." Where there was confidence, citizens would not trouble themselves to turn their paper into gold. Where there was not, convertibility had to be abandoned. The United Kingdom famously returned to gold in 1925 at the prewar parity, precipitating a depression in the export industries. Six years later the United Kingdom abandoned the gold standard. However, gold continued to underpin the global system until President Nixon severed the link between the dollar and gold in 1971. It has taken a long time for the world's monetary systems to free themselves from gold as Wallace had advocated.

Monetary "rules" from the 1970s up to the crisis of 2008 followed Wallace's suggestions to a surprising extent. The new orthodox notion was that unemployment would settle at the so-called "natural rate." The task of the monetary authority was to choose an interest-rate policy that ensured a target rate of inflation. If the market had confidence in the authorities, the target rate of inflation would be the expected rate. There would be no nasty monetary surprises. The main differences between the new orthodox system and Wallace's were (1) that Wallace's inflation target was zero, and (2) that Wallace's control mechanism relied on the issue and withdrawal of notes whereas the new orthodox system made use of the whole range of open market operations. All of this changed after 2008 when it became clear that focusing on the inflation rate had led to a neglect of asset price inflation. The response to the resulting crisis was a "quantitative easing" whereby banks were (electronically) given government debt that they could use as a basis for lending, thus stimulating the economy. What might Wallace have said about this? Three things, possibly. First, the crisis wouldn't have arisen in the first place because his rules on lending would have prevented "casino" banking. Second, if quantitative easing were required, the money should go directly into public expenditure rather than to banks.[68] This alternative is widely advocated today by critics of quantitative easing. Third, there was a danger that monetary expansion would simply push up asset and bond prices.[69] The latter two points are very relevant to the present situation.

Wallace on Depressions

Wallace's little book *Bad Times,* on depressions of trade, was published while he was very busy with land nationalization.[70] It is therefore not surprising to find that land reform was part of his solution to the problem at hand. But there is much else in the essay, and just as in the case of the land nationalization book, it is instructive to look at the full title: *Bad Times: An Essay on the Present Depression of Trade, Tracing It to Its Sources in Enor-*

mous Foreign Loans, Excessive War Expenditure, the Increase of Speculation and of Millionaires, and the Depopulation of Rural Districts; with Suggested Remedies.

Though Wallace dismisses some of the popularly advocated causes such as overproduction, free trade, bad harvests, and monetary policy, he himself can be accused of throwing everything but the kitchen sink at the problem. There are two features for which it is possible to claim some originality, however. One is the international flavor of Wallace's explanation. When exports were doing well (as in the early 1870s), there was an enormous expansion in foreign loans, much commented upon at the time. Much of this lending was unproductive, being wasted on preparations for war. However, the burden of repayments meant that foreign populations were unable to buy British exports, so there was a depression in trade and employment. Wallace assigns his international mechanism an important role in explaining the depression. I shall consider its plausibility below. Equally important was Wallace's linking of the depression to trends in inequality as measured by the incidence of pauperism on the one hand and the number of millionaires on the other. As is typical of Wallace he did not merely allege these facts but gave chapter and verse. But here he encountered a difficulty because he broadly accepted Say's law, which states that income going to millionaires is "spent" just as much as income going to the poor. He could have made an excessive-savings argument here, contradicting Say's law. Instead he produced what we might call a "structural distortion" argument: that redistribution in favor of millionaires causes more luxury items to be produced and fewer wage-goods. As a result there is a shortage of wage-goods: production has been distorted, so that fewer workers can be maintained.

Hardship of both urban and rural populations was, in Wallace's eyes, also linked to the land question. Movement from the countryside into the towns (again well-documented) decreased agricultural output and put downward pressure on urban real wages.

Assessment. By the middle of the nineteenth century it was recognized that capitalist development was accompanied by more or less regular trade cycles with a period of eleven years or so. Often the upper turning point would be characterized by speculation and financial collapse. It was reasonable, therefore, to look to money as a main cause. Until Sir Robert Peel's reforms in the United Kingdom of the 1830s, part of the problem was the overissue of money by private banks. Whether this could cause a crisis was a matter of contention between the "currency" and "banking" schools. The latter taught that there could be no overissue, since bank lending would be based on "real bills" reflecting the needs of trade. As already described, the money supply was more or less regularized with the 1844 act.

Along with this monetary theme ran another much disputed thesis: the impossibility of a general "glut," that is, a collapse in demand. Its alleged impossibility was implicit in Say's law, that supply creates its own demand. This had been an important matter of contention between Malthus (and other dissenters) on the one hand and Ricardo, Mill, and the like on the other. Essentially the doctrine alleges that savings create demand just as consumption does, but it is a demand for investment goods, not consumer goods. On the whole classical economists accepted Say's law and therefore supported the view that a depression could not be due to "over-saving" or "under-consumption."

A typical modern explanation of the business cycle would run as follows.[71] The economy is on a long-term upward trend. From time to time it experiences "shocks" of various sorts that are transmitted over time and may easily be shown to generate more or less modest cycles superimposed on the upward trend. Wages and profits both do better during the boom than during the slump years, so there is apparently little reason for a clash of interest. Inequality seems to play little or no part in explaining the cycle.

Historically, however, there were at least two contributions that flatly contradicted this picture. Marx's theory of capitalist development implied that increasing inequality was inherent in the system with an increasing "organic composition of capital" (capital to labor ratio) and an increasing rate of "exploitation."[72] For Marx inequality is inherent in capitalism rather than a problem for reformers to tackle. Another, much less well-known, economist to link inequality with the cycle was the socialist writer John A. Hobson.[73] Hobson and Wallace both knew Hyndman, and all three were strongly anti-imperialist, so it is very likely that Hobson and Wallace knew of one another (though Hobson's economics was becoming well known only toward the end of Wallace's life). Hobson is now mainly known as an exponent of the "underconsumptionist" explanation of the cycle. He eschewed Say's law, so was able to argue that increasing inequality (measured as in Wallace by the number of millionaires) leads to a restriction of purchasing power and hence consumption. The underconsumptionist aspect of his theory, though not the distributional aspect, was later taken up enthusiastically by Keynes.[74]

So what can be rescued from Wallace's own, relatively brief, contribution? Apart from the great land question, Wallace had emphasized (1) the international nature of the depression and (2) its association with inequality. He was certainly correct in arguing for an international analysis. However, the extent and importance of British lending to Continental governments has been questioned in recent decades, and it has become clear that earlier estimates were too high.[75] Wallace could not have known, of course, that this was the case, and it remains true that excessive debt (whether to Britain or to other lenders) must have had a depressing effect on the demand for British exports. As for 2, capitalism was (and is) characterized by income inequality. But this was, on his own view, mainly due to the private ownership of land and could be remedied, in principle, by nationalization. A major difficulty for Wallace was his belief in Say's law. To overcome this difficulty Wallace offered his own rather idiosyncratic structural theory. For his argument to work, however, the production of luxuries would have to be less labor intensive than the production of necessities. Wallace did not explore this idea further, so his theory, which was essentially a theory of depression rather than of the cycle, remained incomplete.

Wallace on Trade Reciprocity

Wallace was in principle strongly in favor of free trade, except for certain items such as coal and other minerals: "They must be considered to be held by us in trust for the community, and for succeeding generations. . . . [Mining] disfigures the country, diminishes vegetable and animal life, and destroys the fertility (for perhaps hundreds of generations) of large tracts of valuable land. . . . I maintain that it is a wrong to our own population, and a still greater wrong to the next generation, to permit the unlimited export of those mineral products which are absolute necessities of life, but which once destroyed we can never reproduce."[76]

The export of weapons, explosives, alcoholic drinks, and poisons should also be banned. But in line with Wallace's standard position on trade and the price mechanism, he was in favor of free trade (for run-of-the-mill goods). Monopolies were a special case. Writing in *The Malay Archipelago* of the nutmeg trade in Banda, he had defended the Dutch government's monopoly: "Had the Government not kept the nutmeg trade of Banda in its own hands, it is probable that the whole of the islands would long ago have become the property of one or more large capitalists. The monopoly would have been almost the same, since no known spot on the globe can produce nutmegs so cheaply as Banda, but the profits of the monopoly would have gone to a few individuals instead of to the nation."[77]

Apart from these excluded commodities, and what one might call countervailing monopoly power, free trade was beneficial. However, Wallace asserted that it was beneficial only if carried out by all countries. If not, the exports of one country could be hindered by the tariff policies of other countries. Wallace's contributions were brief and to the point. In 1879 he complained in the *Nineteenth Century* that England had become "devoted" to the principle of reciprocity, even if other nations failed to follow its example. He argued, convincingly, that the essence of free trade was mutuality, and its advantages, on a global scale, depended on the whole program being carried out: "the *whole programme* of free trade must be carried out if its advantages are not to be overbalanced by disadvantages."[78] How could this be achieved? By the principle of reciprocity: "It is, to reply to protectionist countries by putting the *very same import duty* on *the very same articles* that they do, changing our duties as they change theirs."[79]

This was an important suggestion because, in the absence of an international agreement, some sort of credible retaliation strategy is required. Wallace briefly restated his position in the *Spectator*, noting the paradox that imposing a duty, if it is a countervailing duty, is nevertheless in the spirit of free trade. Robert Lowe, the politician, among others, had questioned whether Wallace could be a free trader. One could advocate reciprocity, Wallace replied, even if one were a free trader.[80]

Assessment. The "pure milk," as it were, of the classical doctrine was unilateral free trade.[81] Adam Smith thought of exports as a vent for surplus products in which a country had an absolute advantage, and prescribed as wide a market as possible for them: imports should be obtained as cheaply as possible. Ricardo showed that trade was mutually beneficial even without absolute advantage as long as countries had differing "comparative advantage."[82] There was, then, a strong presumption in favor of free trade. The long process of tariff dismantling began under Pitt the Younger and continued steadily under the Huskisson, Peel, and Gladstone administrations. The pure milk was not to everyone's taste, however. Robert Torrens, for example, originally a keen free trader, had come to believe in a policy of tariff reciprocity, that is to say, free trade should only be pursued if other countries also pursued it.[83] Further, he had shown that a strong country could theoretically use tariffs to turn the terms of trade in its own favor. A more practical consideration was that countries such as Germany and the United States were already using tariffs to protect their "infant industries" as part of their development strategies. So it is not surprising that many accepted the reciprocity argument. Indeed the Cobden Free Trade Treaty with France (1860) accepted that unilateralism essentially was dead.

Modern approval of the benefits of free trade relies, in practice, on an implied principle of reciprocity. Indeed Wallace was quite right to question the superiority of free trade when it was not being pursued by all countries. His view is consistent with the so-called "general theory of second best."[84] This states that it might not be optimal to have one efficiency condition (e.g., free trade by one country) in place unless all the other efficiency conditions are in place at the same time. Or, more loosely, the best may be the enemy of the good.

Now consider policy. The experience of the interwar years was that, if left to themselves, countries would use tariffs as a beggar-my-neighbor remedy for unemployment. Consequently, as part of the postwar settlement most countries signed up to the General Agreement on Tariffs and Trade (which subsequently became the World Trade Organization). A series of tariff-reduction "rounds" gradually reduced most tariffs on industrial goods in a reciprocal fashion: it was a multilateral process. Unless there is a binding multilateral treaty, tariff policy takes on the characteristics of a "game." Readers familiar with the prisoners' dilemma game might like to think of "free trade" as the cooperative strategy and "tariffs" as a noncooperative strategy. In a one-off game there is the usual noncooperative outcome. But in a sequential game it is possible for players to "signal" a willingness to play cooperatively, backed by a credible retaliatory strategy if others do not. Wallace's proposed "signal" was the threat of retaliation. His proposal may be seen as an attempt to devise a strategy for a sequential game of trade: similar to the "tit-for-tat" strategy, one of the standard modern approaches.

Wallace on Capital Markets

I have already explained that Wallace, unlike many other critics of capitalism, for example Ruskin, was an advocate of the price mechanism. Writing of Dobbo, in the Aru Islands, he commented, "Here we may behold in its simplest form the genius of Commerce at the work of Civilization. Trade is the magic that keeps all at peace, and unites all these discordant elements into a well-behaved community."[85] This confidence in the price mechanism was especially clear in the case of rent. Wallace, you may recall, was determined that tenants should pay their quit-rents. Like the orthodox economists, he believed that rents and prices were essential for the efficient allocation of resources. But what of interest? Wallace had no difficulty with the charging of interest as such, which he defended rather neatly: "£100 paid a-year in ten years hence is *not* as valuable as £100 pounds paid to-day. To say that it is so is really to say that it has *no value* to-day, for if its payment can be delayed one year without loss it can two, or three, or ten, or a hundred, or a thousand!"[86] Interest was special for Wallace because it directly affected justice between generations. Two examples were the power of trusts and the claims of consol (undated security) holders. Trusts were unjust because they imposed a "slavish adherence to the expressed or implied wishes of the dead." Consols, too, were undesirable because they effected a transfer of resources from workers to rentiers who would become "an ever-increasing class of idle rich."

In addition to these difficulties with interest Wallace was extremely critical of the way in which capital markets were developing. He strongly opposed the principle of limited liability for joint stock companies granted in 1855, as it created an artificial difference between borrowers and lenders that encouraged speculation and gambling. What should be done?

> The true solution to the problem is, I believe, to be found in the proposition that all loans should be *personal*, and therefore *temporary*; and that, as a corollary, the repayment of the capital should be provided for in the annual payments agreed to be made by the borrower, either for a fixed period (if he live so long), or for the term of his life. This would abolish the idea of perpetual interest, which is as impossible in fact as it is wrong in principle, while it would avoid the injustice of compelling one man, or set of men, to pay the debts of a preceding generation from they may have received no real benefit. . . . There would remain no safe investments for money, except in some branch of agriculture, manufactures, or commerce in which either the investor or some relation or friend was personally interested, and thus would be brought about the diminutions and practical abolition of usury as a system, and of whole classes living idle lives on the interest of money derived from the accumulations of previous generations.[87]

Wallace's objection was to the scale and anonymity of capital markets: they should be small scale and personal.

But what of the generational problem? Wallace's treatment of interest was consistent with his position on inheritance. Members of one generation should not be able to bind the freedom of action of future generations.

Hence Wallace's disapproval of trusts. Nor should one generation plunder natural resources: it should, instead, be a custodian on behalf of future generations. But what of the process of capital accumulation itself? How was the capital stock built up by one generation to be passed on to the next? It is true that Wallace was skeptical about the importance of physical capital (innovation and ideas being far more important), but surely he was not arguing that capital accumulation should cease? No, it was simply that the private ownership of capital would gradually be replaced by state ownership. Wallace had relatively little to say about the transfer of private non-agricultural capital, short of that ultimate goal.

Assessment. The issue of consols by governments does not seem to present a special moral problem, for the "present value" of an income stream y forever is simply y/i, where i is the discount rate; and if we know the present value, we can always find a comparable shorter-dated security. So there is nothing unjust about allowing long-undated securities. It is entirely a matter of preference. This argument takes no account of mortality, however. Wallace's point was that in practice the payment of interest on long-dated securities represents the "dead hand" of a previous generation on the lives of the living in a similar manner to the provisions of trusts. His fear was that these payments to the owners of consols would become increasingly oppressive. How likely was this? The ratio of debt to national income increases when the interest rate is greater than the rate of growth of money GDP. For most of the twentieth century, due to inflation, the opposite has been the case.

Wallace's critique of joint-stock and limited liability ran directly counter to the spirit of the age. The great advantage of limited liability was, of course, that investors were only liable for the amount they had invested rather than for the whole of their capital. Since investors were liable only for their investments, without putting house and home at risk, huge amounts of capital could be raised, far more than from personal contacts or local banks. Wallace might have made the point that some industrial empires were built up successfully by "plowing back" private capital and not going to the stock market at all. The dangers, as Wallace saw them, were speculation and increasing inequality, both of which (as we have seen) would decrease stability for the ordinary working man. In a Wallacean "small is beautiful" society there would be less rapid capital accumulation but, perhaps, greater well-being.

Wallace on Progressive Taxes

Wallace's rather brief contribution to capital tax reform had three elements, two of them straightforward and the third rather surprising. One element was progressivity: he took the case for a progressive tax system (in which the proportion of tax rises with capital) for granted. The second element was smoothness: progression, he argued could be achieved by means of a "smooth" function, which he says he had derived "many years" before, rather than a step function.[88] Under his proposal the rate of tax would vary as the square root of the value of assets. He offered a numerical example. Count assets in units of £1,000. Take the square root of the number of units. This gives the percentage tax rate. For example, capital of £4,000 counts as four units. Take the square root. This gives a tax rate of 2 percent. So a capital of £9,000 would be taxed at 3 percent, a capital of £100,000 at 10 percent, and so on. Intermediate capitals would be treated using the same formula. The more unusual element of Wallace's proposal was that the average tax rate (not the merely the marginal rate) could rise to 100 percent at high levels of wealth (in his example £10,000,000). At that level the entire estate would be taxed away. Below that, there would be some level of wealth such that the amount handed on to beneficiaries was maximized, providing a disincentive for the rich to accumulate "too much" wealth.

Assessment. Capital taxation was a live issue in Britain at the time of Wallace's contribution, as various land taxes were, in 1894, being amalgamated into a single estate duty. Wallace's concern with the details of a capital tax schedule indicated, perhaps, that he had accepted that a complete abolition of inheritance of assets other than land might be politically unattainable. Short of abolition, a progressive tax was greatly to be desired. How should it be constructed? Almost all schemes of taxation involve "steps" such that the taxpayer might be confronted by a jump in the marginal tax rate. This, as Wallace realized, can distort behavior as the better-off try to avoid getting into a higher tax band. So a smooth tax function might help reduce tax evasion. It is easy enough to concoct such tax functions: that they are rarely, if ever, used is partly due to practicalities that may admittedly become less important with increasing computerization of tax systems. And it is also true that reasonable degrees of progressivity may be achieved by means of a simple linear schedule with a threshold, which has the advantage of being clear and understood. A finer and more sensitive degree of progression would indeed require something like Wallace's smooth, nonlinear function. His more idiosyncratic idea of a tax rate rising to 100 percent on the whole of income or capital (not just on the marginal amount) has to be understood in terms of Wallace's extreme distaste for large fortunes, which he regarded as morally wrong, as conferring unwarranted privileges on some of the next generation, and as contributing toward persistent unemployment.

Wallace on Immigration

In a letter to the *Clarion* Wallace briefly considered the issue of immigration.[89] To restrict immigration was, he recognized, an evil. But unrestricted "immigration of aliens" was also an evil when the condition of the working people was poor. "When thousands and millions of our own people are struggling for work, and often cannot obtain it, and other thousands are working long hours for barely enough to keep body and soul together, then it may be—and I believe it is—a greater wrong to permit free immigration from every other country."

Assessment. In this fragment, Wallace's main concern is, as usual, with the well-being of ordinary working people. Being Wallace, he added that bad conditions overseas, due to the private ownership of land (!), had probably caused the influx in the first place. As in the case of reciprocity in trade, discussed above, Wallace wished to adhere to the underlying principle of free movement, which he well understood, but was prepared to modify policy so as not to reduce well-being. This resonates today, with resistance to large waves of immigration from North Africa to Europe or from parts of Southeast Asia to Thailand.

The Environment

In Wallace's time environmental economics had not yet been formulated. But the modern environmentalist must surely be impressed by Wallace's frequent comments, especially as he grew older, on the relationship between humankind and the environment. His principal recurring critique of industrial society is to be found mainly in the handbook on nationalization and in his regular presidential addresses to the LNS. The drift of these comments was, of course, that industrialization and the dirt and pollution of city life was bad for people's health and well-being, and that all would be well if land were to be nationalized. Then they would desert the cities, work their plots, and rear their children in the good country air. I have quoted some of his remarks already; in addition the following are typical:

> the labourer has now to pay much higher house-rent, he has generally no garden, and, being usually a weekly tenant, is so dependent on his landlord that he cannot make the most of what he has; the commons and roadside wastes from which he formerly obtained fuel for winter, with food and litter for a cow, a donkey, geese or poultry, have almost all been enclosed; and the result is that he has few means of adding to his scanty wages.[90]
>
> Their hills and valleys become full of furnaces and steam engines; their green meadows are buried beneath heaps of mine-refuse or destroyed by the fumes from copper-works; their waving woods are cut down for timber to supply their mines and collieries; their towns and cities increase in size, in dirt, and in gloom; the fish are killed in their rivers by mineral solutions, and entire hill-sides are devastated by noxious vapours.[91]

Wallace came increasingly to argue that pollution of the atmosphere and of water was the most important issue facing society. Pure air and water were "the one great and primary essential of a people's health and well-being, to which *everything* should, for the time, be subordinate."[92] We have also seen that Wallace opposed overexploitation of natural resources: these should be held in trust by us for future generations, who will almost certainly be damaged by our plunder of the earth and the reckless destruction of the stored products of nature.[93] He was, of course, always skeptical of the alleged superiority of large-scale agriculture and its "artificial manures . . . with which we now pollute our rivers and keep up a full crop of zymotic diseases among our population."[94]

Remarks such as these indicate that, typically, Wallace was thinking not merely of his own and other industrializing countries but of the planet as a whole, embracing the effects of rainfall, sun, wind, etc. and their interaction. Discussing the creation of droughts and deserts due to deforestation, he wrote, "Knowingly to produce such disastrous results would be a far more serious offence than any destruction of property which human labour has produced and can replace; yet we ignorantly allow such extensive clearings for coffee cultivation in India and Ceylon, as to cause the destruction of much fertile soil which generations cannot replace, and which will surely, if not checked in time, lead to the deterioration of the climate and the permanent impoverishment of the country."[95]

And his comments on climate change, not then an issue, were highly insightful: "in all densely-populated countries there is an enormous artificial production of dust . . . chiefly from our enormous combustion of fuel pouring into the air volumes of smoke charged with unconsumed particles of carbon. This superabundance of dust . . . must almost certainly produce some effect on our climate."[96] He even had doubts about the future of tourism, then a negligible concern. Looking up at the great Amazonian canopies, he speculated that "[t]he whole glory of these forests could only be seen by sailing gently in a balloon over the undulating flowery surface above: such a treat is perhaps reserved for the traveller of a future age."[97] But later, in *The Malay Archipelago*, he reflected on tourism's dark side: "should civilized man ever reach these distant lands . . . we may be sure that he will so disturb the nicely-balanced relations of organic and inorganic nature as to cause the disappearance, and finally the extinction, of [the] very beings whose wonderful structure and beauty he alone is fitted to appreciate and enjoy."[98]

Assessment. Though Wallace was familiar with the North American conservation movement—indeed he visited the "ruthlessly destroyed" redwood forests with the conservationist John Muir[99]—he could hardly be described as a "conservationist" in the modern sense. After all, he was a collector by profession and killed (or arranged to have killed) a great many specimens. It was how he made his living. And the pioneer of evolution by natural selection could never argue that the existing stock of creatures was immutable. In any case, the number of specimens, worldwide, must have seemed at the time to be more or less unlimited. The activities of collectors at least enabled people, before the age of photography and film, to appreciate the beauty of creatures they would otherwise have had no chance of experiencing. Moreover, the early Wallace (unlike the later anti-imperialist Wallace) was not opposed to the intrusion of Western man. Thus, in commenting that "an Indian spends a week in cutting down a tree in the forest, and fashioning an article which, by the division of labor, can be made for sixpence,"[100] he implied the superiority of civilized methods. In a similar vein he concluded that "the indolent disposition of the people, and the scarcity of labour, will prevent the capabilities of this fine country being developed till European or North American colonies are formed."[101]

It was the later Wallace who managed to create an intellectual unity from his experiences in the Amazon and Malay Archipelago, his love of nature, and his socialism. This gave him the vision to see that the effects of industrialization and economic development under capitalism could actually reduce the well-being of ordinary people, especially future generations. What could be done? Wallace's implied answer, that all would be well under socialism, was hardly adequate. But, as we have noted, he could not have been expected to invent the then unknown topic of "environmental economics." This required two new concepts: the theory of "externalities" (where social costs are greater than private costs) and the theory of "public goods" (or "bads"). Ironically, and unknown to Wallace, his fellow researcher into spiritualism, Henry Sidgwick, had gone some way toward developing these concepts,[102] but they were not to become part of the mainstream until much later. On these great economic and social issues Wallace is therefore to be seen as a prophet rather than a policy analyst.

Conclusion

This chapter has been concerned with the nuts and bolts of Wallace's contributions to political economy. It has eschewed grand themes such as the linking of natural selection to the alternatives of capitalism and socialism. For what it is worth, I think it can be argued that Darwin's version of natural selection sits more comfortably with capitalism as a struggle for survival than does Wallace's. Moreover, I find it difficult to interweave Wallace's numerous interventions in economic or social policy with his more important work on natural selection. If we take Wallace's view that natural selection had by now done most of its work (our walking upright, development of the brain and use of tools, etc.) the future direction of *Homo sapiens* could be determined by human design rather than by the blind pressures of biological selection. His version of natural selection left him as free as any other Victorian intellectual to speculate on, and more importantly to design, economic and social arrangements for improving the well-being of ordinary people.

Acknowledgments

I am grateful to George Beccaloni for initially stimulating my interest in Wallace, and to Charles Smith for subsequent encouragement.

Notes

1. For discussions of particular aspects of Wallace's work, specifically land nationalization and monetary reform, see Gaffney (1987, 1997) and Patinkin (1993).

2. In this period also, it is claimed, terms such as "sympathy" and "benevolence" disappeared from mainstream economics. See Peart and Levy (2005) and Collard (1978).

3. Wallace (1894f), pp. 177, 184.

4. Whitaker (1996), letter 180.

5. Wallace (1889c).

6. Bellamy (1889).

7. Some of these issues are explored in Benton (2013).

8. For an Aristotelian (as opposed to Benthamite) view of welfare see Nussbaum and Sen (1993).

9. Piketty (2013) charts increasing inequality in the modern period just as Wallace had done for the nineteenth century.

10. Wallace (1898d), p. 377.

11. Wallace (1853b), p. 261.

12. Wallace (1885a) quoted Joseph Arch with approval. Arch was the first president of the National Union of Agricultural Workers. See also Wallace (1885d).

13. Wallace (1882b), p. 216.

14. Engels's (1892) account did much to awaken the English middle classes to the condition of the working class.

15. Wallace (1885b), p. 10.

16. Wallace (1869c), pp. 597–598.

17. Sraffa (1962), chapter 10.

18. Wallace was encouraged by Cairnes's (1874) criticisms of rent. Cairnes has sometimes been described as the last of the classical economists.

19. Wakefield (1849).

20. Mill (1848), book 2.

21. Land Tenure Reform Association (1871), p. 9.

22. Wallace had, however, been exposed to Owenism in his youth.

23. Wallace (1896).

24. Wallace (1905b), vol. 2, p. 236.

25. Wallace (1882b).

26. Spencer (1851).

27. Wallace (1880c), p. 735.

28. Wallace (1892b).

29. Wallace (1905b), vol. 1, p. 157.

30. In a diatribe against so-called scientific agriculture, Wallace suggested the use of the "manurial products" of 40 million people instead of importing artificial fertilizers (Wallace 1889b).

31. Benton (2013), p. 176.

32. Fawcett (1883).

33. Wallace (1883d), p. 490.

34. Wallace (1880c).

35. For a discussion of the later history of the terminable annuity idea, see Dalton (1954) and Peacock and Rizzo (2002).

36. Wallace (1881a, 1881c, 1881d).

37. Wallace (1886c).

38. Wallace (1883c).

39. Wallace (1885b.

40. Wallace (1889b), p. 22.

41. Drucker (2010) refers to the "Wallace insight" that technology was not about tools but about "work," by which Wallace meant ingenuity and application. Wallace, of course, excelled at both.

42. Stigler (1969), pp. 206–207.

43. Wallace (1893b).

44. Wallace (1895a).

45. This section draws heavily on Collard (2001).

46. Fichman (2001), p. 247.

47. George (1912).

48. Marchant (1916b), vol. 1, p. 318.

49. George (1912), p. 326.

50. See Collier (1979).

51. Stigler (1969).

52. Ibid., p. 206.

53. Ibid., p. 207.

54. Wallace (1893b), p. 3.

55. Stigler (1969), p. 183.

56. Giffen (1837–1911) was a distinguished economist and statistician who became chief statistical adviser to the government.

57. Blaug (1996), p. 184.

58. Scottish Government (2014).

59. Wallace (1885c).

60. Wallace (1898d), p. 372.

61. Keynes (1936), p. 172.

62. Lowe (1823).

63. Fisher (1934), p. 42.

64. Fisher (1920).

65. Patinkin (1993), p. 18.

66. See Wallace (1898h) for this and subsequent quotations on money.

67. Law (1705).

68. Wallace (1906a).

69. Wallace (1906b).
70. Wallace (1885c).
71. See, for example, Blanchard and Fischer (1989).
72. Marx (1887).
73. Hobson (1896).
74. Keynes (1936), chapter 23.
75. See, for example, Platt (1980).
76. Wallace (1873c).
77. Wallace (1869c), p. 295.
78. Wallace (1879c), p. 642.
79. Ibid., p. 645.
80. Wallace (1879d, 1879f).
81. O'Brien (1975).
82. Ricardo (1817), chapter 7.
83. Torrens (1833).
84. Lipsey and Lancaster (1956).
85. Wallace (1869c), p. 444.
86. Wallace (1884), p. 150.
87. Ibid., pp. 150–151.
88. Wallace (1894g).
89. Wallace (1904b).
90. Wallace (1892f), p. 112.
91. Wallace (1879c), p. 641.
92. Wallace (1903b), p. 256.
93. Wallace (1898d).
94. Wallace (1889b), p. 22.
95. Wallace (1878c), pp. 20–21.
96. Wallace (1898d), p. 83.
97. Wallace (1853b), p. 36.
98. Wallace (1869c), pp. 448–449.
99. Wallace (1905b), vol. 2, p. 158.
100. Wallace (1853b), p. 172.
101. Ibid., p. 80.
102. Sidgwick (1883).

‹ 9 ›

Physical Geography, Glaciology, and Geology

CHARLES H. SMITH

In the past I have been on record several times observing that, his importance to evolutionary biology notwithstanding, Wallace should perhaps most fundamentally be recognized as a geographer—a geographer who "happened to be interested in the subject of evolution."[1] The reason for this assessment is not only his association with biogeography, or even additionally his land reform work; important as well was his nearly lifelong interest in physical geography, especially glaciology.

This is one of the least known aspects of Wallace's career, though a few of his ideas on geological subjects have met with frequent discussion. Still, it is his writings on glaciology, other perhaps than those on evolutionary biology, that best typify him as a thinker. These have received almost no attention in the general literature.[2]

Wallace's most potent influences in this general realm were the writings of Alexander von Humboldt and Charles Lyell. Humboldt, as noted in chapter 1, provided Wallace with a model of synergistic surface energies: the notion of a physical "general equilibrium of forces." Lyell gave him an understanding of deep history, one amenable to an appreciation of a uniformitarian kind of biological change based on persisting influences (as distinct from catastrophism). Wallace's own background in surveying also was an influence. We begin here by noting Wallace's interest in geodesy, a subject perhaps not familiar to all readers.

Geodesy

Geodesy is the science of the mathematical measure of the size and shape of the earth, and its immediately resultant geophysical characteristics. That Wallace would someday take up related problems was perhaps foreshadowed by his early interest in trigonometry, and his several years as a surveyor apprenticed to his older brother William. The survey work was also his initial exposure to geological subjects, including fossils, and doubtless stimulated his curiosity as to how the various settings within which he plied his trade had taken form. In the mid-1840s he took on surveying work leading to the production of three regional maps of apportionments in the South Wales area.[3] This experience served him well when he voyaged to the Amazon Valley, where he constructed a map of the course of the Rio Negro that remained a standard source for some fifty years.[4]

Wallace did no original work on the subject of geodesy per se, but he did become involved in one of the most celebrated historical episodes concerning it. In 1865 he published a pair of letters replying to the illogical thoughts of a flat-earther; this might have been the end of the matter, but several years later he decided to answer a challenge posed by another flat-earther. This individual, John Hampden, had bet £500 (quite a sum in those days) that no one could produce a proof that the earth was not, indeed, flat. Drawing on his expert knowledge of surveying, Wallace at once came up with such a proof. It involved finding a long stretch of stationary water, setting up markers along it, and using a telescope to show that the markers did not line up vertically as a straight line. It was a foolproof design, if conducted with care, and the experiment was carried out on a six-mile stretch of the Old Bedford Canal in Norfolk on 5 March 1870. Referees and an umpire had been appointed to reach a decision, which went in Wallace's favor. But it turned out that Hampden was an irrational zealot who would not pay up. He eventually turned to harassment and hounded Wallace and his family so severely that he was brought up on charges including libel, and imprisoned for a significant period of time. He remained a thorn in Wallace's side until his death in 1891.

Wallace later recognized his lapse in judgment on this matter, discussing it at length in *My Life* in 1905. *My Life* also contains a second discussion on matters geodesic: Wallace's lucid explanation of the counterintuitive fact that degrees of latitude increase in absolute length from the equator to the poles.[5] Wallace wanted an example to illustrate how he was able to overcome contrary facts once he gained access to clear explanations of them, and it is interesting that he chose this particular problem to make his point.

Geology

Despite his twelve years in the field as a young man, and occasional brief interludes later, Wallace carried out a rather limited amount of original geological investigation. Nevertheless, he did write on geological subjects from time to time, and involved himself in at least three discussions that were significant issues of the day. These concerned the geological age of the earth, the permanence of the oceanic basins and continental masses, and the classification of islands.

The Geological Age of the Earth

By the second half of the nineteenth century geologists were beginning to explore questions extending beyond the descriptive work that had provided a firm foundation for the field. The differing causal forces that distinguished the various rock types were now understood, at least in outline, and fossils were appreciated for what they actually are: the remnant traces of once-living forms. Significantly, however, no way to calculate the absolute ages of rock units or fossils had yet been discovered (this would come only in the twentieth century). Before 1859 and *On the Origin of Species*, this was not a major problem, but once workers began to accept evolution as fact, it became apparent that a reasonable amount of time would be needed to support the development of complex organisms from their originally single-celled progenitors. Most people were estimating that at least several hundred million years would be necessary to accomplish the observed results.

A major challenge to the evolution theory came in the 1860s when the prominent physicist William Thomson (later known as Lord Kelvin), a devout Christian, published calculations estimating the earth's age as being from 20 to 400 million years; later he would restrict his estimate to a mere 20 to 40 million years. Thomson's estimate was based in part on a pre-thermonuclear understanding of the likely age of the Sun and was thus completely off track, but Darwin and his followers, including Wallace, rightly saw this as a threat to their evolutionary model.

Darwin nervously avoided the matter, hoping (correctly, as it turned out) that the great physicist had made some kind of error. Wallace took a different tack: although no direct way to measure absolute geological age had been discovered, perhaps an indirect measurement could be constructed, based on estimates of the rate of deposition of the earth's sedimentary units? In a series of papers and letters to the editor extending from 1870 to the mid-1890s, Wallace set forth, and repeatedly defended, lines of reasoning backing the legitimacy of this approach. His first efforts, addressing various elements of the uniformitarianism model, appeared in his 1869 review of the tenth edition of Lyell's *Principles of Geology*.[6] This was followed by a more specific treatment, "The Measurement of Geological Time," which was printed in two parts in the first volume of the journal *Nature* in 1870.[7] In this work he begins with a discussion of the time measurement problem, alluding to the remarks of Thomson, and quickly draws attention to the relevance of astronomical considerations as possibly affecting recent processes, especially those connected to the Ice Age. This discussion was considerably expanded in following years, as we will see shortly.

The second part of the same paper begins,

> We have now to consider an entirely distinct set of facts which have an important bearing on the probable time elapsed since the last glacial epoch. Messrs. A. Tylor, Croll, and Geikie have shown that the amount of denudation now taking place is much greater than has generally been supposed. The quantity of water discharged by several rivers and the quantity of sediment carried down by those rivers have been measured with tolerable accuracy, and allowing for the difference of specific gravity between sediment and rock, it can be easily calculated, from the known area of each river basin, what average thickness has been removed from its whole surface in a year, since all the matter brought down by the river must evidently have come from some part of its basin. In this way it is found that the Mississippi has its basin lowered 1/6000 of a foot per annum; the Ganges, 1/2358; the Rhone, 1/1526; the Hoang-Ho, 1/1464; the Po, 1/729.

After a good deal more discussion, Wallace eventually concludes that Lyell's estimates of the duration of geological time might be reduced by as much as nine-tenths, leaving the time elapsed since the beginning of the Cambrian period as only 24 million years.

In a subsequent treatment Wallace introduced other observations on possible deposition rates (e.g., of the time it takes to produce stalactites and stalagmites in caves),[8] and several years later provided an updated survey of his thoughts in his book *Island Life*.[9] This produced a number of objections from the scientific establishment to which Wallace responded—satisfactorily, it seemed, for the time being. But when the revised second edition of the book came out in 1892, a flurry of further criticisms arose. Wallace again, however, stood his ground.

Radiometric dating methods were introduced fifteen years later, in 1907, and the discussion of sedimentation rates quickly lost its urgency.

The Permanence of Oceanic Basins and Continental Masses

In retrospect, the discussion on the permanence of oceanic basins and continental masses may be seen as resulting from a lack of information on geological mechanisms and the general characteristics of biodiversity—in terms of both the latter's geographical component and the genetics of individual organisms and populations. To begin with, before the 1960s there was no universal agreement that the continental masses moved with respect to one another; the theory of continent drift was not even seriously put forth until 1915, two years after Wallace's death. Thus, late-nineteenth-century discussion simply revolved around whether, and to what extent, continental and oceanic areas moved up and down relative to one another.

The driving dilemma was a biogeographic one: how populations that seemed to be closely related to each another could sometimes be separated by huge expanses of water, or land (see further discussion in chapter 10). The existence of so-called disjunct populations on land could be attributed to earlier episodes of extinction in the areas in between,[10] but the problem of closely related forms existing on the opposite sides of oceans was more difficult to address. Some naturalists began to propose—some would say "invent"—extensions of continental surface to account for these; these included "land bridges" that ran right across the middle of oceans, or even assuming that continent-sized masses once existed but in more recent periods had presumably disappeared beneath the waves.

Wallace's positions on such hypotheses were generally rather conservative. While in some instances he could imagine sea-level changes and land emergence leading to connections and disconnections between islands or between islands and continents—for example, in the Malay Archipelago, a region with which he was intimately familiar—a basic prerequisite was an ocean depth small enough to permit uniformitarian kinds of events (subsidences, elevations, volcanism, sea-level changes, etc.). Where the evidence suggested large expanses of practically bottomless water, however, it was difficult to imagine how any known process could lead to the emergence of bridges or whole continental masses. His primary discussions of this subject appeared in several places;[11] in one of them, a lecture given at the Rugby School in 1881, he states that "a large body of evidence points to the certainty of the permanence of land and water," and then lists as the main arguments

> 1. The great depth of the ocean compared with the height of the land. 2. The great depth and general flatness of the ocean bed. 3. The nature of the sedimentary rocks, proving the proximity of land from which the sediment was derived, such rocks occurring through all geological time. 4. Over all the continents beds are found containing remains of birds and beasts, many of them of large size, such as would require large areas of continent for their support. 5. Absence of stratified deposits from oceanic islands. 6. The wide-spread oceanic ooze distinct from all known sedimentary rocks.[12]

In 1892, he elaborated on point 1 above:

> The accompanying diagram [see figure 9.1] . . . will better enable the reader to appreciate these proportions, which are of vital importance in the problem under discussion. The lengths of the two parts of the diagram are in proportion to the areas of land and ocean respectively, the vertical dimensions showing the comparative mean height and depth. It follows that the areas of the two shaded portions are proportional to the bulk of the continents and oceans respectively.
>
> The mean depths of the several oceans and the mean heights of the several continents do not differ enough from each other to render this diagram a very inaccurate representation of the proportion between any of the continents and their adjacent oceans; and it will therefore serve, roughly, to keep before the mind what must have taken place if oceanic and continental areas had ever changed places. It will, I presume, be admitted that, on any large scale, elevation and subsidence must nearly balance each other, and, thus, in order that any area of continental magnitude should rise from the ocean floor till it formed fairly elevated dry land, some corresponding area must sink to a like extent. But if such subsiding area formed a part or the whole of a continent, the land would entirely disappear beneath the waters of the ocean (except a few mountain peaks)

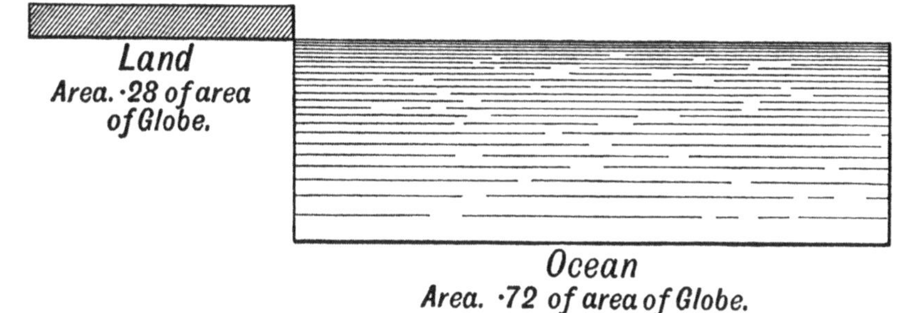

Figure 9.1 Wallace's diagram of the relative mean height of land and depth of oceans. From Wallace (1889a), p. 345.

> long before the corresponding part of the ocean floor had approached the surface. In order, therefore, to make any such interchange possible . . . [w]e must suppose either that when one portion of the ocean floor rose, some other part of that floor sank to greater depths till the new continent approached the surface, or, that the sinking of a whole continent was balanced by the rising of a comparatively small area of the ocean floor. Of course, either of these assumed changes are conceivable and, perhaps, possible; but it seems to me that they are exceedingly improbable, and that to assume that they have occurred again and again, as part of the regular course of the earth's history, leads us into enormous difficulties.[13]

A few months later he responded to criticism of his position with his final words on the matter:

> while admitting with pleasure the growing approximation of views on this subject, I cannot forget that it has been, and still is with many writers, the practice to assume former continental extensions across the great oceans in order to explain difficulties in the distribution of single genera or families; that geologists of repute have claimed the Dolphin bank in the Atlantic trough as the relic of a chain of mountains comparable with the Andes; that oceanic islands have been recently claimed to be merely the tops of submerged mountains, which can only be properly compared with the highest points of continents, and that a geological critic so late as 1879 considered the idea that the oceans had always been in their present positions "a funny one." If such extreme views are now less common than they

> were, I hope that I may, without presumption, claim to have had some share in bringing about the change in scientific opinion now in progress.[14]

Despite these words, the "land bridges" approach to reconciling problems of organic distribution did not go away entirely until the emergence of continental drift theory: that is, until decades after Wallace's death.

The Classification of Islands

Discussions on how to classify islands proceeded largely on the basis of their histories as understood through the geology involved, and the observed kinds of biota they contained. Wallace followed Darwin's lead on this matter, lending additional evidence to the basic understanding:

> Islands, according to Darwin, consist of two radically distinct classes: (1) continental, (2) oceanic.
>
> The oceanic islands have never formed part of continents, though not necessarily very distant from them, but are the result either of volcanic upheaval or coralline formation. They are always surrounded by deep sea, that is, sea of more than 1000 fathoms depth. Of these coralline islands we never can find the foundations, but probably they are founded on old volcanic surfaces worn away by sea action. These exhibit great peculiarities in their natural history, in that they have neither mammals nor amphibia, but contain a small number of reptiles, birds, and other classes, and these are very often distinct from anything found elsewhere in the world.
>
> Continental islands may be subdivided into Recent and Ancient.
>
> Recent continental islands have been separated in recent times so that their geological features are the same, and their natural history nearly the same as those of continents. Examples of these are England, separated by a sea 22 miles wide from Europe, and Borneo, separated by 100 miles from Asia. Though they contain the same animals as the continents near them, the number is smaller and the varieties fewer.
>
> Ancient continental islands are of older date and are always separated by deeper sea, thus while the recent ones are always within the 100 fathoms line, the ancient are outside the 100 but within the 1000 fathoms line. In regard to their natural history, though they contain many forms identical with those of the continent, they contain also many which are totally distinct, so that there must have been a great change in the forms of life,

> and this change implies great lapse of time. Geologically they always agree with the continents, and exhibit a fair sample of the continental geology.[15]

This plan, quite obviously, served a biogeographic (and ultimately evolutionary) purpose. Notwithstanding the later advent of continental drift theory, it remains a largely valid way of ordering the data. Wallace made use of the classification on several occasions as a way of logically organizing his presentations, most notably in the book *Island Life*.[16]

Physical Geography and Glaciology

We now come to the meat of this chapter, on physical geography subjects. Wallace's contributions in this realm feature thoughts on glaciology, climatology, and geomorphology.

Geological Climates and the Cause of the Ice Age

By the time of the publication of the *Origin*, it was evident that the characteristics of animal and plant distribution were intimately tied to historical geological and climatological processes. The climatic aspect of the model was as yet incomplete, however. Since the revelation in the 1840s of glacial periods in earth history, the causes, timing, and extent of glaciation were much debated. Lyell discussed the latest thinking in successive editions of *Principles of Geology*, continuing to maintain that the overall extent and distribution of land and open ocean was the main factor responsible for periods of warming and cooling in the earth's history—what he called "geographical factors." In 1864, however, a paper entitled "On the Physical Cause of the Change in Climate during the Geological Epochs" appeared, and Lyell felt compelled to reassess the causes of "geological climates" ("paleoclimate," in today's terms). This was the first of several related works by the autodidact polymath James Croll (1821–1890), an astronomer and climatologist who was soon to become secretary of the Scottish Geological Survey. In these studies Croll made a compelling case for an astronomical explanation for climate cycles. Building on the work of earlier astronomers, Croll showed how the eccentricity of the earth's orbit (see figure 9.2), obliquity of the ecliptic, precession of the equinoxes, and the distance of the earth from the sun during perihelion and aphelion might together yield cycles of greater and lesser insolation, and in turn dramatic changes in the

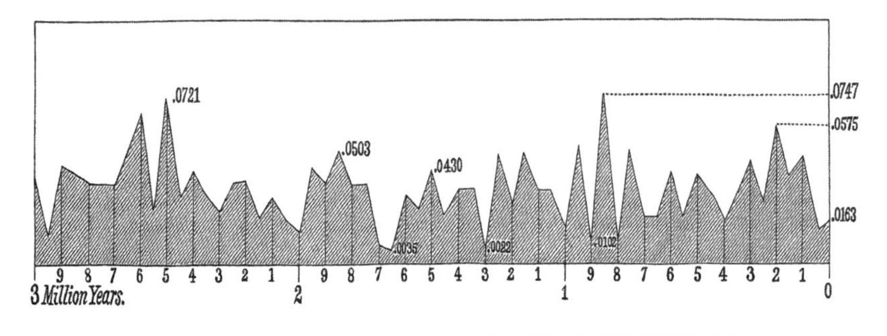

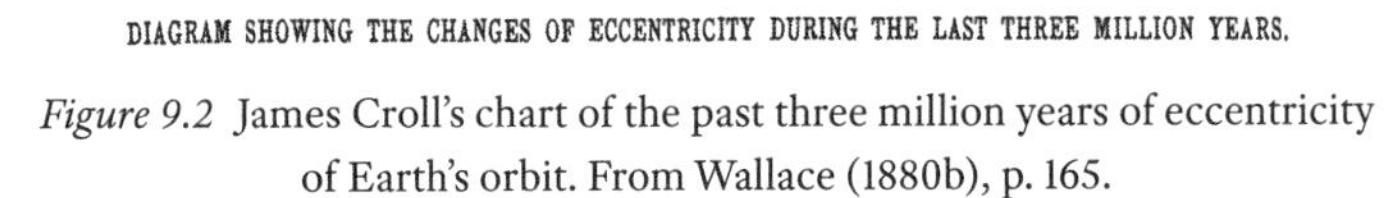

Figure 9.2 James Croll's chart of the past three million years of eccentricity of Earth's orbit. From Wallace (1880b), p. 165.

distribution of heat around the earth via ocean currents. Thus such astronomical phenomena might give rise indirectly to glacial and interglacial periods.[17] Croll's approach was imperfect and largely dismissed by his contemporaries, yet it would later provide a foundation for the work of Serbian mathematician and astronomer Milutin Milanković (1879–1958) in the twentieth century, whose eponymous Milankovitch cycles describe the astronomical basis for glacial and interglacial periods.[18]

In general, Lyell was less enthusiastic than Wallace and Darwin over astronomical explanations of changing climate, insisting that these played a secondary role as compared with geographical factors such as the distribution of land and sea.[19] For Wallace, however, Croll's ideas were a revelation. As he put it in his 1869 review of Lyell's *Principles,* they constituted "some of the boldest speculations of modern science."[20] He thought he recognized common ground between the two positions, applauding Lyell's devotion of space to the subject: "the main vicissitudes of geological climate, and especially the alternations of warm and cold climates of which we have distinct evidence, have been to some extent due to regularly recurring astronomical cause, whose effects were, however, profoundly modified, or sometimes entirely neutralised, by terrestrial changes of surface acting in conjunction with, or in opposition to them."[21] Harkening back to his "more recondite forces" point of view, he could now envision a geologically and climatologically changing earth upon which species evolve, their populations ebbing and flowing subject to the limitations imposed upon them both by other species and the barriers and corridors that open and close dynamically over time. The migration patterns of life were thus related to the inexorably cycling land levels and climate, the evolution and spread of species taking place within the context of more or less permanent continents and ocean basins ultimately giving rise to distinctive and distinguishable Sclaterian biogeographical regions (see chapter 10).

Wallace now set his sights on more detailed explorations of some of the processes suggested by this general understanding. Although most of his efforts featured biological and biogeographical problems, he also continued to thrash out a model of continental glaciation combining geographical and astronomical causal forces.[22] He must have spent much of the 1870s thinking out the relationship, during which he was considering interpretations of glacial processes by reigning experts James and Archibald Geikie; digesting treatises on arctic geology, climate, and paleobotany by the likes of Oswald Heer, Searles Wood, J. W. Judd, James Croll, James Dwight Dana, and Samuel Haughton; and reading up on the investigations carried out by arctic explorers such as Julius Payer and G. S. Nares. By 1879 he had enough put together to publish a lengthy review of the recent literature that featured some of his own conclusions on the matter.[23] These were incorporated into the first sections of *Island Life* a year later.

The treatment of Wallace's model in *Island Life* extends to more than a hundred densely argued pages. He begins with a summary of the evidence supporting the existence of former great ice sheets; he then reviews the various theories that had been introduced to explain the phenomenon, and goes through the evidence for alternating cold and warm periods on the earth at the higher latitudes. Finally, he moves on to his attempt to integrate Croll's cyclical eccentricity thoughts with the competing geographical models of the time. Beyond this brief outline, however, Wallace's discussion defies a quick reduction to the dimensions of an abstract. Complicating matters is the nearly total neglect of his thoughts both by the scientists of his time, and later naturalists and historians. Thus his own summary words, though lengthy, might serve us best in our current mission:[24]

> Looking at the subject broadly, we see that the climatic condition of the northern hemisphere is the result of the peculiar distribution of land and water upon the globe; and the general permanence of the position of the continental and oceanic areas—which we have shown to be proved by so many distinct lines of evidence—is also implied by the general stability of climate throughout long geological periods. The land surface of our earth appears to have always consisted of three great masses in the north temperate zone, narrowing southward, and terminating in three comparatively narrow extremities represented by Southern America, South Africa

and Australia. Towards the north these masses have approached each other, and have sometimes become united; leaving beyond them a considerable area of open polar sea. Towards the south they have never been much further prolonged than at present, but far beyond their extremities an extensive mass of land has occupied the south polar area.

This arrangement is such as would cause the northern hemisphere to be always (as it is now) warmer than the southern, and this would lead to the preponderance of northward winds and ocean currents, and would bring about the concentration of the latter in three great streams carrying warmth to the north-polar regions. These streams would, as Dr. Croll has so well shown, be greatly increased in power by the glaciation of the south polar land; and whenever any considerable portion of this land was elevated, such a condition of glaciation would certainly be brought about, and would be heightened whenever a high degree of excentricity prevailed.

It appears to be the general opinion of geologists that the great continents have undergone a process of development from earlier to later times. Professor Dana says: "The North American continent, which since early time had been gradually expanding in each direction from the northern Azoic, eastward, westward, and southward, and which, after the Palæozoic, was finished in its rocky foundation, excepting on the borders of the Atlantic and Pacific and the area of the Rocky Mountains, had reached its full expansion at the close of the Tertiary period. . . ." A similar development undoubtedly took place in the European area, which was apparently never so compact and so little interpenetrated by the sea as it is now, while Europe and Asia have only become united into one unbroken mass since late Tertiary times.

If, however, the greater continents have become more compact and massive from age to age, and have received their chief extensions northward at a comparatively recent period, while the antarctic lands had a corresponding but somewhat earlier development, we have all the conditions requisite to explain the persistence, with slight fluctuations, of warm climates far into the north-polar area throughout Palæozoic, Mesozoic, and Tertiary times. At length, during the latter part of the Tertiary epoch, a considerable elevation took place, closing up several of the water passages to the north, and raising up extensive areas in the Arctic regions to become the receptacle of snow and ice-fields. . . . [T]he occurrence at this time of a long-continued period of high excentricity necessarily brought on the glacial epoch in the manner already described in our last chapter.

We thus see that the last glacial epoch was the climax of a great process of continental development which has been going on throughout long geological ages; and that it was the direct consequence of the north temperate and polar land having attained a great extension and a considerable altitude just at the time when a phase of very high excentricity was coming on. Throughout earlier Tertiary and Secondary times an equally high excentricity often occurred, but it never produced a glacial epoch, because the north temperate and polar areas had less high land, and were more freely open to the influx of warm oceanic currents. But wherever great plateaux with lofty mountains occurred in the temperate zone a considerable *local* glaciation might be produced, which would be specially intense during periods of high excentricity; and it is to such causes we must impute the indications of ice-action in the vicinity of the Alps during the Tertiary period. . . .

Estimate of the comparative effects of Geographical and Astronomical Causes in producing Changes of Climate.—It appears then, that while geographical and physical causes alone, by their influence on ocean currents, have been the main agents in producing the mild climates which for such long periods prevailed in the Arctic regions, the concurrence of astronomical causes—high excentricity with winter in aphelion—was necessary to the production of the great glacial epoch. If we reject this latter agency, we shall be obliged to imagine a concurrence of geographical changes at a very recent period of which we have no evidence. We must suppose, for example, that a large part of the British Isles—Scotland, Ireland, and Wales at all events—were simultaneously elevated so as to bring extensive areas above the line of perpetual snow; that about the same time Scandinavia, the Alps, and the Pyrenees received a similar increase of altitude; and that, almost simultaneously, Eastern North America, the Sierra Nevada of California, the Caucasus, Lebanon, the southern mountains of Spain, the Atlas range, and the Himalayas, were each some thousands of feet higher than they are now; for all these mountains present us with indications of a recent extension of their glaciers, in superficial phenomena so similar to those which occur in our own country and in Western Europe, that we cannot suppose them to belong to a different epoch. . . .

No doubt a prejudice has been excited against it in the minds of many geologists, by its being thought to lead necessarily to frequently recurring glacial epochs throughout all geological time. But I have here endeavoured to show that this is not a necessary consequence of the theory, because a concurrence of favourable geographical conditions is essential

> to the initiation of a glaciation, which when once initiated has a tendency to maintain itself throughout the varying phases of precession occurring during a period of high excentricity. When, however, geographical conditions favour warm Arctic climates . . . then changes of excentricity, to however great an extent, have no tendency to bring about a state of glaciation, because warm oceanic currents have a preponderating influence, and without very large areas of high northern land to act as condensers, no perpetual snow is possible, and hence the initial process of glaciation does not occur.
>
> The theory as now set forth . . . is in perfect accord with the most recent teachings of the science as to the gradual and progressive development of the earth's crust from the rudimentary formations of the Azoic age, and it lends support to the view that no important departure from the great lines of elevation and depression originally marked out on the earth's surface have ever taken place.
>
> It also shows us how important an agent in the production of a habitable globe with comparatively small extremes of climates over its whole area, is the great disproportion between the extent of the land and the water surfaces. For if these proportions had been reversed, large areas of land would necessarily have been removed from the beneficial influence of aqueous currents or moisture-laden winds; and slight geological changes might easily lead to half the land surface becoming covered with perpetual snow and ice, or being exposed to extremes of summer heat and winter cold, of which our water-permeated globe at present affords no example. We thus see that what are usually regarded as geographical anomalies—the disproportion of land and water, the gathering of the land mainly into one hemisphere, and the singular arrangement of the land in three great southward-pointing masses—are really facts of the greatest significance and importance, since it is to these very anomalies that the universal spread of vegetation and the adaptability of so large a portion of the earth's surface for human habitation is directly due.

At the very least it seems that Wallace deserves some credit for thinking out the problems of global surface change to a degree recognizing synergistic forces among the likely causal elements involved. Astronomical and geographical factors alike remain under scrutiny as an army of investigators continue to examine their implications for both the biological and physical aspects of evolution.

Ice Movement and Glacial Features

Although Wallace's interest in the glacial period centered on how its onset might have affected—and *effected*—episodes of evolutionary change in living things, he also tried to understand some more specific aspects of glacial movement and erosion processes. His first analysis of glacial features came as early as 1867, with an essay titled "Ice Marks in North Wales (with a Sketch of Glacial Theories and Controversies)."[25] The apparent purpose of the work was to provide "readers of this periodical" with "a popular account of such prominent glacial phenomena as are observable in all our chief mountain districts," though as one proceeds through it an additional agenda emerges, one connected to some observations he made "during a month spent near Snowdon and Cader Idris last autumn."

Directly Wallace writes, "We may conveniently consider the chief evidences of a glacial period under the following heads: 1st, The drift; 2nd, Moraines; 3rd, 'Roches moutonnées'; 4th, Grooved and striated rocks; 5th, Boulders and perched blocks; 6th, Alpine lakes."[26] After reviewing the first five features fairly straightforwardly, he sets up his discussion of the sixth evidence rather differently:

> It is only about five years since Professor Ramsay[27] propounded the startling theory that almost all the lakes which form one of the greatest charms of our mountain districts, were actually produced by that comparatively recent irruption of thick-ribbed ice over a great part of the temperate zone, which we can hardly contemplate without a thrill of horror; and that during the preceding warm tertiary epochs they were so scarce as to form no important feature in the scenery of Europe. A short and simple statement of this theory is as follows. In all districts where glaciers have been proved to exist there are numerous lakes. In exactly similar districts where there is no trace of there having ever been glaciers, there are few or no lakes. This holds good all over the globe. Glaciers wear away their beds, as proved,—first, by the immense quantity of sediment in all glacial streams; secondly, by the existence of "roches moutonnées" wherever glaciers have passed. It can almost always be shown that the old glaciers have passed over the exact spots where the lakes now are, and the size of the lakes bears a general proportion to the proved size of the old glaciers. This theory of the glacial origin of Alpine lakes is now the great battleground of physical geologists. In this country Ramsay, Jukes, Geikie, and

> Tyndall are its chief supporters; Sir Roderick Murchison and Sir Charles Lyell, its chief opponents. Every year brings fresh evidence and new combatants; and as it is a question of such great interest, and at the same time one rather of physical than of purely geological science, I shall endeavour to give such an outline of the subject as may enable the general reader to understand the question at issue and form his own judgment upon it.[28]

Wallace ultimately sided with Ramsay, as in following years did an increasing number of other observers. But even twenty-five years later there were still detractors, prompting Wallace to write one of his most successful natural science essays, "The Ice Age and Its Work," in 1893.[29] In this two-part paper, Wallace effectively revisits his earlier treatment; the result, as one recent writer put it, was "a masterly demolition of the glacial protection, or 'no erosion' position."[30] In it Wallace "brought together the whole of the evidence bearing upon the question, and adduced a completely new argument for this mode of origin of the valley lakes of glaciated countries. This is founded on their surface and bottom contours, both of which are shown to be such as would necessarily arise from ice action, while they would not arise from the other alleged mode of origin—unequal elevation or subsidence."[31]

Years earlier Wallace had been embroiled in a discussion concerning the way glaciers move. Glaciologist Doug Benn has very nicely summarized one part of the question in a commentary he provided some years ago on a short Wallace article, "The Theory of Glacial Motion":

> The debate on the causes of glacier motion lasted for many years during the 19th Century. Major sticking points were (1) the fact that ice appears to be a rigid, brittle material, yet the large-scale motion of glaciers requires that the ice flows as a continuous mass; and (2) a mass under a constant force (gravity) should be expected to accelerate downhill (as all falling objects do) whereas glaciers exhibit periodic, often annual, fluctuations in speed. Wallace's note . . . correctly dismisses some of the fallacies current at the time, giving reasons why ice motion cannot be due to freeze-thaw action or "molecular adjustment." His own preferred mechanism, of localised fracture and regelation (healing of fractures by refreezing) is closer to the mark, but still not complete. . . .
>
> Where Wallace does hit true, however, is his recognition of basal sliding as the main mechanism by which Alpine glaciers move. His comments that meltwater reaching the bed will encourage motion by "melting away

its lower surface, and . . . to some extent, buoying it up"[32] are exactly correct. This is indeed the reason why Alpine glaciers flow more rapidly during the day and in summer, when meltwater fluxes are highest. Confirmation of this fact had to await the development of drilling equipment, allowing the real-time measurement of subglacial water pressures simultaneously with accurate velocity measurement.

Wallace was not the first scientist to emphasise basal sliding (de Saussure has this honour), but he certainly had a more lucid view of the problem than many of his contemporaries.[33]

Another glacial movement issue that Wallace considered was whether glaciers could, under the right circumstances, move uphill on a sustained basis (in contrast with the way water flows). In part 1 of the 1893 work he writes,

It is evident that, to have produced such effects as are here described, the glacier must have extended much beyond Soleure, and have been very thick even there. It thus proves to demonstration that a glacier *can* travel for 100 miles over a generally level country, that it *can* pass over hills and valleys, and that, even near its termination, it *can* groove, and grind, and polish rocks, and deposit large masses of hard boulder-clay. And all this was done by a single glacier issuing from a comparatively narrow valley, and then spreading out over an area many times greater than that of its whole previous course. In this case it is clear that such a vast mass of ice, constituting a veritable ice-sheet on a small scale, could not have derived its motion solely from the push given to it by the parent glacier at St. Maurice. Neither could gravitation derived from the slope of the ground have affected it, for it passed mostly over level ground or up slopes, and its termination at Soleure is actually nearly 200 feet higher than its starting-point at the mouth of the valley below St. Moritz! There remains as a cause of motion only the slope of the upper surface of the glacier, the ice slowly flowing downward, and, by means of its tenacity and its viscosity on a large scale, dragging its lower portion still more slowly over the uneven or upward-sloping surface.[34]

An immediate response by Sir Henry Howorth in the pages of *Nature* drew the following exasperated retort from Wallace: "The work of ice on the rocks is as clear as that of palæolithic man on the flints; all the difficulties that may be suggested as to how he lived, or how he shaped the flints do not in the slightest degree affect our conclusion that the palæolithic flint implements are the work of *man*; and there is equally clear evidence that ice *did* march a hundred miles, mostly uphill, from the head of Lake Geneva to Soleure, whatever transcendental qualities it must have possessed to do so."[35]

Wallace's writings on glaciology subjects both small and large are fairly extensive, in total exceeding five hundred pages in print. Consider the preceding review only the proverbial tip of the iceberg.

On the Existence of Southern Glaciations

As a naturalist who was both well known (and thus frequently contacted) and who took a strong interest in the developing literature on his interests, Wallace served a little-credited function by helping keep his colleagues apprised of the potentially important ramifications of foreign or obscure publications. We have already seen this with his notice of Müller's German publication recognizing a new form of mimicry (chapter 1); he also drew attention several times to possible instances of past Southern Hemisphere glaciation episodes. In 1867 and 1870 he discussed evidence presented by Louis Agassiz and Charles Frederick Hartt of possible ice action in the Amazon and southern Brazil; in 1893, with the help of new information provided by geologist J. C. Branner, he showed how these thoughts had been mistaken.[36] In 1874, in a review of Thomas Belt's *Naturalist in Nicaragua*, he passed on Belt's report of possible evidence of ancient glaciations there;[37] years later he introduced findings in Victoria and Tasmania, Australia, suggestive of glacial actions in those places, then contributed to a resulting discussion in the pages of *Nature*.[38] In such efforts he invariably took a conservative position, always calling for additional researches and cautious interpretations.

Climate Synergistics

The extent to which Wallace immersed himself in glaciology studies again reflects his immense debt to the program of Humboldtian science. The modern term "synergistics" is appropriate in this context, for although Wallace was in no sense a founder of such studies any more than he was of cybernetics (as discussed in chapter 1), his efforts may decidedly be viewed as exploratory efforts in these directions.[39] This shows up most clearly in the ways he approached the study of climate.

Wallace made no basic science contributions to climatology, but his particular approach to climate as a causal force was unprecedented in his own time, and barely ever equaled since then. Other scientists have, of course, made climate a deterministic force in their various models of natural process, but Wallace can be singled out for having understood it as contributing interactively to a wide range of natural and social inertias. His applications of related principles extend all the way from his very first writings to his last ones. We can do no more here, unfortunately, than to list some of the most obvious.

- His contemplation of climatic influence as a means of explaining evolutionary modification of adaptations (discussed in chapters 1 and 10).
- His eventual understanding that changes in climate provided the main impetus for recursive adaptive change, via the natural selection model (wherein selection of new traits is akin to the action of a governor on a steam engine, returning the overall system to ecological equilibrium: chapters 1 and 4).
- His integration of the effects of climate into his model of human racial differentiation, wherein natural selection is superseded by our ability to invent clothing and other devices of protection from the elements (chapters 2 and 6).
- His enlistment of climate as an explanation of biological diversity gradients (chapter 10).
- His understanding that climate can be changed to our detriment anthropogenically, through the disruption of natural cycles by human activities (chapters 8 and 11).
- His model of the interplay of astronomical events with climate, causing, among other things, the great Ice Ages (chapters 9 and 12).
- His essay on the importance of atmospheric dust, in which he links its presence to a number of important physical processes fundamental to our presence on this planet.[40]
- His anticipation of the factors underlying climate on distant worlds, affecting their capacity for hosting life (chapter 12).

Other basal synergisms can easily be identified in other aspects of Wallace's work (e.g., in his approach to spiritualistic development, his economics, his understanding of social good, and his approach to adaptation), but nowhere does his penchant for this kind of thinking show up more clearly than it does with his writings on climate. The story of this newly advanced ecological point of view, and its relation to the changing world of science in the nineteenth century, could be an excellent dissertation project for some enterprising graduate student.

Perspective

In Wallace's time the pioneering studies of Louis Agassiz and others on glaciation processes were still new, and many details of the phenomenon were still under debate (as indeed they still are). Wallace obviously reveled in the discussion; here was the ultimate integration of complex natural forces into observable physical results! The clues were here, there, and everywhere, and Wallace was always up for a good puzzle. One of the main reasons for his undertaking the North American lecture tour in 1886–1887 was almost certainly a desire to observe firsthand a variety of glacier-influenced landscapes, and indeed his travel journal for that period is filled with related observations.[41]

Wallace undertook a considerable amount of observational fieldwork in North America, but during that ten-month period he was giving his attention to more than just strictly physical, or even biological, features. In chapter 3 we mentioned Wallace's examination of museums in and near Boston and Washington, DC, and subsequent writings on them, but it was more than just the organization of the facilities themselves that held his interest. Much of his time in those cities, and in other places such as Cincinnati and San Francisco, was spent examining, reading studies on, and visiting the sites of glacial and postglacial human activities. He took particular interest in the collections of the Museum of Prehistoric Archaeology at the Smithsonian Institution, and the excavations being carried out near Cincinnati on the remnants of the mound-builders culture. He also spent time reading up on the evidence of early Pacific coast humans during his visit to California in 1887. These efforts produced a review paper, "The Antiquity of Man in North America," published shortly after his return to England later that same year. In typical Wallace fashion, he ended this study with the following words:

> the proper way to treat evidence as to man's antiquity is to place it on record, and admit it provisionally wherever it would be held adequate in the case of other animals; not, as is too often now the case, ignore it as unworthy of acceptance or subject its discoverers to indiscriminate accusations of being either impostors themselves or the victims of impostors.

> Error is sure to be soon detected, and its very detection is often a valuable lesson. But facts, once rejected, are apt to remain long buried in obscurity, and their non-recognition may often act as a check to further progress.[42]

Taking his own advice, a few years later Wallace printed a notice in *Nature* of an obscure government report by a James Terry that among other things discussed the discovery of "three rude, yet bold, characteristic, and even life-like sculptures of simian heads, executed in basalt" near the John Day River in Oregon.[43] Wallace eventually sides with the first of Terry's two attempts to explain these figures, that perhaps "the animals which these carvings represent once existed in the Columbia valley." Wallace's discussion of this report provides a fascinating glimpse into the intellectual world he inhabited, combining elements of his interests in glaciology, geomorphology/geology, archeology, ethnology, and cultural and ecological/evolutionary biogeography.[44]

Acknowledgments

Thanks to my co-contributors and two referees for their constructive remarks on earlier manuscript versions of this essay.

Notes

1. Smith (1991b), p. 7. For an extensive article on Wallace as geographer, see Smith (2010).

2. The only substantial treatment I am aware of is Tinkler (2008).

3. These maps, covering the areas of Neath, Briton-Ferry Demesne, and Llantwit Lower, are now held by the National Library of Wales.

4. This map was printed in Wallace (1853a, 1853b).

5. Wallace (1905b), vol. 2, pp. 40–41. Geographer Yi-Fu Tuan (1963, p. 260) once wrote that "Wallace's account has not been surpassed in clarity by modern textbooks on geography."

6. Wallace (1869d).

7. Wallace (1870a). As an aside, it is sometimes forgotten that Wallace was one of this famous journal's founders; over the years he contributed more than 150 letters, reviews, and articles to its pages, as well as serving an unknown amount of time behind the scenes as an uncredited editor.

8. Wallace (1873d).

9. Wallace (1880b).

10. Wallace produced a key paper on this subject in 1879 (1879b).

11. See especially Wallace (1877a, 1880b, 1881b, 1892d).

12. Wallace (1881b), p. 3.

13. Wallace (1892d), p. 421.

14. Wallace (1892g), p. 718.

15. Wallace (1881b), p. 4.

16. Wallace (1880b).

17. Reviewed by Fleming (2006).

18. See the historical overview by Imbrie and Imbrie (1979).

19. Lyell nevertheless incorporated much of Croll's thinking into the tenth *Principles of Geology* edition (1867-1868)—indeed, more than doubling the space devoted to "geological climates" from the ninth edition, including a new chapter on the astronomical hypothesis with thirty-seven pages discussing Croll's work. See Fleming (1998, 2006).

20. Wallace (1869d), p. 372.

21. Ibid., p. 379.

22. If this trajectory sounds familiar to the reader, it should: to a remarkable extent, Wallace's efforts closely paralleled his attempts to reconcile the views held by the monogenist and polygenist camps on human evolution. See chapter 6.

23. Wallace (1879g).

24. Wallace (1880b), pp. 197–202.

25. Wallace (1867a).

26. Ibid., pp. 34–35.

27. Sir Andrew C. Ramsay (1814–1891).

28. Wallace (1867a), p. 43.

29. Wallace (1893f).

30. Tinkler (2008), p. 193.

31. Wallace (1905b), vol. 2, pp. 389–390.

32. Wallace (1871a), p. 310.

33. Benn (2000).

34. Wallace (1893f), p. 628.

35. Wallace (1893h), p. 52.

36. Wallace (1867a, 1870d, 1893e).

37. Wallace (1874b).

38. Wallace (1892e, 1893c, 1893g).

39. I was recently contacted by an investigator who regarded Wallace, in similar fashion, as effectively a pioneer in the field of biosemiotics, the study of the meaning and interpretation of biological signs and codes. This is not a difficult argument to make, considering, for example, his attention to the meaning of the communication inherent in biological coloration patterns, most notably his theory of organismal recognition marks.

40. Wallace (1898f).

41. Smith and Derr (2013).

42. Wallace (1887e), p. 679.

43. Wallace (1891b).

44. Cryptozoology might also be added to this list, as in recent years the Terry report has been mentioned as some of the earliest scholarly literature bearing on the alleged existence in this area of the apelike creature known as Sasquatch (or Bigfoot). We are reminded of Wallace's words, noting how "facts, once rejected, are apt to remain long buried in obscurity," with "their non-recognition" often acting "as a check to further progress." See Eberhart (2002).

‹ 10 ›

Historical and Ecological Biogeography

JAMES T. COSTA

> It is evident that, so long as the belief in "special creations" of each species prevailed, no explanation of the complex facts of distribution *could* be arrived at or even conceived.
>
> —WALLACE, *Island Life*

Alfred Russel Wallace's contributions to the field of historical and ecological biogeography were as incisive as they were varied. After evolutionary biology itself, this is the second field in which he is regarded as a "founding father," an appropriate label resonant with US "founding father" Thomas Paine's declaration that "a share in two revolutions is living to some purpose."[1] Even without the plethora of papers Wallace wrote over the course of half a century on such topics as the geographical distribution of various faunal groups, systematic regional classification, analysis of biotas, island biogeography, theoretical biogeography, and the interrelationship of evolution, paleogeography, geological history, physical geography, and geographical distribution, Wallace's landmark works *The Geographical Distribution of Animals* (1876) and *Island Life* (1880) alone would have secured his stature as founder of this rich discipline.

Wallace's interest in geographical distribution was of a piece with his interest in the question of species origins and earth history, as he understood that earth processes (geological and climatological) and the evolution and distribution of life on earth go hand in hand. Like these evolutionary processes themselves, Wallace's thinking on some key issues changed over time in light of new perspectives and evidence. Accordingly, I first give a brief overview of the scientific context, or state of biogeographical understanding, as Wallace entered the field. I then take a roughly chronological approach in treating the primary biogeographical subjects Wallace investigated, and their status in light of current thinking. These include (1) the riverine barrier hypothesis; (2) land bridges and continental extensionism; (3) biogeographical realms and the permanence of continents and ocean basins; and (4) latitudinal biodiversity gradients and other aspects of ecological biogeography.

Foundations

The history of biogeography can be succinctly described as a quest to infer process from pattern: specifically, pattern in the geographical distribution of species discerned from empirical observation. Depicting, if not mapping, the geographical occurrences of species is a centuries-old convention, where identifying representative animals and plants of different regions was a natural application of cartography that went hand in hand with the mapping of geographical features and human settlements and boundaries. The database of regional flora and fauna grew exponentially, beginning with the voyages of exploration and discovery of the fifteenth century, inevitably confusing the actual with the fabulous or mythical in some cases. By the late eighteenth century sufficient (though still incomplete) information on regional flora and fauna and other aspects of natural history had accumulated, making it possible to draw thematic maps depicting distribution. The earliest, from the 1770s and 1780s, featured but scattered names and a few distributional limits. Generalizations, however, were beginning to arise, such as "Buffon's law": a label later given to the comte de Buffon's observation in 1761 that the mammalian fauna of the Old and New World tropics differed significantly despite environmental similarity.[2] Most attempts to explain species distributions in this period had been informed by theological considerations, but Buffon (1707–1788) posited a more material process of dispersal and transmutation, believing in an Old World origin of major animal groups followed by subsequent migration to the New World, where transmutational "degeneration" ensued under environmental influence.

The framework for making sense of species distributions in which Wallace worked—what these distributions can teach us about ecological and historical processes, to use modern terms—was established in the years just prior to his birth in 1823. Based on explorations of South America between 1799 and 1804, the renowned German naturalist Alexander von Humboldt generalized Buffon's "law" to birds, reptiles, insects, spiders, and plants. The first publication to stem from the travels of Humboldt and his fellow explorer, French botanist Aimé Bonpland (1773–1858), was *Essai sur la Geographie des Plantes* (*Essay on the Geography of Plants*) of 1807, a work that had the greatest impact on the early study of biogeography. The *Essay* was the first detailed analysis of plant distribution in relation to environment and adaptation: "The geography of plants does not merely categorize plants according to the various zones and altitudes where they are found; it does not consider them merely in relation to the conditions of atmospheric pressure, temperature, humidity, and electrical tension in which they live; it can discern, just as in animals, two classes having a very different kind of life, and, so to speak, very different habits."[3] The significance of plant distribution for inferring historical processes was clearly acknowledged as well:

> In order to determine the ancient link between neighboring continents, geology bases itself on the analogy of the coastal structures, on the ocean beds, and on the types of animals living there. The geography of plants can furnish precious materials for this kind of research: up to a point, it can show how islands that were previously linked are now separated; it can show that the separation of Africa from South America occurred before the development of organized forms of beings. . . . The geography of plants can assist us in going back with some degree of certainty to the initial state of the earth: it can determine if, after the initial retreat of the waters which were abundant and agitated as attested by rocks filled with sea shells, the entire surface of the globe was covered at once with various plants, or whether, as traditional accounts of various peoples say, the earth, now stilled, produced plants only in one area, from which they were transported by sea currents in a progressive march to far-flung regions over the course of centuries.[4]

At the heart of the matter was the essential question of "centers of origin" or "centers of creation"—that is, why and how flora and fauna are distributed as we see them, and whether this provides clues to the very origin of species (albeit in a theological context). Did they have a single regional point of origin (creation) and subsequently spread far and wide across the globe, or were multiple centers of origin possible? In a subsequent essay published a decade later, Humboldt observed that "studies of the law of the distribution of forms lead naturally to the question whether there exist plants common to both [Old and New World]—a question that inspires all the more interest, for it relates directly to one of the most important problems of zoology."[5] He was referring to the "centers of origin" problem.

Swiss botanist Augustin de Candolle (1778–1841) built upon Humboldt's work in his classic *Essai Élémentaire de Géographie Botanique* (1820), in which a distinction was made between environmental and historical factors shaping species distributions, more or less articulating modern concepts of ecological and historical biogeography. Focusing on plants, which he divided into twenty botanical provinces worldwide, de Candolle developed the concept of species "station" (habitat, in modern terms), as something "determined uniquely by physical causes actually in operation" such as rainfall, soil, and climate. He distinguished these from larger-scale "habitations," the regional distributions or ranges of species that are "determined by geological causes that no longer exist today." Proximate environmental factors thus determine species habitat, while regional distribution patterns reflect earth history. Ultimately, de Candolle argued, "All of the theory of geographical botany rests on the particular idea one holds about the origin of living things and the permanence of species."[6]

De Candolle's ideas were championed by geologist Charles Lyell in the second volume (1832) of his watershed work *Principles of Geology*. Referring to the "luminous" *Essai Élémentaire* in chapter 5, the first of four chapters devoted to geographical distribution, Lyell emphasized the central importance of the subject: "Next to determining the question whether species have a real existence, the consideration of the laws which regulate their geographical distribution is a subject of primary importance to the geologist." He went on to describe the phenomenon of regionalization that de Candolle had pioneered: the "parcelling out of the globe amongst different *nations*, as they have been termed, of plants and animals,—the universality of a phenomenon so extraordinary and unexpected, may be considered as one of the most interesting facts clearly established by the advance of modern science."[7] Lyell readily accepted de Candolle's botanical provinces, recognizing that these were, by definition, sufficiently distinct and stable as to suggest that the boundaries delineating them represent barriers to dispersal. How and why such distinctiveness was maintained were thus key questions, and this is why Lyell immediately followed his introduction of de Candolle's botanical regions with an extended discussion of means of dispersal and the geographical dispersion of species. In the following two chapters Lyell took a parallel approach with respect to animal species, treating the "migrations and facilities of diffusion" of various groups of animals, including humans.

Building upon "Buffon's law," Lyell speculated that species originate as single (specially created) pairs in some one locale, from which they multiply and expand—albeit not all species at once, but in complementary sets in succession over geological ages. He acknowledged that a complex of factors must over time shape observed distributions: natural barriers, interspecific dependencies, predation and other interactions, environment and other physical conditions, gradual extinction, etc. Barriers in particular would be expected to give rise to distinct botanical and zoological provinces. It is noteworthy that Lyell discussed the vicissitudes of species, their populational ebb and flow, in relation to "continual strife," quoting de Candolle in an early expression of what became known as the struggle for existence: "'All the plants of a given country,' says Decandolle in his usual spirited style, 'are at war one with another. The first which establish themselves by chance in a particular spot, tend, by the mere occupancy of space, to exclude other species—the greater choke the smaller, the longest livers replace those which last for a shorter period, the more prolific gradually make themselves masters of the ground, which species multiplying more slowly would otherwise fill.'"[8] This was likely Wallace's introduction to the concept of the struggle for existence.

In the second volume of *Principles*, Lyell thus articulated the complexities of *history* (successive origin, increase, powers of dispersion, limits placed by geographical barriers, decline, and extinction) and what is now called *ecology* (climatic and other abiotic factors as well as myriad biotic interactions) in not only shaping the distribution and relative abundance of species in the world today, but over geological time. This Lyellian vision of dynamically changing assemblages of species in time and space very much informed Wallace's thinking. It was likely in reading Lyell that Wallace first gained an appreciation of the important factors shaping distribution, becoming aware of the broad philosophical implications of biogeography. He apparently first read *Principles* in the mid-1840s (in one letter to Henry Walter Bates dated April 1846, Wallace commented that he was "much pleased" to find that his friend "so well appreciated 'Lyell'"[9]). Importantly, however, around the same time Wallace was reading other notable works discussing the significance of geographical distribution, some of them also drawing on Lyell. Perhaps the most important of these was Robert Chambers's *Vestiges of the Natural History of Creation*, published anonymously in 1844, which Wallace encountered in 1845.

It is clear from the tenor of Wallace's letters to Bates on the subject that he was deeply impressed with *Vestiges*.[10] Perhaps his enthusiasm lay mainly with the book's support of transmutationism, but this went hand in hand with Chambers's treatment of the problem of distribution, to which a dozen pages were dedicated. Among the most important concepts discussed by Chambers were (1) the idea that the same species are not necessarily found in the same environment elsewhere; (2) there are multiple, not single, centers of origin; and (3) the idea that species are related in both time and space. "Hence arises an interesting question," Chambers wrote. "Are the plants of the various isolated regions which enjoy a parity of climate and other conditions, identical or the reverse? The answer is—that in such regions the vegetation bears a general resemblance, but the *species* are nearly all different. . . . [P]arity of conditions does not lead to a parity of productions so exact as to include identity of species, or even genera."[11] Barriers, or isolation, were responsible. Like Lyell, Chambers discussed de Candolle, noting that his "well-marked" botanical regions would be greatly increased in number "if remote islands and isolated mountain ranges were to be included." He saw these as so many "distinct foci of organic productions throughout the earth" that have developed (evolved, in modern terms) independently in their respective areas in response to local conditions. Moreover, like Lyell, Chambers pointed out the relationship between the distribution of species geographically and over geological time: "We must now call to mind that the geographical distribution of plants and animals was very different in the geological ages from what it is now." He rejected the idea of a single center of origin on the basis of plant and animal species now found in isolated locales. Speculating that earth's climate was uniformly warm at one time, which gave rise to uniform and widespread plant and animal groups, Chambers suggested that the highly varied climates arising in more recent times gave rise to more varied species, and more endemism. "It may have only been when a varied climate arose, that the originally few species branched off into the present extensive variety."[12] His use of the word "branched" is telling, as Chambers explicitly advocated a branching pattern of ancestor-descendant relationship of species.

Humboldt, Lyell, Chambers, and other authors (e.g., James Cowles Prichard in *Researches into the Physical History of Man* [1813] and William Swainson in *Treatise on the Geography and Classification of Animals* [1835]) thus introduced Wallace to the overarching philosophical questions concerning species origins, species distribution, and the relationship between the two, convincing him that "there is no part of natural history more interesting or instructive than the study of the geographical distribution of animals."[13] Investigating geographical distribution was thus very much a part of Wallace's general plan to investigate the "theory of the origin of species," as he put it in a letter to Bates in October 1847.[14] As centrally important as distribution remained to Wallace even after solving the mystery of species origins in 1858, we shall see that his views on and inferences from geographical distribution changed over time.

The Riverine Barrier Hypothesis

Several of Wallace's writings stemming from his four years in Amazonia reflect his interest in biogeography, but none is more explicit in this regard than his paper "On the Monkeys of the Amazon." Read before the Zoological Society of London on 14 December 1852, two and a half months after his return home, this paper is noteworthy as Wallace's first expression of his philosophical interest in distribution, boundaries, and barriers. He introduced the subject by criticizing the practice of natural history collectors of providing only general, imprecise, locality information for their collections. Owing to this, he laments, "there is scarcely an animal whose exact geographical limits we can mark out on the map." This matters because, he continued,

> [o]n this accurate determination of an animal's range many interesting questions depend. Are very closely allied species ever separated by a wide interval of country? What physical features determine the boundaries of species and of genera? Do the isothermal lines ever accurately bound the range of species, or are they altogether independent of them? What are the circumstances which render certain rivers and certain mountain ranges the limits of numerous species, while others are not? None of these questions can be satisfactorily answered till we have the range of numerous species accurately determined.[15]

Wallace then articulated what has become known as the "riverine barrier hypothesis." He noted that the Amazon and its great tributaries the Rio Negro and Rio Madeira formed natural boundaries—or barriers—defining the ranges of the monkey species of the region. Collectively these rivers delineate four "districts"—Guiana and Ecuador, north of the Amazon and separated by the Rio Negro; and Brazil and Peru, south of the Amazon and separated by the Rio Madeira. Each district has its complement of monkey species not found in the others, though Wallace observed that "[o]n approaching the sources of the rivers they cease to be a boundary, and most of the species are found on both sides of them."[16] Wallace noted that the effects of these rivers are not restricted to monkeys, as they also seem to limit the distribution of some birds and insects.[17] He later emphasized the importance of this observation to Bates, saying that he wished he had paid closer attention to their distribution patterns while in Amazonia:

> There is however another fact I think of equal interest & importance, which you have barely touched upon, & yet I think your own materials in this very paper [Bates's paper on papilionid butterflies] establish it: viz. that the river in a great many cases limits the range of species or of well marked varieties. . . . In mammals this fact was not so much to be wondered at, but few persons would credit that it would extend to Birds & winged insects. Yet I am convinced it does & I only regret that I had not collected & studied birds there with the same assiduity I have here, as I am sure they would furnish some most interesting results.[18]

Wallace did not suggest that the rivers play a causal role in delineating the ranges of the monkeys, but as he was a transmutationist when he delivered this paper he likely saw the riverine range boundaries as more than coincidental, especially as he perceived that several groups of organisms were apparently all affected in a similar way.

The matter is still under active discussion; it is now posited that large rivers restrict gene flow and promote speciation by either (1) splitting ancestrally continuous populations (vicariance), (2) isolating founder populations colonizing one side of a river from another, or (3) preventing the recolonization of one side of a river from another following the local extinction of a population. It has been proposed that one or more such processes, all of which involve allopatric speciation over time, may help explain the megadiversity of the Amazonian rain forest,[19] in view of the number and size of river systems throughout the Amazon basin. The hypothesis received increasing attention with the advent of genetic and analytical procedures to compare relationships of taxa in a geographical context, across river systems.[20] Reviews of efforts to test the riverine barrier hypothesis in relation to alternative processes that may shape tropical rain-forest diversity and species distribution (e.g., Pleistocene refugium hypothesis, disturbance-vicariance hypothesis, and gradient hypothesis) reveal mixed support, based on a diversity of studies of vertebrate and invertebrate taxa and river systems.[21] Some phylogeographic studies of Amazonian birds, felids, primates, and butterflies found patterns of species relationship consistent with the riverine barrier hypothesis,[22] in terms of either congruence of river systems with areas of endemism or gene-flow rates that vary as predicted between upriver and downriver sites. Some of the most comprehensive and explicit tests of the hypothesis, however, yielded weak or no support.[23] Similarly, the first test of the riverine barrier hypothesis looking at bird assemblages in an Afrotropical river basin found only partial support.[24] The mixed findings of these studies point to the obvious: no single process can explain the rich tropical rain-forest diversity and its distribution; as ecologist Robert Colwell said in one review, "In short, the answer to the riverine barrier question, like so many in evolution, ecology, and biogeography, turns out to be 'It depends.'"[25]

Land Bridges, Continental Extensionism, and Dispersal

The prevailing model for long-term changes in landscape and geological formations that Wallace would have learned when he became interested in the subject in the 1830s and 1840s was Huttonian/Lyellian "actualism," later termed "uniformitarianism" by William Whewell. The uniformity model projected what we would now term a "steady-state" understanding of earth history: slow and steady geological evolution over vast periods of time, driven by such material phenomena in evidence today as volcanism, wave action, earthquakes, running water, glacial ice, and so on. Continual and gradual uplift of some regions of the world was viewed as balanced by equally gradual subsidence of other regions, combined with cycles of climatic change. Continents and ocean basins were broadly permanent features except for the endless gradual cycling of uplift and subsidence. The principles of uniformity of natural law and natural processes were foundations for the principles of uniformity of *rate* of change and, in Lyell's earliest formulation in *Principles of Geology*, even a controversial uniformity in a *state* of change that extended to the species populating the earth. Lyell held that there was no evidence of sustained "progressive" change in the fossil record, and that since species are designedly adapted to their environment, when particular environmental conditions cyclically re-emerge (such as tropical conditions in Britain), presently extinct species suited to those conditions might once again appear.

Wallace was among the majority who disagreed with Lyell's argument that the fossil record shows no evidence of progressive change, let alone the idea that extinct species might reappear. But he very much accepted the Huttonian/Lyellian concepts of gradual uplift and subsidence of the earth's crust, occurring continually on a regional to continental scale. The evidence from marine incursions onto the land was clear, in the form of extensive marine deposits of fossiliferous limestones and sandstones far inland and high in mountains, and unconformities and alternating marine-derived strata of different kinds suggested repeated or cyclical submersion and exposure. Gradual Lyellian uplift and subsidence was corroborated by Darwin's observations of stair-step fossiliferous terraces on the west coast of South America, and was central to his model of coral reef and island atoll evolution, which he first reported in two papers read at the Geological Society of London in January and May 1837.[26] In the second of these, on elevation and subsidence as deduced from coral formations, Darwin pointed to the importance of such observations for understanding geographical distribution in relation to the origin of species: "certain coral formations acting as monuments over subsided land, the geographical distribution of organic beings (as consequent on geological changes as laid down by Mr. Lyell) is elucidated, by the discovery of former centres whence the germs could be disseminated. . . . [S]ome degree of light might thus be thrown on the question, whether certain groups of living beings peculiar to small spots are the remnants of a former large population, or a new one springing into existence."[27]

Edward Forbes and Continental Extensionism

A few years later, however, Edward Forbes (1815–1854), paleontologist with the Geological Survey of Great Britain and professor of botany at King's College in London, took the uplift and subsidence model to an extreme in an effort to explain apparent anomalies in geographical distribution. Forbes is perhaps best known today for having stimulated Wallace to write the 1855 "Sarawak law" paper, in opposition to Forbes's "polarity theory." Wallace was apparently more receptive to Forbes's 1846 theory of continental extensions or land bridges, wherein he proposed the former existence of a vast continental landmass occupying much of the North Atlantic basin as a means of explaining the occurrence of mainland European plants on Atlantic islands such as Madeira and the Azores.[28]

The idea of now-sunken landmasses or land bridges was generally in accordance with Lyell's theory of oscillating land levels—indeed, Lyell criticized Forbes for not acknowledging his own work suggesting that a land bridge once connected Britain and France, an isthmus that Lyell posited to explain differences in the fossil marine deposits on either side of the hypothesized barrier, and he speculated on the possibility of former continents to explain the origin of extensive sedimentary formations: "If asked where the continent was placed from the ruins of which the Wealden strata were derived, we might be almost tempted to speculate on the former existence of the Atlantis of Plato as true in geology, although fabulous as an historical event. We know that the present European lands have come into existence almost entirely since the deposition of the chalk [Cretaceous period] . . . and the same period may have sufficed for the disappearance of a continent of equal magnitude, situated farther to the west."[29] Lyell was less willing than Forbes and his followers to postulate lost continents and extensive land bridges spanning ocean basins in the recent geological past to explain puzzles of geographical distribution. He was, however, sufficiently tolerant of those views that Darwin—a staunch dispersalist who decried what he saw as the too-ready invocation of vast land bridges on the flimsiest of evidence—took him to task:

> Forbes made a continent to N. America and another (or the same) to the Gulf weed [Sargasso Sea]. Hooker makes one from New Zealand to S. America & round the World to Kerguelen Land. Here is Wollaston speaking of Madeira and P. Santo "as the sure & certain witnesses" of a former continent. Here is Woodward writes to me if you grant a continent over 200 or 300 miles of ocean-depths (as if that was nothing) why not extend a continent to every island in the Pacific & Atlantic oceans! . . . If you do not stop this, if there be a lower region for the punishment of geologists, I believe, my great master, you will go there. Why your disciples in a slow & creeping manner beat all the old catastrophists who ever lived.—You will live to be the great chief of the catastrophists![30]

In was in this same spirit of opposition to land bridges, part and parcel of Forbesian continental extensions, that Darwin registered his disagreement with Wallace over the issue.

Wallace and Darwin: Land Bridges versus Dispersal

Wallace arrived in the Aru Islands, in the far east of modern-day Indonesia, on 8 January 1857. In the subsequent half-year spent exploring and collecting in that small archipelago, Wallace made a number of geological and biological observations that suggested to him that the Aru Islands were once physically linked to New Guinea, about 150 miles away. He laid out his argument in a paper published in the *Annals and Magazine of Natural History* at the end of that year, pointing out the faunal similarities between Aru and New Guinea, the shallow depth of sea between the two, and the peculiar winding saltwater channels dissecting the Aru Islands, the constant width, depth, orientation, and meanders of which suggested that they are relictual riverbeds representing the terminus or near terminus of rivers that once coursed from the New Guinean highlands across an intervening plain that once connected New Guinea and Aru. The biogeography of Aru and environs was especially important to Wallace, and he made it a centerpiece of his argument for the existence of a former land bridge linking Aru to New Guinea.

The structure of his argument reveals much about Wallace's use of faunal similarities and differences to reconstruct historical relationships at the time. In the case of Aru, he enumerates the bird and mammal fauna, highlighting the broad overlap with Australia and New Guinea. Fully one-half of his New Guinean birds also occurred on Aru; considering how imperfectly the New Guinean avifauna was known, he argues, all the birds of Aru would likely be found in New Guinea, and he declares that "[s]uch an identity occurs, I believe, in no other countries separated by so wide an interval of sea."[31] He points to several islands (Sardinia, Tasmania, and Sri Lanka) closer to their respective mainlands than Aru is to New Guinea, yet more "marked" by the (endemic) elements in their fauna. Only Britain and Sicily as islands possessed a rich fauna identical with that of the nearest mainland—a circumstance "held to prove that they have been once a portion of such continents, and geological evidence shows that the separation had taken place at no distant period."[32]

Wallace thus uses degree of endemism as an indication of degree or duration of isolation, and therefore of time for new and distinct species to arise—presumably by transmutation, though he does not state this. It is an idea he expressed in a different way in his Species Notebook, in an entry on the Galápagos Islands made around the same time: "In a small group of islands not very distant from the main land, like the Galápagos, we find animals & plants different from those of any other country but resembling those of the nearest land. If they are special creations why should they resemble those of the nearest land? Does not that fact point to an origin from that land?"[33] In the Aru paper Wallace observes that the deficiencies of the islands' fauna "teach as much as what it possesses."[34] Putting his Aru observations into broader context, Wallace was building upon the central observation of his "Sarawak law" paper, published in 1855 just a year after his arrival in Southeast Asia.

This so-called law states that "every species has come into existence coincident both in space and time with a pre-existing closely allied species."[35] The Aru fauna provided a case in point, and Wallace used his paper as an opportunity to highlight a fundamental flaw in Lyell's reasoning in *Principles* on this count. Wallace paraphrases Lyell as asserting that in the wake of physical changes like the upheaval of a new mountain chain in, say, the Sahara, "the animals and plants of northern Africa would disappear, and the region would gradually become fitted for the reception of a population of species *perfectly dissimilar in their forms, habits, and organization*."[36] He then refutes Lyell's claim by demonstrating that environmental similarity or dissimilarity has little to do with species relationships in the Malay Archipelago: Borneo and New Guinea are comparably large and topographically varied equatorial islands in the same region, yet differ dramatically in their faunas. New Guinea and Australia, on the other hand, differ in latitude, size, topography, and climate, yet have a high degree of congruence in their respective fauna. "We can hardly help concluding, therefore, that some other law has regulated the distribution of existing species than the physical conditions of the countries in which they are found, or we should not see countries the most opposite in character with similar productions, while others almost exactly alike as respects climate and general aspect, yet differ totally in their forms of organic life."[37]

Wallace provides a plausible scenario for the observed distribution: when land areas such as New Guinea and Australia, or Aru, are physically connected, species largely range throughout the whole. After separation, as when an intervening low-lying area subsides and becomes a strait or channel, new species somehow "appearing" in each now isolated area are related to the pre-existing species of that country. Over time more and more groups "peculiar" (endemic) to each would arise, but always bearing a resemblance to their immediate forebears. Wallace was explicit in drawing a parallel between Aru and New Guinea on one hand and New Guinea and greater Australia on the other: a progressive process of separation and continued introduction of species gives rise to sets of allied yet distinct species, first differing at the species level, then generic, and later familial and ordinal level as time and isolation have their effect. The taxonomic level of similarity or difference between two areas is related to the time since separation of those areas; there is no Lyellian discontinuity, only diversification with allied species over time. Circling back to geology, he concluded, "It is evident that, for the complete elucidation of the present state of the fauna of each island and each country, we require a knowledge of its geological history, its elevations and subsidences, and all the changes it has undergone since it last rose above the ocean. This can very seldom be obtained; but a knowledge of the fauna and its relation to that of the neighbouring countries will often throw great light upon the geology, and enable us to trace out with tolerable certainty its past history."[38]

In September 1857 Wallace wrote to Darwin about his Aru hypothesis. Darwin replied that he was "extremely glad to hear that you are attending to distribution in accordance with theoretical ideas," and pointed out that others had been thinking along similar lines—namely, navigator and geographer George Windsor Earl (1813–1865), who had invoked uplift and subsidence to explain the anomalous distribution of species in the Malay Archipelago in an 1845 paper. In 1853 Earl wrote that the similarities in the mammalian fauna of peninsular Malaysia, Sumatra, Borneo, and Java proved that these areas "continued attached to the continent of Asia at a comparatively recent epoch," and pointed out that the zoological connection with Asia "is as distinct as that of Timor, Ceram, and New Guinea, with the continent of Australia"[39] He did not invoke Lyell, but his discussion of distribution in the context of volcanism, uplift, and subsidence in the region is a purely Lyellian model. Such land bridges were acceptable to Darwin, but he argued that Forbes and his followers took the idea too far. He continued in his December 1857 letter to Wallace,

> I can see that you are inclined to go much further than I am in regard to the former connections of oceanic islands with continent: Ever since poor E. Forbes propounded this doctrine, it has been eagerly followed; & Hooker elaborately discusses the former connections of all the Antarctic islds & New Zealand & S. America.—About a year ago I discussed this subject much with Lyell & Hooker (for I shall have to treat of it) & wrote out my arguments in opposition; but you will be glad to hear that neither Lyell or Hook[er] thought much of my arguments: nevertheless for once in my life I dare withstand the almost preternatural sagacity of Lyell.[40]

Darwin concluded his letter jokingly, wishing that "all your theories succeed, except that on oceanic islands, on which subject I will do battle to the death."

In the summer of 1859 Wallace sent Darwin his manuscript "On the

Zoological Geography of the Malay Archipelago," which Darwin passed on to the Linnean Society (it was read on 3 November 1859 and published the following year). In this paper Wallace thoroughly described the faunal discontinuity that would later become known as "Wallace's Line" (a phrase coined by Huxley in 1868), and it occasioned another letter from Darwin registering his disagreement with Wallace over land bridges and transoceanic dispersal: "I differ wholly from you on colonisation of oceanic islands, but you will have everyone else on your side. I quite agree with respect to all islands not situated far in ocean. . . . I wish I had given fuller abstract of my reasons for not believing in Forbes's great continental extension."[41] Darwin found Lyell and Hooker's endorsement of the land-bridge idea especially irritating, and was unhappy that in his paper Wallace wrote how Hooker had "convincingly applied this principle" of land bridges in explaining the disjunct plant groups scattered throughout the southern polar region. Wallace further maintained not only that "though many naturalists are inclined to regard all such views as vague and unprofitable speculations, we are convinced they will soon take their place among the legitimate deductions of science."[42] Worse yet, he also cited Darwin's own coral reef work in support of the doctrine:

> Geology can detect but a portion of the changes the surface of the earth has undergone. It can reveal the past history and mutations of what is now dry land; but the ocean tells nothing of her bygone history. Zoology and Botany here come to the aid of their sister science, and by means of the humble weeds and despised insects inhabiting its now distant shores, can discover some of those past changes which the ocean itself refuses to reveal. They can indicate, approximately at least, where and at what period former continents must have existed, from what countries islands must have been separated, and at how distant an epoch the rupture took place. By the invaluable indications which Mr. Darwin has deduced from the structure of coral reefs, by the surveys of the ocean-bed now in progress, and by a more extensive and detailed knowledge of the geographical distribution of animals and plants, the naturalist may soon hope to obtain some idea of the continents which have now disappeared beneath the ocean, and of the general distribution of land and sea at former geological epochs.[43]

Darwin may have despaired over Wallace's vigorous application and defense of Forbesian land bridges in the "Zoological Geography of the Malay

Archipelago" paper, but ironically it was also there that Wallace planted the seeds for his own reversal of opinion on the land-bridge issue, as Fichman has shown.[44] This was Wallace's endorsement of Sclater's proposed system of six biogeographical regions or realms, which was to become the centerpiece of Wallace's global biogeographical vision.

Thesis, Antithesis, Synthesis

In 1857 the ornithologist Philip Lutley Sclater (1829–1913) proposed making sense of bird distributions by dividing the globe into six more or less well-defined biogeographical regions, or realms: Neotropical, Nearctic, Palearctic, Ethiopian, Indian, and Australian.[45] Insofar as regional development of distinctive bird faunas must involve historical geological and climatic processes that affect other groups of organisms at the same time, Wallace believed that Sclater's system should be broadly applicable. "With your division of the earth into six grand zoological provinces I perfectly agree," Wallace wrote Sclater, "and believe they will be confirmed by every other department of zoology as well as by botany."[46] Very importantly, at the same time Wallace realized that for these biogeographic regions to be stable—that is, for them to be recognizable as regions of largely distinctive assemblages of plants and animals—extensive land bridges in the recent geological past were untenable. He had previously argued persuasively that it was the high degree of congruence between Aru and New Guinea that pointed to the likelihood of a former connection. The same would have to be true between any areas connected by large-scale land bridges or continental extensions of the kind postulated by Forbes. In short, Wallace realized that land bridges may make sense on a fairly localized scale, but on a Forbesian scale they would only serve to homogenize floras and faunas, erasing the distinctiveness that Sclater pointed out. In the early 1860s Wallace thus reined in his endorsement of land bridges, but did not throw the baby out with the bath water: "in all cases where we have independent geological evidence, we find that those islands, the productions of which are identical with those of the adjacent countries, have been joined to them within a comparatively recent period, such recent unity being in most cases indicated by the very shallow sea still dividing them; while in cases where the natural productions of two adjacent countries is very different, they have been separated at a more remote epoch—a fact generally indicated by a deeper sea now dividing them."[47] These remarks are found in Wallace's 1863 paper "On the Physical Geography of the Malay Archipelago," read at the Royal Geographical Society in June of that year—a paper that Darwin called "an epitome of the whole theory of geographical distribution."[48]

"Independent geological evidence" on a case-by-case basis was the criterion for accepting the possibility of a land bridge—the kinds of lines of evidence he had elucidated in his paper on the Aru Islands in 1857. The tide was beginning to change in the view of other naturalists too. Hooker exemplifies this struggle; having been a vocal supporter of land bridges, in his 1866 presidential address to the British Association for the Advancement of Science he made it clear he had changed his mind. The difficulties of dispersalism were preferable on principle to the untenable claims of lost continents. Both hypotheses "are as yet unverified and insufficient; neither geological considerations, nor botanical affinity, nor natural selection, nor all these combined, have yet helped us to a complete solution of this problem, which is at present the *bête-noir* [*sic*] of botanists."[49]

An alternate hypothesis was beginning to gain favor, one accepting a general permanence of the continents and ocean basins, overlain with transoceanic dispersal and climatic cycles to facilitate movement and admixture. Wallace's championing of dispersalism came with his increasingly strong support for the reality of biogeographic regions: limited (local) land bridges and a dynamic process of origin, diffusion, and barriers' checking diffusion. Some barriers may be more or less permanent, and others arise and disappear over time through gradual elevation and subsidence of the land and climatic cycles. The change in Wallace's biogeographical thinking is clear in his papers on the subject between 1863 and 1869, beginning with "On Some Anomalies in Zoological and Botanical Geography" (read in 1863, published in 1864). There he praised Sclater as having succeeded in providing "the foundation for a general system of Ontological regions," and set out to arrive at "a division of the earth into Regions, which shall represent accurately the main facts of distribution in every department of nature"[50] (see figure 10.1). Wallace laid out five principles that he argued could explain both the broad patterns and anomalies of geographical distribution:

> 1st, The tendency of all species to diffuse themselves over a wide area, some one or more in each group being actually found to have so spread, and to have become, as Mr Darwin terms them, dominant species.
>
> 2d, The existence of barriers checking, or absolutely forbidding that diffusion.
>
> 3d, The progressive change or replacement of species, by allied forms, which has been continually going on in the organic world.

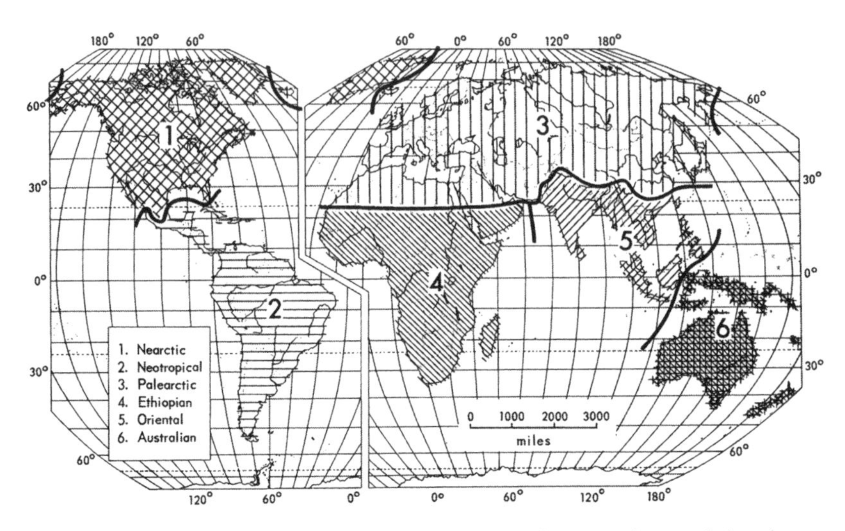

Figure 10.1 The Sclater/Wallace faunal regions classification. After Smith (1983). Copyright Charles H. Smith.

> 4th, A corresponding change in the surface, which has led to the destruction of old and the formation of new barriers.
>
> 5th, Changes of climate and physical conditions, which will often favour the diffusion and increase of one group, and lead to the extinction or decrease of another.[51]

Thus by the mid-1860s Wallace had a model of geographical distribution intimately tied to historical geological and climatological processes. He spent a good part of the next fifteen years working out related details (see chapter 9), but only a few years later he was encouraged by an 1864 paper by James Croll entitled "On the Physical Cause of the Change of Climate during Geological Epochs." This work argued that astronomical phenomena indirectly give rise to glacial and interglacial periods (see chapter 9).[52]

Sclater, Darwin, and the naturalist Alfred Newton now encouraged Wallace to produce a comprehensive treatment of distribution. He took up the challenge, publishing *The Geographical Distribution of Animals* in 1876. This watershed work built on Sclater, subdividing the six Sclaterian regions into twenty-four subregions. It opens with a section called "The Principles and General Phenomena of Distribution," laying out the means of dispersal and migration of animals (chapter 1) and the effects of both environment and historical changes in the earth's surface (chapter 2) as jointly key to understanding the earth's six zoological regions (chapter 3). The work was regarded as a masterpiece by the scientific community—Darwin called it "grand and memorable"[53]—and indeed today is still widely regarded as perhaps the founding document of modern evolutionary biogeography. One reviewer, writing in the *Academy*, declared that with this book, "[n]ow all is changed." Before, "whether some petty weed or noxious insect was a native of this or that country, or whether it should be arranged in this or that family, were questions which might interest the specialist, but were beneath the attention of the philosopher." No more. "Upon the minutest comparisons, upon the most diligent collection of seemingly trivial facts, generalisations and deductions are advanced and combated which will not yield in importance to the profoundest problems of the physicist, the geologist or the astronomer." In the view of this reviewer, "Among the chief causes of this reform in Biology have been the wonderful advance of our knowledge of the history of extinct life and of the development of living beings, the scientific study of the geographical distribution of animals and plants, and—springing from these—the promulgation of the doctrine of Evolution."[54]

The *Academy* reviewer put his finger on the key innovations of Wallace's treatise, among the most important of which was Wallace's incorporation of fossil data, treating the distribution of extinct animals "in order that we may see the meaning and trace the causes of the existing distribution of animal forms."[55] This gave rise to the interrelated subdisciplines of geobiology and paleobiogeography.[56] Another was the complementary ways in which Wallace presented global zoogeographical data, what he termed "zoological geography" on one hand and "geographical zoology" on the other. "In the first of these we divide the earth into regions and subregions, study the causes which have led to the differences in their animal productions, [and] give a general account of these, with the amount of resemblance to and difference from other regions. . . . [In the second, we] have given for every family of mammals, birds, and reptiles a diagram, which in a single line exhibits its distribution in each of the four subregions of the six regions."[57] There had been nothing like it in the history of biogeography. Wallace himself saw the work as bearing "a similar relation to the eleventh and twelfth chapters of the 'Origin of Species' as Mr. Darwin's 'Animals and Plants under Domestication' does to the first chapter of that work."[58]

Describing Wallace's goal in *The Geographical Distribution of Animals* of producing "a rigorous and coherent science of zoogeography," historian of science Martin Fichman notes that Wallace succeeded in achieving "a unified theory which explained both the existing and past zoological features" of the earth in terms of geological history and evolutionary history.

In so doing, Fichman concludes, Wallace "freed zoogeography from its dependence on *ad hoc* hypotheses" and "showed that the known agents of animal dispersal were sufficient to determine the probable birthplace and subsequent geographical history of the most important genera and families *without* positing vast major changes in physical geography."[59] Note, however, that although Wallace had by then banished ad hoc land bridges, he continued to entertain the possibility of lost continents. Sclater himself promoted the idea beginning in 1864 with a paper entitled "The Mammals of Madagascar," in which a lost southern continent occupying the Indian Ocean was postulated to explain what Sclater thought (erroneously) were disjunct lemur distributions, with species in Madagascar, Sri Lanka (in fact lorises), and the Malay Archipelago (in fact dermopterans). Sclater dubbed this continent "Lemuria." On the basis of the apparently anomalous lemurid distribution, Wallace concurred with Sclater in *The Geographical Distribution of Animals*: "This is undoubtedly a legitimate and highly probable supposition, and it is an example of the way in which a study of the geographical distribution of animals may enable us to reconstruct the geography of a bygone age," he declared; and "[the remarkable groups of Madagascar] speak to us plainly of enormous antiquity, of long-continued isolation; and not less plainly of a lost continent or continental island, in which so many, and various, and peculiarly organized creatures, could have been gradually developed in a connected fauna, of which we have here but the fragmentary remains."[60] He soon became disenchanted with Lemuria, however, and by 1880 was criticizing the "weakness and inutility" of the hypothesis, "in the hope that a check may thus be placed on the continued re-statement of this unsound theory as if it were one of the accepted conclusions of modern science."[61]

This condemnation is found in Wallace's final biogeographical masterwork, *Island Life*. Despite the success of *The Geographical Distribution of Animals*, he felt that the work was incomplete. By necessity (in terms of space), he had not said much about islands, despite having "become aware of the great interest attaching to them, and the great light they threw upon the means of dispersal of animals and plants, as well as upon the past changes, both physical and biological, of the earth's surface."[62] Four years later Wallace published *Island Life*, a tour de force that he saw as a companion volume to *The Geographical Distribution of Animals*, calling it the "completion of that work" in the opening sentence of the preface.[63] Dedicated to Joseph Hooker, *Island Life* was Wallace's last grand biogeographical synthesis (notwithstanding later summaries such as the chapter dedicated to the topic in *Darwinism* in 1889). As such, it was his definitive statement on the key issues: evolution as key to understanding distribution, reaffirmation of the permanence of continents and oceans, the manifold means of dispersal, the distinctions between and insights from oceanic and continental islands, and a bold "revision and attempted solution" to the problem of geological climates (glacial epochs), building on James Croll's astronomical theories (see chapters 9 and 12).

Mammalogist and biogeographer Lawrence R. Heaney provides an excellent overview of these themes of *Island Life* in his introduction to the 2013 facsimile reissue of the book.[64] Heaney points out that Wallace at times anticipated phenomena and processes shaping island biota that were only formally elucidated by others much later—for example, ideas recalling the immigration-extinction dynamic, the taxon cycle, and the very ecological idea of seeing occupancy as a barrier (discussed below). The factors that Wallace recognized as shaping species richness, such as available area, island age, and habitat heterogeneity, were similar to those treated in MacArthur and Wilson's equilibrium model of island biogeography some eighty years later,[65] though Wallace's view of species and communities was fundamentally one of dynamic nonequilibrium: "with changes in community composition due to extinction and colonization and species changing because of ongoing natural selection"—a concept resonant with modern metapopulations and patch dynamics theory.[66]

The two parts into which Wallace divided *Island Life* say a great deal about his biogeographical thinking at the time. The first part is entitled "The Dispersal of Organisms; Its Phenomena, Laws, and Causes," giving Wallace's overarching argument about the permanence of continents and oceans, and the concomitant stability of Sclaterian regions, which boils down to dispersal (or lack thereof). The second is simply entitled "Insular Faunas and Floras," a systematic overview of the biotas of islands worldwide, which illustrate the different means and timescales of dispersal: degrees of endemicity and disharmony with the nearest continent representing a kind of signature, indicating degree of isolation and likely origin of islands. Wallace's (and Darwin's) theory of geographical distribution became in essence a grand theory of dispersal and barriers to dispersal in space and time, dominating thought for nearly a century. More modern treatments of species spread and movement distinguish between three modes of dispersal that Wallace would recognize: *corridor* or *land-bridge* dispersal consisting of sizable land bridges exposed when sea level drops (e.g., Beringia), *filter-bridge* dispersal consisting of stepping-stone-like archipelagoes that require island hopping (e.g., Wallacea), and *sweepstakes* or *jump* dispersal involving trans-oceanic transport by carriage or rafting.[67] Over time, however, the over-reliance on dispersal and insistence on static continents would prove increasingly problematic for the Wallace-Darwin model, until dispersalism itself was attacked as weak ad hoc hypothesizing (which Wallace and Darwin themselves were so critical of in the extensionists). These developments are considered in the next section, in the context of later thinking on the biogeographical region concept.

The Biogeographic Region Concept

While Wallace's dual magnum opuses of biogeography seem quite modern in many respects—precisely why they are seen as founding documents of the modern science—it is important to contextualize their limitations in the modern view: continents and ocean basins are *not* permanent, fixed features of the earth, especially, and on shorter timescales it is largely the level of the sea that rises and falls, not the land (and it does so in response to extensive continental glaciation; Wallace made the connection between glacial epochs and sea level, but because he was unsure of the extent of sea-level change during glacial periods, he tended to emphasize land-bridge formation via uplift). A better understanding of these concepts came decades later. There are a number of other errors of fact or interpretation very much of the time (e.g., concerning the age of the earth, and the time since the last glacial period). Indeed, the reality of deep time and continental movement help shed light on some anomalous distributions that Wallace was otherwise forced to explain in terms of dispersal scenarios—for example, ratite birds, the circum-Antarctic flora, and other distributions that we call "Gondwanan" today.[68]

It is interesting to speculate on how Wallace might have received Alfred Wegener's (1880–1930) argument for the "drift theory," lateral continental movement. Published in German as *Die Entstehung der Kontinente und Ozeane* (*The Origin of Continents and Oceans*) in 1915, two years after Wallace's death, Wegener used the same kind of consilience argument that Wallace himself applied in the 1850s to argue for the reality of transmutation, and for the former connection of the Aru Islands with New Guinea.[69] Wegener also argued that distributional fossil data were informative, much as Wallace argued in his biogeographical works. The drift theory, which gave rise to modern plate tectonics, was contentious until its predictions were confirmed in the 1950s and 1960s by analysis of paleomagnetic signatures and age sequences of ocean crustal rocks across ocean basins, among other lines of evidence—all made possible by technological advances in marine geology, radiometric dating, and ferromagnetic mineral analyses.[70]

Plate tectonics theory, together with advances in biological systematics, reinvigorated a vicariance model of biogeography,[71] leading some biogeographers to go to the extreme of viewing continental movement as the sole or dominant process shaping modern distributions and rejecting dispersal altogether (as reflected in Nelson's famous dismissal of the Wallace-Darwin dispersalist tradition as "a science of the improbable, the rare, the mysterious, and the miraculous"[72]). In retrospect it seems illogical to maintain that only one or the other process explains extant geographical distributions.[73] Related schools of biogeographical thought, such as the "panbiogeography" of the Venezuela-based biogeographer Leon Croizat (1894–1982), rejected the dispersalism of Wallace (and Darwin), yet are essentially Wallacean in embracing the concept of the earth and its biota evolving in concert—Croizat's mantra that "Earth and Life evolve together" is simply a restatement of a view long promoted by Wallace, Darwin, and even, later in life, Lyell, albeit involving a far more dynamically changing earth than those nineteenth century naturalists could imagine.[74] Oddly, however, the eccentric Croizat rejected plate tectonics, the very discovery that seemed to vindicate his "track" method of reconstructing the historical origin and movement of species.[75]

What does plate tectonics and the reality of continental movement in geological time mean for the biogeographical region concept, given Wallace's argument that the very existence of such recognizable regions must result from a historical stability of continents and ocean basins? Insofar as Wallace based his global biogeographical analysis on animals (particularly birds and mammals), are they relevant to the full range of taxa on earth? Do, say, plants, protists, or fungi recognize the same general Sclaterian boundaries as animals? Not always: the Sclaterian regions were criticized as arbitrary or of limited applicability (in taxonomic terms) even in the nineteenth century, but as late as 1894 Wallace was still defending them: "The subject which I now propose to discuss, is the purport and use, and therefore the essential nature, of what are termed zoological regions. This seems necessary because, although such regions have been more or less generally adopted for more than thirty years, there has of late grown up a conception as to their nature and purport which seems to me to be altogether erroneous, and which, if generally adopted, is calculated to lead to confusion, and to minimise, if not to destroy, whatever advantages may be derived from their use."[76]

This raises the question of whether the division of the earth into biogeographic regions is simply a convenient means of bookkeeping. It is, but not merely so. Such biogeographical organization is not arbitrary in the way that, say, the constellations are arbitrary but convenient means of imposing order on otherwise confusing star fields. With species distributions we are identifying associations that are not spatially and temporally related by chance the way two distant and altogether unconnected stars might be associated by the chance occurrence of appearing side by side in our line of sight. Some biogeographers frame the question of biogeographic regions in terms of "natural units," which may be defined as areas of endemism (groups of endemic species) with a shared geological history (which New Zealand biogeographer Bernard Michaux proposed calling "Wallacean biogeographical units" in Wallace's honor),[77] as opposed to areas of endemism consisting of composite geological formations, amalgamated by plate collisions.

At different spatial scales there is a practical need to identify and categorize more or less distinctive assemblages of species (ecozones, communities, biogeographical provinces, etc.), in terms of (1) better understanding the ecology of communities and ecosystems, (2) testing hypotheses regarding historical relationships or such phenomena as evolutionary convergence, and (3) informing management and policy decisions in conservation. If we grant that there is a biological reality to such "units" as biogeographic regions, provinces, or ecozones, by which I mean they reflect a deterministic set of factors both ecological and historical and are thus nonrandom groupings in space and time, what is the best way to define these units? There may be no consensus, stemming from a belief that there can be no "one size fits all" with regard to identifying biogeographic regions and subregions, even if their broad delineation has more than just heuristic value. Yet it is surely possible, and desirable, to identify a hierarchical set of regions and subregions that reflect both ecological, evolutionary, and historical-geological processes. In 2002 the Mexican biogeographer Juan Morrone put forth a comprehensive effort at such a general system of regionalization, dividing the globe into three kingdoms or realms, Holarctic, Holotropical, and Austral (which largely map onto the paleogeography of Laurasia and Gondwana), plus twelve regions.[78]

More recently, in 2013, a team headed by biologists at the University of Copenhagen proposed a new statistical approach to the study of biogeographical realms, analyzing pairwise phylogenetic beta diversity for over 21,000 vertebrate species worldwide to quantify turnover in phylogenetic composition among global species assemblages, and using this to statistically delineate zoogeographical realms and regions.[79] This novel approach is made possible by the availability of phylogenetic and distributional data. The results, yielding eleven broad-scale realms divided into twenty finer-scale zoogeographical regions, agree in some respects with Wallace's 1876 treatment of zoogeographical realms, but important differences emerged on the basis of which the authors proposed five new zoogeographical realms. They argued that their statistical approach helps delineate realms and regions based on evolutionary history, and also helps us to better understand the historical relationships of realms and regions themselves. But while this approach may have great potential in principle, critics have pointed out that the variable quality of input phylogenetic data and reliance on algorithm-based methodology with arbitrary assumptions undermines the quality of this analysis. Significantly, it was shown that altering some of the assumptions of the initial analysis only slightly changed the outcome, collapsing several of the proposed new realms into groupings essentially congruent with Wallace's.[80]

The Problem of Wallacea

Because the boundary area between the Oriental (also called Indo-Malayan) and Australian Sclaterian regions constitutes a vast archipelago, the faunal discontinuities are especially striking where islands representing the eastern and western extremes of their respective regions approach one another—as they do at Bali and Lombok. Wallace recognized the significance of the shallow continental shelves underlying the eastern (Sahul) and western (Sunda) portions of the archipelago, making sense of the similarities and differences between the eastern and western faunal assemblages in terms of periodic dry-land connection with Asia in the west and Australia/New Guinea in the east. In between, however, was a puzzle. In 1928 this region between the Sahul and Sunda shelves was dubbed "Wallacea."[81] (See figure 10.2.)

Wallace struggled to assign the Philippines, Celebes (Sulawesi), and several other central Indonesian islands of Wallacea to one or the other faunal region. In *The Geographical Distribution of Animals* he came down on the side of placing the Philippines in Asia and Sulawesi in Australasia, while Huxley (in the lecture in which he coined the term "Wallace's Line") reassigned the Philippines to the Australian realm. Later authors (Weber,

Figure 10.2 Map showing the Malay Archipelago and Wallace's Line (with alternative placements thereof). From Mayr (1944). By permission, University of Chicago Press.

Lydekker, Mayr) redrew the line of demarcation between the regions according to their taxonomic group of study, leading George Gaylord Simpson to despair that there were "too many lines" in a well-known 1977 paper.[82] Following Simpson, British biogeographer Barry Cox cut the Gordian knot by arguing that "it seems quite clear that we should no longer view the islands between the continental shelves of Southeast Asia and Australia as belonging to either the Oriental or the Australian zoogeographic regions."[83]

The confusion over Wallacea stems from its complex geological history, with several central islands in the region consisting of amalgams of very different geological terranes.[84] Michaux identified those areas of endemism within Wallacea exhibiting a shared geological history (Wallacean biogeographical units), showing that these often correspond not to entire islands, but to distinct regions of islands. He further pointed out that Wallacea as a whole may not be a natural (Wallacean) biogeographical unit, yet it is not altogether artificial either, "because all the areas currently recognized as parts of Wallacea are tectonically derived, predominantly Australasian fragments caught up in a complex collision zone. Although it may not meet the strict criteria for 'naturalness,' Wallacea is composed of a number of Wallacean biogeographical units that are linked by geological processes resulting from the interplay between the Eurasian and Australasian continents and the Philippine Sea Plate."[85]

Ecological Biogeographical Ideas Explored by Wallace

Wallace is best known for his extensive work on historical/evolutionary biogeography, but his far-ranging interests included subjects that fall under the heading of ecological biogeography in modern terms. Of course, some ecological factors and processes are part and parcel of Wallace's historical thinking, as we have seen in chapter 9. I will limit my discussion here to five interrelated ideas that Wallace discussed, bearing on the generation and geographical distribution of species diversity: latitudinal diversity gradients, the "tropics as museums" concept, taxon cycles, the intermediate disturbance hypothesis, and succession.

Latitudinal Diversity Gradients

Probably since the earliest era of exploration, naturalists have noted the richness of plant and animal species at lower latitudes in comparison with the temperate and polar regions. *Why* that is the case has been, and still is, a topic of considerable debate, and numerous papers, books, and reviews have been dedicated to the issue.[86] One reviewer enumerated some thirty vying hypotheses proposed over more than a century to explain the pattern![87] Space precludes a comprehensive treatment of the subject here; instead I will restrict my treatment to Wallace's thinking and the current status of his and other hypotheses.

Humboldt gave what is perhaps the first explicit description of the latitudinal species diversity gradient in his *Ansichten der Natur*,[88] in which he offered an explanation for the pattern: "In the frigid zone, Nature periodically becomes frozen. . . . On the other hand, the nearer one comes to the tropics, the greater the variety of shapes, the gracefulness of forms and color combinations, and the perpetual youth and power of organic life. . . . [W]hoever is able to comprehend Nature with a single look and knows to abstract localized phenomena will see how, with the increase in invigorating heat from the poles to the equator, there is also a gradual increase in organic power and abundance of life."[89] To Humboldt, it all came down to climate. Wallace, perhaps following Humboldt, concurred. In chapter 2 of *Tropical Nature and Other Essays*, which first appeared in 1878, Wallace described the primeval forests of the equatorial zone as "grand and overwhelming by their vastness, and by the display of a force of development and vigour of growth rarely or never witnessed in temperate climates." Like Humboldt he commented on the variety of tropical growth forms, including "the extent to which parasites, epiphytes, and creepers fill up every available station with peculiar modes of life," and he pinpointed climate as the ultimate causal factor:

> Atmospheric conditions are much more important to the growth of plants than any others. Their severest struggle is against climate. As we approach towards regions of polar cold or desert aridity the variety of groups and species regularly diminishes; more and more are unable to sustain the severe climatal conditions, till at last we find only a few specially organized forms which are able to maintain their existence. . . . In the equable equatorial zone there is no such struggle against climate. Every form of vegetation has become alike adapted to its genial heat and ample moisture, which has probably changed little even throughout geological periods; and the never-ceasing struggle for existence between the various species in the same area resulted in a nice balance of organic forces, which give the advantage, now to one, now to another, species, and prevents any one type of vegetation from monopolizing territory to the exclusion of the rest.[90]

Later in the book Wallace gave a more nuanced explanation:

> The equatorial zone, in short, exhibits to us the result of a comparatively continuous and unchecked development of organic forms; while in the temperate regions, there have been a series of periodical checks and extinctions of a more of less disastrous nature, necessitating the commencement of the work of development in certain lines over and over again. In the one, evolution has had a fair chance; in the other it has had countless difficulties thrown in its way. The equatorial regions are then, as regards their past and present life history, a more ancient world than that represented by the temperate zones, a world in which the laws which have governed the progressive development of life have operated with com-

> paratively little check for countless ages, and have resulted in those infinitely varied and beautiful forms . . . which delight and astonish us in the animal productions of all tropical countries.[91]

In these few sentences Wallace articulates the most enduring explanation for the latitudinal diversity gradient: that is, climate plays an overarching role. But note his use of the phrases "periodical checks and extinctions" in reference to high latitudes, and "continuous and unchecked" in reference to the tropics. He envisions that high-latitude species are periodically knocked back, even driven to extinction, by extreme cold, necessitating repeated repopulation, while in the tropics unchanging warmth permits long-term persistence and fosters intense biotic interaction and mutual adaptation. As he expresses it in his 1910 book *The World of Life*, "It is this long-continued *uniformity* of favourable conditions within the tropics, or more properly within the great equatorial belt about 2000 miles in width, that has permitted and greatly favoured ever-increasing delicacy of adjustments of the various species to their whole environment. Thus has arisen that multiplicity of species intermingled in the same areas, none being able, as in the temperate zone, to secure such a superior position as to monopolise large areas to the exclusion of others."[92]

Wallace's thinking about the prevailing evolutionary dynamic at high versus equatorial latitudes clearly stems from the then new and exciting developments in understanding the glacial periods; he had Pleistocene glaciation in mind in thinking about the destructive effects of the far-northerly climate. Later work has shown that the latitudinal diversity gradient far predates the Pleistocene, likely extending back to the Paleozoic.[93] Wallace's ideas on the subject have resonated with an important and long-standing school of modern ecological and evolutionary thought holding that abiotic factors hold sway at high latitudes while biotic interactions (competition, predation, parasitism, etc.) dominate in equatorial regions, leading to a greater diversity and more specialized forms at low latitudes.[94] Inevitably, however, there is far more to it than that.

Wallace recognized that the pattern involves an underlying evolutionary process, but do the tropics experience higher rates of speciation or simply lower rates of extinction than the temperate and polar zones? Or if speciation rates are similar at high and low latitudes, perhaps the *time* available for diversification is greater in the tropics? Perhaps the most widely held view relates the pattern to antiquity (the "time and area" model), as expressed by ecologist Kevin Gaston in one review: "Ultimately, the latitudinal gradient in species richness must be a consequence of a greater period of net diversification in the tropics, likely following the origins of life there, or higher speciation rates and/or lower extinction rates at low latitudes compared with other regions."[95] Recall that Wallace suggested that the tropics are a more "ancient world" than other regions, with more time for speciation. This idea, later expressed as the "tropics as cradle" model (with accumulation of species owing to intrinsically higher speciation rates), versus the "tropics as museum" model (where accumulation of species stems from lower extinction rates), has also been an important line of investigation,[96] and the evidence suggests that the tropics are both cradle *and* museum, depending on taxonomic and temporal scale of analysis.

Of course, diversification is just one factor relevant to geographical distribution over evolutionary time. In ecological and shorter timescales, biotas also show a great propensity for movement and adaptive response to environmental conditions, and some of these might contribute to the latitudinal diversity gradient. This may happen relatively quickly in ecological time. In one study it was found that realized annual evapotranspiration, a proxy for primary production and therefore energy availability, explained significantly more of the observed distributional variation than latitude alone.[97] Further, these authors pointed to the influence of factors such as average climatic conditions and topography as more important than historical factors. Energy available to communities and ecosystems may thus be the dominant factor in shaping distribution on a continental scale—an idea that resonates with the well-known relationship between plant community architecture more generally and the available energy and water budget (expressed in terms of evapotranspiration) of communities.

The idea of energy availability as a major determinant of structure at the community level and the distribution of species richness at higher spatial scales is essentially the position of both Humboldt and Wallace. The complexities of the tropics are not, however, fully explained by any one or a few factors: a multitude of processes both ecological and evolutionary, contemporary and historical, play a role, including stability, antiquity, net productivity, intra- and interspecific competition and predation.[98] Wallace would have found that exciting, no doubt.

Competition: Exclusion, Relictualism, and Taxon Cycles

Another aspect of Wallace's "ecological biogeographical" thinking concerns the role of biotic interactions, especially competition, in shaping geographical distribution of species. Although Darwin emphasized competition far more than Wallace, to the point of making it central to his vision of evolutionary diversification,[99] competition was nonetheless a central factor in Wallace's approach to how selection shapes distribution, even in his early writings. One expression of this is found in Wallace's Species Notebook, the most important of his Southeast Asian field notebooks, in which a dynamic reminiscent of competitive exclusion is described. In an entry dated 20 January 1858, Wallace speculated on the origin of the dense grassy expanses he encountered on the island of Halmahera (Gilolo) in central Indonesia, a puzzle in areas otherwise dominated by dense tropical rain forest. His explanation combined recent uplift with rapid colonization by grass seeds, which he posited were dispersed farther and more rapidly than those of trees. Once established, the grasses dominate and prevent encroachment by the trees; in effect, they competitively exclude the woody plants:

> Plains in the tropics. Why are some covered with lofty forests,—others with grasses only? This for a long time puzzled me, but I think I have found the explanation. It depends upon the more or less rapid rise of the plain from the ocean. Where a forest covered plain gradually increases by deposit, as in a delta &c. the seeds of trees vegetate in the mud & each spot is occupied as it is formed,—but if a rapid elevation should convert a shallow sea into a muddy plain, the seeds of grasses carried by the wind or by birds will vegetate immediately & cover it, forbidding the advance of forest. Thus may be explained the open plains of Celebes compared with the forest plains of Borneo,—& the small flat grassy valleys in Gilolo. Ground once taken possession of by grasses cannot be reconquered by forest even if surrounded by it. A clearing for a few years only, will if left become forest, from roots & seeds left in the earth, but if once covered with grass all woody growth is kept down. Even the "llanos" of the Orinooko as compared with the forest valley of the Amazon may perhaps be accounted for in this manner. Climate or soil will not account for it—these vary in both districts & in parts of both are similar.[100]

Competitive exclusion came up in other contexts in the Species Notebook: for example, as part of Wallace's critique of Lyell's claim of "balance" or "harmony" in nature, he pointed out that some species "exclude all others in some tracts."[101] It is perhaps surprising therefore that Wallace did not appear to realize that competition might explain the maintenance of the striking faunal discontinuity that he discovered between the Oriental and Australian regions. Observing that only a handful of species of various groups had successfully "co-mingled" across the Wallace line, Wallace attributed this to recent subsidence of the continental shelves (and conversion of continental areas to archipelago). In several of the papers and books where he described the discontinuity, Wallace emphasized barriers to dispersal (maintaining, for example, that "birds never pass over"), but also acknowledged that "gradual co-mingling" was taking place as the occasional bird or insect species from one region becomes established in the other.[102] Wallace would not be surprised that some groups show evidence of considerable success in "jumping" his line (e.g., a number of plant groups, fanged frogs, and some hawkmoths), while others have not despite having strong dispersal abilities (e.g., honeyeaters and other hawkmoth groups).[103] Nor, presumably, would he be surprised at the potential importance that biotic interactions appear to hold in determining successful dispersal for many groups, even the birds.[104]

In other contexts Wallace showed an understanding of this process, for example invoking it to explain a biogeographical anomaly cited as a criticism of Sclaterian regions: How is it that the snakes of Japan have an Indo-Malayan affinity while the birds and batrachians there are Palearctic? His explanation is premised on the order of colonization of the Japanese islands by these groups. Wallace hypothesized that Japan was first connected to Asia to the north, during which time birds and batrachians colonized the islands, but not snakes or certain butterflies, simply owing to the paucity of these groups in the north. Then, he suggests, Japan later became connected to south Asia via an island chain, allowing southerly species-rich groups like snakes and butterflies to colonize. They would have no problem doing so, finding the islands largely unoccupied by these groups, but any *new* species of birds or batrachians attempting to colonize from the south would have a hard time succeeding because the area was already occupied—that is, would-be colonists "would find a firmly established Palaearctic population ready to resist the invasion of intruders, and it is therefore not to be wondered at that but few, if any, Indian forms of these groups should have been able to maintain themselves."[105] This is an argument based on competitive exclusion. Ultimately Wallace would come to agree with Darwin on the overwhelming importance of competition in limiting distribution.[106]

Competition had a leading role in Wallace's vision of geographical distribution in other ways too. In *The Geographical Distribution of Animals*, *Tropical Nature*, *Island Life*, and other writings of the 1870s and 1880s he often invoked the idea of relaxed competition pressure in isolated environments in explaining (1) the persistence of relictual taxa and (2) wide disjunctions in distribution. Relictualism associated with isolation (as on islands, or in extreme island-like habitats such as caves) is a well-established principle discussed at length by Wallace and Darwin, where forms (e.g., lemurs of Madagascar, the lizard-like *Sphenodon* of New Zealand, and even the marsupials of Australasia) that would be outcompeted in species-rich continental areas can persist in the relative safety that isolation affords. Wallace built upon this idea to propose an explanation for disjunct distributions, describing a process reminiscent of the modern taxon cycle concept, which describes a process of expansion, diversification, contraction, and extinction of species over time in archipelagos.[107]

In Wallace's version, well-adapted, highly competitive species increase in number and spread over a wide area, supplanting inferior competitors and diversifying (speciating) as they move into new environments. At some point the species of the group begin to (rather deterministically) wane as other, superior competitors that have similarly spread from other centers come into competition with them. As Wallace wrote in one article, the species making up the formerly dominant genus go extinct one by one, the last remaining few holding on in isolated locales where competition is less severe:

> First one species and then another will dwindle away and become finally extinct, and by so doing will necessarily leave gaps in its area of distribution. This process going steadily on, the time will at last come when two or three species only will remain, most likely in widely separated parts of its former area; their position being determined either by the competition being there somewhat less severe, or by some speciality of conditions which are exceptionally favourable to the dying-out group. Then one and then another of these species will die out, and the once extensive genus will only be represented by a single species inhabiting a very restricted locality. This will become rarer and rarer, the necessary preliminary to that final extinction which we know to be the fate, sooner or later, of every group of living things.[108]

Biogeographical disjunctions thus arise in this model in the final stages of increase, spread, and decline of groups of allied species: the disjunct species are remnants. In *Island Life* Wallace further pointed out that this process explains why disjunctions of congeneric species are common, while intraspecific population disjunctions are rare.

Succession and the Intermediate Disturbance Hypothesis

Wallace's far-ranging musings on the interplay of geological and climatic history, evolution, dispersal, and species interactions in shaping geographical distribution led him to speculate on several other processes in ways that seem strikingly modern. Two ecological concepts expressed by Wallace late in life include ecological succession and the intermediate disturbance hypothesis. The succession concept has long been understood by naturalists—the pioneering ecologist Frederic Clements (1874–1945) provided a historical overview some twenty-four pages long in his magnum opus *Plant Succession*, tracing the earliest expression of the idea to 1685 in the context of bogs.[109] Clements can be forgiven for overlooking Wallace, whose most succinct expression of the succession concept was tucked away in the 1910 volume *The World of Life*, a book whose subtext can be described as "intelligent design" in modern terms:

> A modified form of the same general law is seen when any ground is cleared of all vegetation, perhaps cultivated a year or two, and then left fallow. A large crop of weeds then grows up (the seeds of which must have been brought by the wind or by birds, or have lain dormant in the ground); but in the second and third years these change their proportions, some disappear, while a few new ones arrive, and this change goes on till a stable form of vegetation is formed, often very different from that of the surrounding country. . . . All these phenomena . . . are manifestly due to that "struggle for existence" which is one of the great factors of evolution.[110]

In the same volume Wallace leads into the succession concept by articulating a state of dynamic ecological equilibrium maintained by periodic disturbance:

> But besides these inorganic causes—soil, climate, aspect, etc.—which seem primarily to determine the distribution of plants, and, through them, of many animals, there are other and often more powerful causes in the organic environment which acts in a variety of ways. Thus, it has been noticed that over fields or heaths where cattle and horses have free access seedling trees and shrubs are so constantly eaten down that none ever grow to maturity, even although there may be plenty of trees and woods around. But if a portion of this very same land is enclosed and all herbivor-

> ous quadrupeds excluded, it very quickly becomes covered with a dense vegetation of trees and shrubs. Again, it has been noticed that on turfy banks constantly cropped by sheep a very large variety of dwarf plants are to be found. But if these animals are kept out and the vegetation allowed to grow freely, many of the dwarfer and more delicate plants disappear owing to the rapid growth of grasses, sedges, or shrubby plants, which, by keeping off the sun and air and exhausting the soil, prevent the former kinds from producing seed, so that in a few years they die out and the vegetation becomes more uniform.[111]

This is a fair expression of what was later dubbed the "intermediate disturbance hypothesis," following the work of ecologists J. Philip Grime, Henry S. Horn, and Joseph Connell in the 1970s.[112] The hypothesis posits that maximal species diversity is maintained under conditions of intermediate rates of ecological disturbance; if disturbance is too rare, then one or a few dominant species eventually take over (competitive exclusion), and if it is too frequent, diversity is suppressed because there is never time enough for species accumulation between disturbance events. Wallace's example of too-frequent disturbance—"fields or heaths where cattle and horses have free access seedling trees and shrubs are so constantly eaten down that none ever grow to maturity"—is likely based on Darwin's account in *On the Origin of Species* of an apparently barren heath in Surrey with freely ranging cattle, in which he found, looking closely, "a multitude of seedlings and little trees, which had been perpetually browsed down by the cattle." Darwin further observed that "as soon as the land was enclosed, it became thickly clothed with vigorously growing young firs,"[113] as Wallace noted.

Conclusion

This overview by necessity provides but an imperfect picture (to echo Wallace's Victorian prose) of the remarkable depth and breadth of Wallace's historical and ecological biogeographical thinking and contributions. It omits, for example, his myriad specific treatments of invertebrate, avian, and mammalian distribution and faunas, plant distribution and systematics, the statistical measure of biotas, and cultural biogeography. It is not an overstatement to say that geographical distribution was always a primary scientific interest of Wallace's. In his autobiography, *My Life*, Wallace rather immodestly listed a dozen of the most important "ideas, suggestions, or solutions of biological problems, which I have been the first to put forth," of which only three are biogeographical.[114] But in a life as long, astonishingly productive, and intellectually rich as Wallace's, for 25 percent of his top dozen scientific contributions to bear on biogeography is saying something.

Acknowledgments

I am grateful to Charles Smith for his helpful editorial comments on earlier versions of this chapter and for kindly providing figure 10.1, and to the University of Chicago Press for permission to republish figure 10.2. Special thanks to Andrew Berry and George Beccaloni for numerous stimulating conversations on Wallace's life and thought, and Lynda Brooks and Elaine Charwat at the Linnean Society of London for their kind assistance in facilitating my study of Wallace's Species Notebook. My Wallace studies were greatly facilitated by a year-long scholarly leave as fellow of the Wissenschaftskolleg zu Berlin, with additional support from Western Carolina University and Highlands Biological Station.

Notes

1. Thomas Paine letter to George Washington, 16 October 1789; see Paine (1995), p. 370.

2. Buffon (1761), excerpted in Lomolino et al. (2004), pp. 16–18.

3. Humboldt and Bonpland (1807/2009), p. 65.

4. Ibid., p. 67.

5. Humboldt (1816), p. 450.

6. De Candolle (1820), p. 417; translated by Nelson (1978), p. 285.

7. Lyell (1830–1833), vol. 2, p. 66.

8. Ibid., p. 131.

9. Wallace letter to Henry Walter Bates, 11 April 1846 (WCP340).

10. Wallace letters to Henry Walter Bates, 9 December 1845 (WCP345) and 20 December 1845 (WCP346).

11. [Chambers] (1844), pp. 252–253.

12. Having made a case for species being "bound together in development, and in a system of both affinities and analogies," Chambers then shows that the facts of geographical distribution are consistent with these patterns. See [Chambers] (1844), pp. 251–262.

13. Wallace (1890f), p. 326.

14. Wallace letter to Henry Walter Bates, 11 October 1847 (WCP348).

15. Wallace (1852b), pp. 109–110.

16. Ibid., p. 110.

17. Wallace (1890f), pp. 328–329.

18. Wallace letter to Henry Walter Bates, 10 December 1861 (WCP377).

19. Gaston (1996, 2000).

20. Moritz et al. (2000).

21. Colwell (2000), Moritz et al. (2000), Haffer (1997, 2008), Michaux (2008).

22. For example, Capparella (1988, 1991), Ayres and Clutton-Brock (1992), Peres et al. (1996), Eizirik et al. (1998), Hall and Harvey (2002), Racheli and Racheli (2004), Cheviron et al. (2005), and Fernandes et al. (2012).

23. Gascon et al. (1998, 2000), Lougheed et al. (1999), Link et al. (2015).

24. Voelker et al. (2013).

25. Colwell (2000), p. 13471.

26. Darwin (1837a, 1837b); see also Darwin (1839), pp. 201–206, 362, 423; and Darwin (1842).

27. Darwin (1837b), p. 554.

28. Forbes (1846). Forbes even identified this landmass as "Old Atlantis" on a map in his memoir, invoking the ancient legend of the lost continent as suggestive evidence! My treatment of Wallace and the land-bridge issue benefited from the lucid treatment of the subject by Fichman (1977), whom I acknowledge with thanks.

29. Charles Lyell letter to Edward Forbes, 14 October 1846 (Lyell 1881, vol. 2, pp. 106–110; Lyell 1839, 1841; Lyell 1830–1833, vol. 4, p. 316). Other leading geologists were far less willing to speculate on such matters. In an extensive volume on the geology of the Pacific, published in 1849 just a few years after Forbes's memoir, American James Dwight Dana (1813–1895) attempted to more precisely map areas of subsidence as indicated by coral atolls, recognizing, like Darwin, that coral islands are records of movement: "Had there been no growing coral the whole would have passed without a record. These permanent registers . . . exhibit in enduring characters the oscillations which the 'stable' earth has since undergone." "Yet," Dana cautioned, "we should beware of hastening to the conclusion that a continent once occupied the place of the ocean, or a large part of it, without proof. To establish the former existence of a Pacific continent is an easy matter for the fancy; but Geology knows nothing of it, nor even of its probability" (p. 400).

30. Darwin letter to Charles Lyell, 16 June 1856 (DCP1902).

31. Wallace (1857d), p. 478.

32. Ibid., pp. 478–479.

33. Costa (2013b), p. 120; Species Notebook (Linnean Society manuscript 180), p. 46.

34. Wallace (1857d), p. 479.

35. Wallace (1855b), p. 186.

36. Wallace (1857d), p. 480, quoting Lyell (1835), p. 154; emphasis in the original.

37. Wallace (1857d), p. 481. As an aside, after nearly two years of collecting in the western part of the Malay Archipelago (1854–1856), Wallace had been serendipitously delayed in voyaging to Aru in the far east, leading to an unintended visit to the islands of Bali and Lombok, where he was struck by astonishing faunal differences between the two, despite being separated by a mere fifteen-mile-wide strait. This was Wallace's

first introduction to the faunal discontinuity—Wallace's Line—that he would elaborate in his paper "On the Zoological Geography of the Malay Archipelago" (1860a). In the meantime, a brief account was published as a letter (Wallace 1857b), so Wallace's comparison of Borneo and New Guinea in the Aru paper is his second published mention of this phenomenon.

38. Wallace (1857d), p. 483.

39. Earl (1845, p. 359; 1853, pp. 37, 39).

40. Darwin letter to Wallace, 22 December 1857 (DCP2192).

41. Darwin letter to Wallace, 9 August 1859 (DCP2480).

42. Wallace (1860a), p. 181. This is possibly a reference to Darwin's determined opposition to land bridges and continental extensions.

43. Wallace (1860a), pp. 181–182.

44. Fichman (1977), pp. 51–52.

45. Sclater (1858).

46. Wallace (1859), p. 449.

47. Wallace (1863b), p. 227.

48. Darwin letter to Wallace, 29 January 1865 (DCP4757, WCP1861).

49. Hooker (1867), p. 50; see also Williamson (1984) and Berry (2009). Describing Darwin's dispersal experiments in some detail, Hooker (1867, p. 75) declared the dispersal hypothesis a more "rational solution" than continental extensions: "to my mind, the great objection to the continental extension hypothesis is, that it may be said to account for everything, but to explain nothing; it proves too much: whilst the hypothesis of trans-oceanic migration, though it leaves a multitude of facts unexplained, offers a rational solution of many of the most puzzling phenomena that oceanic islands present: phenomena which, under the hypothesis of intermediate continents, are barren facts, literally of no scientific interest—are curiosities of science, no doubt, but are not scientific curiosities."

50. Wallace (1864a), pp. 1, 2.

51. Ibid., p. 4.

52. Reviewed by Fleming (2006).

53. Darwin letter to Wallace, 5 June 1876 (DCP10531).

54. Alston (1876a), p. 63.

55. Wallace (1876a), vol. 1, p. 107.

56. Lieberman (2005).

57. Wallace (1905b), vol. 2, pp. 95, 96.

58. Wallace (1876a), p. xv.

59. Fichman (1977), pp. 62–63; emphasis in the original.

60. Wallace (1876a), vol. 1, pp. 76, 278.

61. Wallace (1880b), p. 399. Note on this page Wallace's somewhat disingenuous disavowal of his previous support for the Lemuria hypothesis in *The Geographical Distribution of Animals*. His comments in that volume, however, suggest otherwise.

62. Wallace (1905b), vol. 2, p. 99.

63. Wallace (1880/2013), p. lxxvii.

64. See Heaney (2013), pp. xi–lxxi.

65. MacArthur and Wilson (1963, 1967).

66. Heaney (2013), p. xx.

67. Morrone (2009), p. 15; Kricher (2011), p. 42.

68. Gondwanaland (named for the Gondwana region of north-central India) is the name given in 1861 by Austrian geologist and geographer Eduard Suess (1831–1914) to the Southern Hemisphere supercontinent, whose existence he inferred based on the distribution of *Glossopteris* fern fossils.

69. Wegener (1912/2002, 1966). For Wallace's consilient approach, see Costa (2015).

70. For general reviews see Nield (2007), pp. 152–156, 195–199; and Cox and Moore (2005), pp. 31–35.

71. Reviewed by Wiley (1988).

72. Nelson (1978), p. 289.

73. Tectonic-related vicariance is evident in some groups, while chance long-distance dispersal surely has occurred in others (perhaps most obviously in remote oceanic islands geologically too young to have ever been connected or even near a continental landmass, yet populated with endemic species closely related to species of the nearest mainland). In recent years genetic evidence has vindicated dispersal as an important historical process in shaping modern geographical distribution of species, while other distributional patterns are explicable in the light of plate tectonics. See Cox and Moore (2005), pp. 371–374; Morrone (2009), p. 20; and A. de Queiroz (2005, 2014) for historical overviews of the dispersal-vicariance schism, and A. de Querioz (2005, 2014) on the vindication of transoceanic dispersal as a valid historical process shaping some modern geographical distributions.

74. Croizat (1964), p. 605. For a synopsis of Croizat's ideas, see Cox and Moore (2005), pp. 30–31; and Lomolino et al. (2004), pp. 653–656.

75. Cox and Moore (2005), p. 31.

76. Wallace (1894b), pp. 610–611.

77. Michaux (2010), p. 195.

78. Morrone (2002, 2009), pp. 174–176.

79. Holt et al. (2013).

80. Kreft and Jetz (2013).

81. Dickerson et al. (1928), p. 101.

82. Simpson (1977).

83. Cox (2001), p. 514.

84. Whitmore (1982); Michaux (1991, 2008), pp. 182–185.

85. Michaux (2010), p. 204.

86. Early overviews were provided by Fischer (1960), Simpson (1964b), and Pianka (1966); for recent reviews see Chown and Gaston (2000), Willig et al. (2003), Brown and Sax (2004), Davies et al. (2004), Jablonski et al. (2006), Gaston (2007), Mittelbach et al. (2007), and Mannion et al. (2014).

87. Willig et al. (2003).

88. *Ansichten der Natur* was first published in German in 1808; one of the first English editions of 1849, published as *Views of Nature* (another one, published the same year, was given the title *Aspects of Nature*), is a translation of the second German edition of 1826.

89. Humboldt (1849/2014), pp. 158, 159.

90. Wallace (1878c), pp. 65, 66.

91. Wallace (1878c), p. 123.

92. Wallace (1910c), p. 96.

93. Mittelbach et al. (2007).

94. Dobzhanzky (1950, p. 220) wrote, "Where physical conditions are easy, interrelationships between competing and symbiotic species become the paramount adaptive problem." This idea was also expressed and refined by MacArthur (1972), chapters 7 and 8, and lies at the heart of the "Red Queen hypothesis" of Van Valen (1973).

95. Gaston (2007), p. R574.

96. Chown and Gaston (2000), Kricher (2011), pp. 124–130.

97. Currie and Paquin (1987).

98. Kricher (2011), pp. 132–151.

99. Intraspecific competition in particular was absolutely central to both Darwin's "Principle of Divergence" and his tendency to favor (in modern terms) sympatric versus allopatric speciation: Browne (1980), Tammone (1995), Schweber (1980), Beddall (1988). See overview in Costa (2014), pp. 225–227, and annotations in Costa (2009a), pp. 111–126.

100. Costa (2013b), pp. 244, 246; Species Notebook (Linnean Society manuscript 180), pp. 108–109.

101. Costa (2013b), p. 126; Species Notebook (Linnean Society manuscript 180), p. 46.

102. Wallace (1860a, p. 175; 1863b, p. 228; 1869c, p. 25n.).

103. For plants: Welzen, Slik, and Alahuhta (2005), Welzen, Parnell, and Slik (2011), Muellner et al. (2008), Bacon et al. (2013). For sphingid moths: Beck et al. (2006). For fanged frogs: Evans et al. (2003). See overview by Whitmore (1982).

104. Ornithologist and evolutionary biologist Ernst Mayr (1904–2005) explored the "efficiency" of the barriers posed by the straits between the Lesser Sunda islands, from Sumatra in the west to Kambing in the east, in limiting the intermingling of Oriental and Australian-region birds. The island chain is rather like a filter bridge, attenuating the numbers of eastern and western species penetrating in opposite directions. The Bali-Lombok strait marks the strongest apparent barrier in the chain by far, as Wallace observed, but for birds it is a barrier that stems more from ecology than mere distance (which is trivial for birds): Mayr held that ecological factors (competition most likely foremost among them) have "as much or more to do with the limits of the [birds'] ranges than age or width of the straits between [the islands]" (Mayr 1944; and see figure 10.2). Simply put, robust populations of species occupying available niches pose a formidable barrier to invasion, particularly since chance immigrants are almost by definition likely to be few.

105. Wallace (1864a), p. 5.

106. Wallace highlighted the importance of competition in two encyclopedia articles: in the first (Wallace 1875a, p. 84) he wrote that "[i]t has been shown by Mr Darwin that, in the case of most animals and plants in a state of nature, the competition of other organisms is a far more efficient agency in limiting their distribution than the mere influence of climate." In the second (Wallace 1878b, p. 268) he noted that "the actual power of dispersal is by no means the only factor in determining the distribution of a species or a group. It is no use to bring a creature to a new country if it cannot live and maintain itself there. Whether it can do so depends upon many causes. It must be able to adapt itself to a different climate, and generally to different physical conditions; it must be able to live upon whatever food it may find in its new abode; and, most important of all, it must be able to defend itself against new kinds of enemies and to live in successful competition with allied organisms which are already in possession of the soil."

107. See Cox and Moore (2005), p. 186; Ricklefs and Bermingham (2002); and references therein.

108. Wallace (1879b), p. 249; see extended discussion in Wallace (1880/2013), pp. 59–63.

109. Clements (1916), pp. 8–32.

110. Wallace (1910c), pp. 13–14. Henry David Thoreau (1817–1862) gave perhaps the earliest American expression of the concept in an 1860 essay, based on a lecture entitled "The Succession of Forest Trees," given to the Middlesex Agricultural Society. The lecture was posthumously published in a volume of collected works under that title, with a biographical sketch by Ralph Waldo Emerson; see Thoreau (1887), pp. 33–52.

111. Wallace (1910c), p. 13.

112. See Kricher (2011), pp. 166–169; and Wilkinson (1999).

113. Darwin (1859/1976), p. 72.

114. Wallace (1905b), vol. 2, pp. 386–388.

‹ 11 ›

Wallace at the Foundations of Biogeography and the Frontiers of Conservation Biology

MARK V. LOMOLINO

"The vastness of the tropical archipelagoes also provided the knowledge Wallace needed to conceive the biological discipline of biogeography, which has expanded during the late twentieth century into a cornerstone of ecology and conservation biology."[1] These words, from one of the most influential scientists of the twentieth century, E. O. Wilson, both acknowledge the fundamental importance of biogeography for understanding biological diversity, and position Wallace—if only implicitly—at the foundation of what Wilson and others eventually articulated as the field of conservation biology. Wallace's contributions to these fields of study, while vastly underappreciated in the past, are gaining more recognition in recent years: not only for establishing the foundations, but in applying the patterns and principles of biogeography to advance the frontiers of conservation biology as well.

Charles Smith has summarized Wallace's insights on conservation into six general areas where his influence was felt many decades before the discipline of conservation biology was formally articulated:[2]

- *The importance of nature.* His quasi-poetic musings on the place of nature include the plea that humans better familiarize themselves with nature's workings if they expect to plot a responsible trajectory into their future. This line of thinking speaks directly to the moral-ethical underpinnings of the current biodiversity conservation movement.[3]

- *Collections and systematics.* His collecting and systematics work brought to professionals and laypersons alike a wealth of knowledge on the basic characteristics of biodiversity. In addition to being one of the most prolific specimen collectors of any time, he is increasingly being recognized both as history's pre-eminent tropical naturalist, and as one of its leading field biologists (see chapter 3).
- *The conservation of natural resources.* On several occasions Wallace wrote about the conservation of nonrenewable resources, especially in the context of international trade (see chapter 8).
- *The degradation of landscape and climate.* Wallace's knowledge of physical geography was very extensive, and he was a pioneer in recognizing how various synergies might result in the deterioration of surface processes (see chapter 9).[4]
- *The protection of technologically unadvanced peoples.* No one in Wallace's time was more vocal in insisting that culturally unsophisticated native peoples were yet our equals so far as intelligence and morality go. In practice, this implicitly became a plea for their fair and humane treatment (see chapter 6).
- *Land planning.* Wallace's ideas and plans for land nationalization and other land conservation programs were ahead of their time, featuring suggestions for "rural renewal" such as the preservation of ancient monuments and common grounds, the conservation of nonrenewable resources, geographically graded rate plans for property rental, and the construction of greenbelts.[5]

Beyond these influences on the early programs in biodiversity and conservation, Wallace's writings and insights continue to influence modern conservation biology. This will be my focus here: the enduring influence of Wallace's insights on the frontiers of conservation biology, as evidenced in three emerging concepts and syntheses in biogeography and biodiversity research: the Wallacean shortfall, the Anthropocene, and conservation biogeography.

The Wallacean Shortfall

In an important paper published in 1992, Peter Raven and E. O. Wilson described the "Linnaean shortfall" as what then seemed to be a cavernous gap between the number of described species and the actual total number of species believed to inhabit the planet.[6] As such, the Linnaean shortfall was viewed as a challenge but also a great opportunity for naturalists and systematists, with many thousands of species awaiting discovery and description—some of these proving so distinct that they were destined to be recognized as representatives of undescribed families, orders, or even phyla.[7] The last two decades have witnessed great progress in addressing the Linnaean shortfall,[8] but to develop effective plans for conserving biological diversity we require much more than taxonomic designations and the identifying characteristics of species. Essential information, but lacking for most of the described species (and, in particular, for those of conservation concern), is accurate descriptions of species distributions.

First articulated in 2006 in the text *Biogeography*,[9] the term "Wallacean shortfall" was a conscious effort to pay homage to the contributions and lasting legacy of Wallace, whose singular career of discovery and description of new species was surpassed only by Linnaeus and a select few naturalists. His voyages across the Malay Archipelago totaled some 14,000 miles, during which he collected more than 100,000 insects, 8,000 bird skins, and 410 mammal and reptile specimens, including thousands of species that were new to science. In addition to his contributions to the world's biodiversity collection, the homage is also acknowledgment that Wallace's life served to inspire generations of naturalists and systematists to continue to explore and describe the fascinating natural menagerie of species across the planet and, perhaps most important, for his continual emphasis on the geographical context of biological diversity. Borrowing the assertion of Theodosius Dobzhansky, that "*nothing in biology makes sense except in the light of evolution*,"[10] the central mantra of biogeography—one clearly evident in the writings of Wallace throughout his professional career—is that patterns in biological diversity over time and across all levels of complexity (from genes to entire ecosystems and regional biotas) make little sense unless placed within an explicit geographical context. After all, the main inspirations for both Wallace's and Darwin's epiphanies on evolution by natural selection were the differences in species and biotas among islands and geographic regions. Plants and animals dispersed to isolated regions, and in the absence of the homogenizing effects of interbreeding (and what would later be known as gene flow), they diverged in response to different selection regimes. Wallace's appreciation for the transformative value of the geographical context of life is clear in the final paragraph of his seminal volume *The Geographical Distribution of Animals*: "And it is a study which will surely lead them [naturalists] to an increased appreciation of the beauty and the harmony of nature, and to a fuller comprehension of the complex relations and mutual interdependence, which link together every animal and vegetable form, with the ever-changing earth which supports them, into one grand organic whole."[11]

Equally important is the realization—evident in the earlier quotation from E. O. Wilson—that many of the most important questions and challenges for conserving biological diversity have a geographical context. Where does the species occur? What sets the limits to its geographical range? To what degree do geographical distributions of different species overlap, creating hotspots of diversity and high-priority regions for conservation? Where are the key threats to biological diversity, and how might these threats or extinction forces spread to other regions in the future? As Wallace realized more keenly than most other scientists of his era (and those in more recent times as well), understanding and conserving what we call biological diversity requires that we map it. Thus it seemed entirely appropriate for my colleagues and me to advance the term "Wallacean shortfall" to describe the paucity of information on species distributions and on the geographical dynamics of extinction forces, especially those associated with the dynamics of human civilizations.[12] Again, this homage to Wallace was not only from our appreciation for his contributions to biogeography. These were, of course, without question fundamental, transformative, and arguably without rival, but we also wished to acknowledge that he was thinking and writing about the "plundering" of the natural world many decades if not over a century before the issues and discussions coalesced into one of the most important applications of the natural sciences to be undertaken by humanity: conservation biology.[13]

Emergence of the Anthropocene

Throughout the history of the natural sciences and, in particular, during the ages of Enlightenment and European exploration (fifteenth through eighteenth centuries), few if any individuals possessed a knowledge of the diversity and geography of life in its natural state that rivaled that of Wallace's. This wealth of information was derived from a lifetime of working in and studying the natural world, including his own field excursions conducted at local and regional scales, across two continents (Europe and South America) and throughout the vast, complex, and biologically rich set of archipelagoes of Indonesia. As chronicled earlier in this book, his years as a young man were spent as an assistant and apprentice to his brother

William, surveying lands for the railroads across England and Wales. These experiences were likely foundational to his appreciation of the geographical context of life and are reminiscent of the earlier experiences of William Smith (often viewed as the father of British geology and the creator of "the map that changed the world"[14]) during his surveys of the British landscape for the empire's expanding canal system during the late eighteenth century. For both men, these early years included hundreds of hours spent patiently walking across the land, surveying its underlying soils and rock formations and, more especially in Wallace's case, wondering about the interrelations of all of this with the diversity and distributions of plants and animals. Smith's grand map, first published in 1815, demonstrated how the distributional patterns of rock strata and their associated fossils could be used to reconstruct the evolution of the earth, including that of its ancient, now extinct, life-forms. This may well have served as a template of sorts for Wallace's 1876 map of zoogeographical regions, which demonstrated how species distributions among the continents could be used to reconstruct the evolutionary histories of isolated biotas.

Smith's surveys and explorations were limited to Great Britain, and although they eventually produced his famous map, he was soon to fall tragically into a life of ignominy and destitution. Wallace on the other hand sought to rise from his modest beginnings by following the lead of great natural explorers—in particular Alexander von Humboldt, whose narratives of exploration in the New World tropics were essential reading for all ambitious young scientists, including Wallace and Darwin. Despite the loss of many of the specimens he collected during his four years along the Amazon and Rio Negro, the contrast to the landscapes he had surveyed across Great Britain served to emphasize the altered state of "civilized" lands, even those of the "rural" countryside. Wallace also came to realize that the disturbing effects of humanity were penetrating deeper and deeper into the pristine forests of the tropics.

Within two years of his return to England in 1852, Wallace would set out on his defining eight-year voyage of discovery across the Malay Archipelago. His net experience as a naturalist in South America, while unsuccessful economically, had nevertheless primed him for one of the most extensive natural history surveys ever conducted, and for tackling and ultimately unlocking the mysteries of the geography and origins of life. Just as the geographical signature of finches, tortoises, and other life-forms across the Galápagos led Darwin to his epiphany, so too would the stark contrast in species assemblages across the Malay Archipelago lead to the coalescence of Wallace's theory of evolution by natural selection. We may wonder, in fact: Had the geographical signature of nature been less apparent to Wallace (or to Darwin, for that matter), would the great revelations destined to produce one of the most revolutionary theories of all science have occurred so directly?

Wallace's eight years of exploration across the Malay Archipelago also revealed that the undeniable footprint of humanity had traveled far into the most diverse, and previously presumed pristine, tropical Edens on the planet. In the final chapter of his *Wonderful Century*, Wallace demonstrated his understanding of both the functioning and the fragility of tropical ecosystems in the face of humanity's "rush to obtain wealth":

> In tropical countries many valuable products can be cultivated by means of cheap native labor, so as to give a large profit to the European planter. . . . The rich soil, the product of thousands of years of slow decomposition of the rock, fertilized by the humus formed from decaying forest trees, being no longer protected by the covering of dense vegetation, was quickly washed away by the tropical rains, leaving great areas of bare rock or furrowed clay, absolutely sterile, and which will probably not regain its former fertility for hundreds, perhaps thousands of years. The devastation caused by the great despots of the Middle Ages and of antiquity, for purposes of conquest or punishment, has thus been reproduced in our times by the rush to obtain wealth.[15]

This virtually unrivaled wealth of knowledge on biological diversity, and in particular, its geographical signature and context, meant that Wallace was not only positioned to transform our understanding of the natural sciences and lay the foundation for one of its most holistic disciplines—biogeography—but that he was also sensitive to, if not incapable of ignoring, the impacts of human civilization on the natural world.

The Plundering of Lands and Natural Resources

Wallace was struck by the contrasts in condition of the areas he saw as he traveled across Great Britain, and even more so by the stark differences between the state of these long-inhabited, heavily impacted lands and those he saw during his explorations of the Amazon and Rio Negro and islands of the Malay Archipelago. In each region, albeit to varying degrees, he had witnessed the misguided power of civilization to—in Wallace's words—"plunder" natural resources, and to transform, degrade, and homogenize natural landscapes wherever human colonization took hold.

The following is from "The Plunder of the Earth—Conclusion," a chapter in Wallace's *Wonderful Century*:

> The struggle for wealth, and its deplorable results, as sketched in the preceding chapter, have been accompanied by a reckless destruction of the stored-up products of nature, which is even more deplorable because more irretrievable. Not only have forest-growths of many hundreds of years been cleared away, often with disastrous consequences, but the whole of the mineral treasures of the earth's surface, the slow products of long-past eons of time and geological change, have been and are still being exhausted, to an extent never before approached, and probably not equaled in amount during the whole preceding period of human history.[16]

David Collard (chapter 8) provides a thorough discussion of Wallace's objections to problematic practices of land use by advanced civilizations, and his proposal for land nationalization. Wallace's views on the morality of land use—its plundering, and alternative more ethical and ecologically sound means of managing the land—were likely formative for developing conservation views such as Aldo Leopold's land ethic, itself soundly based on the foundations of ecology, natural selection, and evolution. This similarity in viewpoint is obvious, for example, in Leopold's description of his land ethic in the 1949 work *A Sand County Almanac*:

> There is as yet no ethic dealing with man's relation to land and to the animals and plants which grow upon it. Land, like Odysseus' slave-girls, is still property. The land-relation is still strictly economic, entailing privileges but not obligations. . . .
>
> This extension of ethics, so far studied only by philosophers, is actually a process in ecological evolution. Its sequences may be described in ecological as well as in philosophical terms. An ethic, ecologically, is a limitation on freedom of action in the struggle for existence . . .
>
> In short, a land ethic changes the role of *Homo sapiens* from conqueror of the land-community to plain member and citizen of it. It implies respect for his fellow-members, and also respect for the community as such.[17]

Natural and Anthropogenic Extinctions

Wallace, of course, did not limit his incisive studies and insights to extant species, but also viewed an understanding of fossils and ancient biotas as invaluable clues to the origins and diversification of life over space and time. Wallace and his more astute colleagues were led to two seemingly undeniable deductions: (1) extinctions have occurred, and (2) especially in recent times, many of them resulted from human activities. Not only have human colonizations led to a plundering of natural resources and transformation of native lands, but overexploitation of plants and wildlife and introductions of exotic predators and competitors to islands have driven a startling number of species to extinction in recent times. The effects of recent extinctions on native biotas (whether anthropogenic or natural) became clear to Wallace when he compared the extant wildlife of North and South America—in particular, their large mammals—to the lists of mammals in the fossil record and to those of extant biotas of Africa and other relatively pristine regions.

> It is clear, therefore, that we are now in an altogether exceptional period of the earth's history. We live in a zoologically impoverished world, from which all the hugest, and fiercest, and strangest forms have recently disappeared; and it is, no doubt, a much better world for us now they have gone. Yet it is surely a marvellous fact, and one that has hardly been sufficiently dwelt upon, this sudden dying out of so many large mammalia, not in one place only but over half the land surface of the globe. We cannot but believe that there must have been some physical cause for this great change; and it must have been a cause capable of acting almost simultaneously over large portions of the earth's surface, and one which, as far as the Tertiary period at least is concerned, was of an exceptional character. Such a cause exists in the great and recent physical change known as "the Glacial epoch."[18]

Wallace returned to this subject—of glacial-epoch extinctions—in his entry "Distribution" for a new edition of *The Encyclopaedia Britannica* in 1878. After summarizing the lists of large mammals and birds that are now extinct, but that once inhabited Europe, North America, South America, and Australia, Wallace inferred, "It appears then that in all parts of the world where we have been able to obtain the requisite information, the period which immediately preceded that in which we live was characterized by great movements or migrations of the higher animals where that was possible; and everywhere, by the extinction of a variety of huge animals belonging to almost every order of Mammalia and to several orders of birds, many of which are now totally unrepresented on the globe."[19]

Thus, Wallace appreciated the dramatically altered state of native biotas across much of the world, what we now call the megafaunal extinctions during the late Pleistocene and early Holocene. Although at first attributing these extinctions to climate change during the glacial epoch, Wallace later rejected this idea for at least three lines of reasoning.[20]

- "[S]pecies which undoubtedly survived that event [the Ice Age] have since become extinct."
- "This great climatic catastrophe did undoubtedly produce extensive migration . . . but . . . is not held to be a sufficient cause for so general a destruction of the larger forms of life."
- "[I]n two very remote parts of the earth, both enjoying a warm or even a sub-tropical climate—Australia on the one hand, and Brazil to Argentina on the other,—exactly the same phenomena [megafaunal extinctions] have occurred."

Instead, Wallace attributed the relatively recent, global wave of megafaunal extinctions to the same force that is now identified by an emerging consensus of scientists studying this phenomenon: "the rapidity of the extinction of so many large Mammalia is actually due to man's agency, *acting in co-operation with those general causes* which at the culmination of each geological era has led to the extinction of the larger, the most specialised, or the most strangely modified forms. . . . [T]he fact that man should everywhere have helped to exterminate the various huge quadrupeds, whose flesh would be a highly valued food, almost becomes a certainty."[21] Wallace further explains this conclusion:

> It is therefore certain, that, so soon as man possessed weapons and the use of fire, his power of intelligent combination would have rendered him fully able to kill or capture any animal that has ever lived upon the earth; and as the flesh, bones, hair, horns, or skins would have been of use to him, he would certainly have done so even had he not the additional incentive that in many cases the animals were destructive to his crops or

> dangerous to his children or to himself. The numbers he would be able to destroy, especially of the young, would be an important factor in the extermination of many of the larger species.[22]

We now realize that rather than "the" glacial period, there were some twenty or so periods of glaciation that dominated the Pleistocene, interrupted by relatively short interglacials such as the one we are currently experiencing. The now much more detailed fossil record reveals that most of the megafaunal vertebrates survived all the previous climatic upheavals; so why did they perish during the later ones? Also, instead of there being a simultaneous collapse in megafaunal assemblages across the globe—which would have been consistent with a glacial (climatic) cause—these extinctions occurred in a succession of waves across the globe (beginning in Australia about 50,000 years BP, North America about 13,000 years BP, and South America about 11,000 years BP, and cascading across other continents and islands to most recently claim the megafauna of Madagascar and New Zealand around 1,000 and 700 years BP, respectively). How could a glacial event, or any other major climatic shift of the Pleistocene, account for these asynchronous waves of extinctions, some occurring prior to the last glacial maximum, some during, and others well after? Although inconsistent with the glacial-period extinction hypothesis, the global succession of extinctions described above is entirely consistent with the hypothesis that they were driven by ecologically significant humans, with each episode of megafaunal extinctions coming on the heels of colonization as humans migrated out of Africa to inhabit and dominate most regions across the globe.

Again, Wallace was keenly aware of the role humanity had played, and would continue to play, in causing recent and imminent extinctions. In his 1863 paper "On the Physical Geography of the Malay Archipelago," he speaks persuasively to the mission of the naturalist, the importance of geographical distributions in understanding the origins and diversification of life, of his appreciation for the irreplaceable value of species, and the responsibility of government and science in conserving these treasures:

> It is for such inquiries the modern naturalist collects his materials; it is for this that he still wants to add to the apparently boundless treasures of our national museums, and will never rest satisfied as long as the native country, the geographical distribution, and the amount of variation of any living thing remains imperfectly known. He looks upon every

> species of animal and plant now living as the individual letters which go to make up one of the volumes of our earth's history; and, as a few lost letters may make a sentence unintelligible, so the extinction of the numerous forms of life which the progress of cultivation invariably entails will necessarily render obscure this invaluable record of the past. It is, therefore, an important object, which governments and scientific institutions should immediately take steps to secure, that in all tropical countries colonised by Europeans the most perfect collections possible in every branch of natural history should be made and deposited in national museums, where they may be available for study and interpretation.
>
> If this is not done, future ages will certainly look back upon us as a people so immersed in the pursuit of wealth as to be blind to higher considerations. They will charge us with having culpably allowed the destruction of some of those records of Creation which we had it in our power to preserve; and while professing to regard every living thing as the direct handiwork and best evidence of a Creator, yet, with a strange inconsistency, seeing many of them perish irrecoverably from the face of the earth, uncared for and unknown.[23]

Some seventy years later, Aldo Leopold would echo Wallace's imagery of species serving as precious elements in the grand story of Earth's natural history when he observed that "[t]o keep every cog and wheel is the first precaution of intelligent tinkering."[24]

Taken together, Wallace's observations on the plundering of natural resources, transformation of native lands, and recent and imminent anthropogenic extinctions made it clear to him that we had long ago entered into an entirely singular age of life, one influenced to an ever-increasing degree by humanity.

Origins of the Anthropocene

As I observed earlier, one of Wallace's enduring legacies is how he inspired countless naturalists over many generations to catalogue, describe, and compare local and regional biotas in a quest to unlock the forces influencing the diversity and dynamics of the natural world. The accumulated knowledge available to Wallace eventually made it apparent to him that we had indeed entered a period that is quite distinct from those of previous ages, one dominated by the activities of one species: *Homo sapiens*. While there is and will no doubt be continuing debate over when this period of anthropogenically altered life—the Anthropocene—began,[25] it is likely that Wallace would agree with those asserting that its onset was quite early: arguably as early as the time anatomically modern (and ecologically advanced) humans migrated out of Africa, some 100,000 years ago. Many hundreds if not thousands of native species suffered extinctions following the arrival of ecologically significant populations of humans in places where the native biota had no prior experience with these novel hunters, competitors, and ecosystem engineers.

Beyond his recognition of anthropogenic influence and other work establishing the foundations of biogeography and biodiversity research, Wallace's efforts also serve to demonstrate the value of a holistic approach in science. One such demonstration of this approach for advancing science is an emerging synthesis among the fields most central to Wallace's life's work—one that combines the insights from biogeography to address the challenges of conserving species in the face of continuing "advances" of humanity.

The "New" Synthesis: Conservation Biogeography

The past few decades have witnessed tremendous advances across multiple lines of biodiversity research. This includes not only an exponential increase in information on the morphological, physiological, behavioral, genetic, and ecological traits of an impressive diversity of species, but also impressive advances in our ability to store this information and visualize and analyze it in evolutionarily and geographically meaningful ways. These transformative advances not only validate the legacy of Wallace and his contemporaries in laying the foundation for the modern science of biodiversity research, but they extend and amplify Wallace's emphasis on distributions and the geography of life—the central theme of his life's work. Many if not most of the challenges for conserving biological diversity in the face of continuing and seemingly incessant "advances" of humanity have an explicit geographical context. *Where* should we search for undiscovered species? *Where* are the range boundaries of endangered species, and how will they shift in response to climate change or the advances of land transformation and invasive species? *Where* should we locate nature reserves, and *how large* should they be and *how isolated* from other reserves or from centers of human activity?

In the latter part of the twentieth century, the term "conservation biogeography" was introduced to draw attention to what Wallace (and, more recently, an ever-growing body of concerned citizens) understood as an intensifying crisis in biodiversity: the need to apply and advance the principles and tools of biogeography to discover, describe, and develop effective strategies for preserving species before they are "plundered" by humanity.[26] The term may have first appeared in the title of an article by John Grehan that emphasized the potential utility of panbiogeography for biodiversity research.[27] The research program, however, was not clearly articulated until a volume stemming from the founding of the International Biogeography Society,[28] *Frontiers of Biogeography: New Directions in the Geography of Nature*, identified conservation biogeography as one of the emerging frontiers of the field.[29] Conservation biogeography became the explicit theme of the second meeting of the society, held in 2005 at the US National Conservation Training Center in at Shepherdstown, West Virginia. At that meeting, the society also established the Alfred Russel Wallace Award, which recognizes a lifetime of outstanding contributions by an eminent scholar in any subdiscipline of biogeography. The *Frontiers* volume elaborated the fundamental assertions of conservation biogeography, including (1) that success in conserving biological diversity depends heavily on our understanding of the geography of nature; and (2) that in order to conserve the true nature of native species (including their distinct physiologies, morphologies, behaviors, and ecological capacities), we need to conserve their distributions, thus conserving their evolutionary as well as geographical context.[30] This theme, that the evolutionary history of a lineage (and its evolutionary potential as well) is inseparable from its geographical history—that is, that evolution occurs across space as well as over time—is the essence and central mantra of biogeography, and one clearly visible throughout Wallace's major works on the subject.

In sum, through his research, writing, and deep ethical concerns, Wallace's lifelong dedication to understanding the geography of life not only established the foundations of one of science's most holistic disciplines, but continues to serve as an exemplar and inspiration for those attempting to advance our understanding of, and develop effective means for conserving, biological diversity.

Acknowledgments

My thanks go to our three editors for their constructive remarks on earlier manuscript versions of this essay.

Notes

1. Wilson (1999), p. xii.

2. Smith (1998-b).

3. Wallace's example has inspired many writers over the years (most famously Joseph Conrad). For treatments of his influence on novelists and poets seeking to "engage" the natural world, see Krasner (1992), Houston (1997), Hampson (2000), Cluysenaar (2008), Bignami (2009), and Christensen (2011).

4. A brilliant example of this attention is provided by a sequence from an 1887 interview Wallace gave in Sioux City, Iowa, during his North American lecture tour: "The doctor alluded . . . to the systematic tree planting in the western states and territories. The relation of forests to climate, he said, is one to which he has given much study. 'Tree planting,' said he, 'cannot fail to be of the greatest value to the forestless prairies of the northwest. It certainly modifies winds, and it will probably modify rainfall, though it is not so certain about this. One fact is very important about this matter of tree planting. The trees should be planted near the streams. In every country in which the trees have been cut down there have been great floods, and this is especially so where the hillsides have been denuded of the forests. Systematic and extensive tree culture along the minor streams of the northwest will certainly benefit the land. Wherever there is a forest growth moisture is necessarily retained in the soil. Moreover, the mold and the debris which collect under the trees tends to obstruct drainage and thus counteracts the effects of drought. There is no reason why all this benefit cannot be secured to the prairie territory of the northwest by a few years' attention to tree culture.'" See http://people.wku.edu/charles.smith/wallace/S735B.htm.

5. For example, see Smith (2003), Clark and York (2007), and Collard (chapter 8 in this volume). For other important studies of Wallace's influence on the foundations of conservation biology, see Stepan (2001), Clark and York (2007), Knapp (2008), and McDowall (2010).

6. Raven and Wilson (1992).

7. See Blackwell (2011), Mora et al. (2011), and Costello et al. (2013).

8. See IISE (2012) and Shuker (2012).

9. Lomolino et al. (2006), pp. 714–715, 747.

10. Dobzhansky (1973), p. 125.

11. Wallace (1876a), vol. 2, p. 553.

12. Lomolino et al. (2006), p. 714–715.

13. For treatments of this and related shortfalls challenging conservation biologists, see Beck et al. (2013), Diniz-Filho et al. (2013), and Hortal et al. (2015).

14. See Winchester (2001).

15. Wallace (1898d), pp. 369–373.

16. Ibid., p. 369.

17. Leopold (1949), pp. 202–204.

18. Wallace (1876a), vol. 1, pp. 150–151.

19. Wallace (1878b), p. 275.

20. Wallace (1910c), p. 244.

21. Ibid., pp. 246–247.

22. Ibid., p. 249.

23. Wallace (1863b), p. 234.

24. Leopold (1953), p. 147.

25. See Crutzen (2002), Zalasiewicz et al. (2008), and Steffen et al. (2011).

26. See Lomolino (2004), Whittaker et al. (2005), Ladle and Whittaker (2011), and Richardson (2012).

27. Grehan (1993).

28. www.biogeography.org.

29. Lomolino and Heaney (2004).

30. Ibid., pp. 293–360.

‹ 12 ›

Wallace and Extraterrestrial Life

ROBERT W. SMITH

Wallace played a very significant part in the often intense and hard-fought debates on extraterrestrial life in the first decade of the 1900s.[1] In so doing, he argued that the only advanced life to be found in the solar system is on Earth, and he dismissed the possibility of intelligent beings of the same order as humans existing anywhere else in the universe. Yet throughout Wallace's lifetime it was widely accepted by British astronomers, science popularizers, and likely the general public that the universe is populated by extraterrestrials. In this chapter, then, the central issues will be why Wallace rejected extraterrestrials so emphatically and how he came to regard humans as being, in his words, "the unique and supreme product of this vast universe" and that "the universe was actually brought into existence for this very purpose."[2]

Wallace and Astronomy

As a surveyor in Wales and parts of England early in his life, Wallace gained a familiarity with optical instruments. He also made his own simple telescope by mounting lenses in a paper tube, an activity that prompted him to learn more about the construction of larger devices. This small telescope "also led me throughout my life to be deeply interested in the grand onward march of astronomical discovery,"[3] and he remained a close student of astronomy. We find him in 1864, for example, urging Charles Darwin to read

Herbert Spencer's essay on the nebular hypothesis and the formation of the solar system as well as more broadly the history of the universe itself. Wallace told Darwin that it was "the most masterly astronomical paper I have ever read," clearly implying Wallace had read many others.[4]

In his writings on the Ice Age as well as in a chapter of his 1880 book *Island Life*, he drew extensively on James Croll's ideas on possible astronomical causes of climate change. Wallace was impressed by, and examined closely, Croll's writings and told Croll in 1880 that "my theory of geological climate . . . is really founded almost wholly on your own researches."[5] Wallace, drawing on Croll, developed a theory in which changes in the eccentricity of Earth's orbit led to variations in the climate that drove fluctuations in the pace of natural selection, in effect speeding up the rate of organic change, and so reducing the disagreement with the physicists on the age of Earth.[6]

The Extraterrestrial Life Debate

Robert Chambers's anonymously published *Vestiges of the Natural History of Creation* of 1844 was, as we have seen elsewhere in this volume, a major influence on the young Wallace. Chambers began his account in the remote past with matter spread throughout space, a "universal Fire Mist" as he put it. From that Fire Mist, stellar and planetary bodies were slowly born. Over time, as Earth cooled, life began to form and then develop through natural processes until humans appeared. Chambers also insisted that the plurality of worlds was an accepted doctrine, with every one of the numberless planets "either a theatre of organic being, or in the way of becoming so." As all planets are subject to the same physical laws and contain the same chemical elements, then "[w]here there is light there will be eyes."[7] If Wallace had not encountered writings on extraterrestrial life before, he certainly had done so in *Vestiges*.

One of the great scientific controversies of the Victorian era erupted after the publication of William Whewell's 1853 book *Of the Plurality of Worlds*, in which he launched a robust attack on claims of extraterrestrial life.[8] In making his case Whewell—and Wallace was a very close student of Whewell's writings on this topic, although he termed *Plurality of Worlds* an "able if rather vague and diffuse work"[9]—seized on what Michael Crowe has characterized as "dubious observations, overly extended analogies, questionable theoretical claims, and problematic methodological assumptions."[10] But Whewell also regretted that the question of extraterrestrial life would be settled less on scientific grounds than on moral, metaphysical, and theological considerations. For Whewell, however, pluralism was, as he by then viewed matters, religiously dangerous, as it undermined what he regarded as the most remarkable of God's providential actions: the incarnation and redemption.[11] As Wallace would put it in 1903, siding with Whewell on the absence of extraterrestrials, "[T]he belief that other planets are inhabited has been generally entertained, not in consequence of physical reasons but in spite of them."[12]

The leading writer on extraterrestrial life in the 1870s and 1880s was the English astronomer and popularizer Richard Proctor (1837–1888). Wallace studied his writings very carefully and had a high opinion of them. In his treatments of pluralism, Proctor was able to bring to bear much more astronomical evidence than Whewell had in the 1850s. After 1860, increasing numbers of astronomers began to pursue what would come to be called the "new astronomy" (later widely termed "astrophysics"). This featured the analysis of the light from distant bodies with the aid of spectroscopes attached to telescopes. Astronomers were in effect handed remarkable new powers, as they could now gather previously inaccessible information on the chemical and physical properties of celestial bodies. Spectroscopes could provide astronomers with information on, for example, the possible presence of water vapor in the atmosphere of Mars. The new astronomy thereby took the debate on extraterrestrial life in new directions.[13]

In part due to the writings of Proctor, the closing decades of the nineteenth century also saw both an increase in interest in extraterrestrial life and a narrowing of the ground over which the debate on life in the solar system was fought. More specifically, as we will see later, it narrowed to Mars, as lively and sometimes ill-tempered discussions swirled around the possibility of life on the Red Planet. Wallace was not, however, drawn to the extraterrestrial life debate by the Mars furor.

Wallace's Connected Argument

In 1898 Wallace published *The Wonderful Century*, in which he examined both what he regarded as the crucial scientific advances of the nineteenth century and its chief failures. The first edition of the book contained a chapter called "Astronomy and Cosmic Theories." For a later expanded edition, published in September 1903, Wallace wrote four chapters on astronomy and described the astronomical progress of the second half of the century. As he worked through his sources, Wallace became aware that leading astronomers placed the Sun close to the center of the Milky Way, and that they also believed the latter was the only such star system that could be sighted in even the biggest telescopes.

He was very surprised that this understanding, as well as the associated one that our stellar system is finite (although the cosmos itself was universally regarded as infinite), was little commented on. To the great majority of astronomers, it appeared to Wallace, these observations were curiosities. Wallace, however, saw meaning in them. He displayed a profound sense of place and boundaries in his scientific practice, and his emphasis on the Sun's location may be an example of that sense.[14] He also later recalled that when he considered, and rejected, life on the other planets of the solar system besides Earth, "there flashed upon me the idea that it was only near the centre of this vast material universe that conditions prevailed rendering the development of life, culminating in man, possible."[15] Around this time, the London agent of the New York newspaper the *Independent* asked him whether he might like to write an article for the *Independent* on any scientific subject of his choosing. As was often the case throughout his life, Wallace, who never held a salaried position, was again beset with financial concerns, this time largely because of debts incurred in a move to a new house. At first Wallace declined the *Independent*'s invitation, but when the agent reported he could name his own fee, Wallace relented. He took for his subject this notion of a possible relationship between the central position of the Sun and intelligent life on Earth.[16]

In "Man's Place in the Universe," a title with an obvious link to T. H. Huxley's famous 1863 study "Man's Place in Nature," Wallace reported on knowledge gained on stars and stellar systems in the previous quarter century. In so doing, he made clear his teleological and anthropocentric commitments and adopted the position "of those who believe in some Intelligent Cause at the back of this universe, some Creator or creators, Designer or designers."[17] He was therefore explicitly offering a discourse of design, as Proctor had done earlier (though Proctor had arrived at different conclusions). Even though Wallace's religious views were far from an orthodox theism, he was very much writing in the manner of natural theology.

Wallace discussed briefly the evidence that bore on four astronomical questions: "Are the stars infinite in number?" "What is the distribution of stars in space?" "What is the nature of the Galaxy or 'Milky Way'?" and "What is the nature of the star cluster of which our Sun is a member?" For him, the evidence pointed to the Sun's being located in a nearly central position, "if not actually at, the center of the whole visible universe, and therefore in all probability in the *center* of the *whole material universe*."[18] The nub of Wallace's argument was that it was only in this central or very nearly central position that conditions were stable enough over a sufficiently extended period for life to emerge and thrive. Away from the center, the actions of different radiations (e.g., X-rays) as well as gravitational and electrical forces would make for variable conditions, so that "while some of these necessary radiant forces may be wanting," others might be overabundant or highly irregular and "antagonistic to the delicate and nicely balanced forces which are essential to the orderly development of life."[19] He concluded that no other planet in the solar system could offer the same sorts of conditions over the millions or perhaps even hundreds of millions of years that would be required.

One feature Wallace stressed was that there had to be an uninterrupted supply of atmospheric dust for the generation of rain clouds, and in turn beneficial rains and mists. Without these, "the whole course of meteorological phenomena would be so changed as to endanger the very existence of a large portion of the life upon the earth." Adequate amounts of dust called for a sufficient number of deserts and active volcanoes. Constant winds were needed to distribute the dust. These three elements—dust, active volcanoes, and wind—were therefore essential for the development of intelligent life, and he reckoned that he was the first to point out this utility of deserts and volcanoes.[20] Further, he sided firmly with those who "[hold that] the universe is a manifestation of Mind, and that the orderly development of Living Souls supplies an adequate reason why such a universe should have been called into existence, believe that we ourselves are its sole and sufficient result, and that nowhere else than near the central position in the universe which we occupy could that result have been attained."[21]

Wallace's article, which appeared in both London's *Fortnightly Review* and New York's *Independent*, caused a splash, but he was taken aback by the antagonism and what he termed "rather contemptuous criticism" of some astronomers and physicists, criticisms to which he responded in a follow-up article.[22] Wallace was also attacked for moving beyond his own sphere of competence as a scientist. As a review in the *Monist* put it, a scientist working in his own area of expertise might be the right man in the right place—"a round peg in a round hole"—but take him out of his area, and he would be "a round peg in a square hole. It needs only common sense to perceive that he is out of place."[23]

Wallace's agent nevertheless thought they were onto a winning subject even before the *Independent* article was published. He proposed that Wallace write a 70,000-word book on the topic.[24] This Wallace duly did, and it involved him in what he regarded as six months of the stiffest reading and study he had ever undertaken. The article for the *Independent* was published in February 1903; he started the book right away, and it was completed in July. It had been a rush job. As he raced to finish, he sent chapters to his son. As Wallace conceded, of "course this book has been written much too quickly to be first rate. I ought to have had a full year about it—but neither I nor the publishers could wait so we must make it as good as we can, & then when it is well cut up improve it for a new edition."[25]

In preparation for the book, Wallace did not undertake original researches with a telescope or other astronomical instruments, but in addition to his reading he corresponded with various astronomers as well as, most importantly, the popularizer and historian of astronomy Agnes Clerke, who also practiced natural theology in her own works. The result was *Man's Place in the Universe*, published in late 1903. Wallace's agent was right: the book *was* a financial success and brought Wallace a new circle of readers. His ideas again proved highly controversial, however.

Rather than plunge into discussions of Intelligent Causes, Creators, or Designers at the back of this universe, as he had done near the start of his article in the *Independent*, Wallace, although he was again writing for a general educated audience, adopted a quite different narrative strategy for his book. Wallace also stressed in the preface of *Man's Place in the Universe* that he had gone beyond the earlier literature on the plurality of worlds. He had founded his work on the findings and conclusions of the "New Astronomy" "together with those reached by modern physicists, chemists, and biologists. Its novelty consists in combining the various results of these different branches of science into a connected whole, so as to show their bearing upon a single problem," the problem of whether Earth is the only inhabited planet in the entire stellar universe.[26]

In the first chapter, Wallace briefly reviews earlier works on the plurality of worlds and in so doing discusses religious elements of these writings, but he does so as a commentator—not a proponent—in the debate. Five chapters of general astronomical background follow. Wallace then provides three chapters in which he discusses, "Are the stars infinite in number?" "Our relation to the Milky Way," and "The uniformity of matter and its laws throughout the stellar universe." In these chapters he again underlines how contemporary astronomers were placing the Sun very close to the center of our own system of stars, the galaxy, with no distant galaxies being visible in even the most powerful telescopes. The overall stellar system is spherical

DIAGRAM OF STELLAR UNIVERSE (Plan)

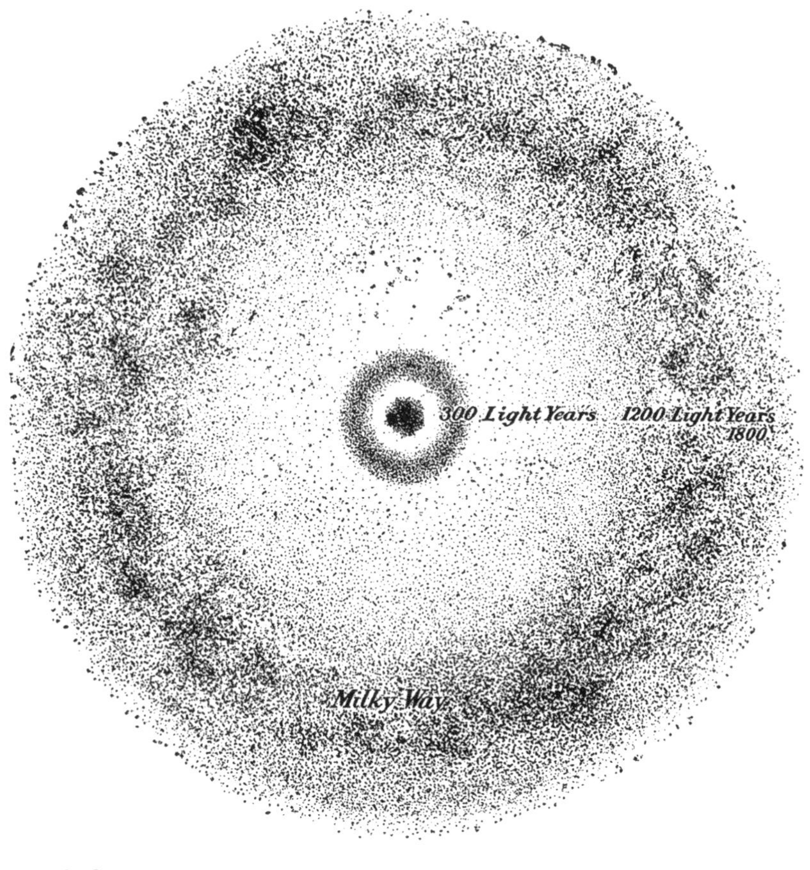

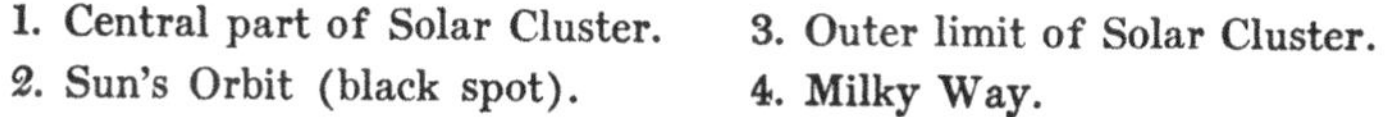

Figure 12.1 Wallace's diagram of the Milky Way. From Wallace (1903b), p. 300.

and the cluster of stars to which the Sun belongs, the solar cluster, is part of the larger system and contains anywhere from several hundred to several thousand stars. The Sun is not at the very center. Rather, it is at the outer edge of the solar cluster's very dense central portion. The larger system is itself some 3,600 light-years across (see figure 12.1).

In making these points, Wallace was broadly in line with the generally accepted astronomical thinking of the day. For example, Agnes Clerke had advocated a single-galaxy universe in *The System of the Stars* of 1890, and the leading American astronomer Simon Newcomb had done likewise in his *Popular Astronomy*, which went through numerous editions in the late nineteenth century. Wallace read both Clerke and Newcomb very closely.

The leading investigators of the size and structure of the galactic system were practitioners of what has been called "statistical cosmology," and they plied their trade with highly sophisticated mathematical tools to analyze large bodies of information on the distances and properties of stars. The scientific papers they produced were far beyond Wallace's technical competence, but their findings were consistent in placing the Sun quite close to the center of a single system of stars that was perhaps lens shaped and some thousands of light-years in diameter.[27] As he had already emphasized in his *Independent* article, that the Sun is in the center or near the center of this system was crucial for Wallace, "the very heart of the subject."[28]

In the early 1900s, there were various theories as to the source of the Sun's energy, none widely accepted (general acceptance of the idea that stars provide their own fuel supply via nuclear fusion was decades away).[29] Wallace claimed that the solar cluster acted via gravitation and perhaps electrical forces to draw in matter from the outer regions of the Milky Way. The inrushing matter fell into stars and fueled them. This process worked most efficiently for the stars at the center of the stellar system, that is, for those stars, like the Sun, within the solar cluster. However, the very inner circle of the solar cluster would receive too much matter, and so the stars there could not be stable. In contrast, the outer edge of the solar cluster, the location of the Sun, was ideal, as a star in such a place would receive a steady and continuing flow of fuel. In regions beyond the solar cluster, the stars, as they received an insufficient inflow of material, would not be able to shine long enough for life to develop on the planets orbiting them. The number of stars that might have planets capable of supporting life could therefore, Wallace reckoned, be put at around several hundred.

Wallace then sets about a process of what he called "successive eliminations." Time was a critical consideration, as a planet would need a relatively steady supply of heat for a very long period for the development of life. But some stars would either be too small or too large to be capable of providing the appropriate amount over the eons required. Other stars had only just formed or were in the process of formation, so there had not been enough time for life to emerge on the planets traveling around them. Stars having one or more companion stars would subject the planets in such systems to wide differences in the amount of heat they received. Life could not develop on such planets, as the changes in conditions were too big. Even if there *were* planets traveling around some suitable single stars, these planets might not be of the right mass or at the required distance from their star for appropriate and stable conditions. If a planet were too far away from its central star, for example, it would be too cold.[30]

Wallace also emphasizes the remarkably complex basis of organic life. T. H. Huxley had identified protoplasm as the physical basis of life, and for Wallace the staggering intricacy of, and chemical reactions in, protoplasm was crucial.[31] Yet protoplasm was just the starting point from which "the infinitely varied structures of living bodies are formed."[32] Wallace's earlier studies of climatic change had also impressed on him that mild climates and generally steady conditions persisted throughout all geological epochs on Earth, and that the maintenance of these outcomes was the result of an extremely delicate balance of conditions.[33] Further, and this was crucial for him, Wallace saw no reason to suppose organic life could develop unless the same conditions and laws that applied on Earth applied elsewhere in the universe too. Through successive eliminations, combined with his profound sense of the astonishing nature of protoplasm, Wallace concludes that the probability is completely against life beyond the Earth.[34]

Wallace also advances a short argument from purpose, but does so only at the end of the book and after more than three hundred pages of what he presents as strictly scientific analyses. Now, Wallace asks, "[w]hat if the Earth is *really* the only inhabited planet in the universe?" In his view, one sizable group, including probably the majority of scientists, would explain it as coincidence. He nevertheless expected there was probably a much larger body of opinion, people for whom "mind is essentially superior to matter and distinct from it." And "when they see life and mind apparently rising out of matter and giving to its myriad forms an added complexity and unfathomable mystery," Wallace contends, "they see in this development an additional proof of the supremacy of mind . . . and when they are shown that there are strong reasons for thinking that man *is* the unique and supreme product of this vast universe, they will see no difficulty in going a little further, and believing that the universe was actually brought into existence for this very purpose."[35] Moreover, with "infinite space around us and infinite time before and behind us, there is no incongruity in this conception," Wallace explains. "A universe as large as ours for the purpose of bringing into existence many myriads of living, intellectual, moral, and spiritual beings, with unlimited possibilities of life and happiness," he argues, "is surely not *more* out of proportion than is the complex machinery, the lifelong labour, the ingenuity and invention which we have bestowed upon the production of the humble, the trivial, *pin*."[36] To presume that all the planets are inhabited would also "introduce monotony into a universe whose grand character and teaching is endless diversity."[37] Wallace goes further yet, claiming that there probably are other universes "perhaps of other kinds of matter and subject to other laws, perhaps more like our conceptions of the ether, perhaps wholly non-material, and what we can only conceive of as spiritual."[38]

For the fourth edition of *Man's Place in the Universe*, Wallace added an appendix with another argument for the uniqueness of intelligent life on Earth. His basic point was that humans are the result of an extremely long series of modifications of organic life that have occurred under particular circumstances. The chance of the very same such circumstances occurring elsewhere must be tiny. Intelligence in any other form was therefore so exceedingly unlikely that if the connected astronomical argument put the chances of life at a million to one, then the evolutionary improbabilities would be at least a hundred million to one, so that, given the known laws of evolution, the total chances against the evolution of humankind, or what he would regard as an equivalent moral and intellectual being, were "a hundred millions of millions to one."[39]

Man's Place in the Universe was very widely read. In addition to seven editions in five years, it was later published in two "cheap" editions; German and French editions also appeared quickly (in 1903 and 1907, respectively). Together with the original article in the *Independent*, it drew more than forty reviews.[40] As was the case for the article, the book version proved highly controversial. A review in the science weekly *Nature* reckoned the book needed to "be treated with considerable respect" and complimented Wallace for the "masterly way" in which he had marshaled the available subject matter as well as the very clear and succinct summary of the necessary astronomical knowledge. Nevertheless, the suggestion that Earth constitutes the only inhabited body in the universe, the reviewer argued, was a reversion to prehistoric ideas. More powerful arguments were needed before humanity's isolation in the cosmos could be accepted as a fact.[41]

At the end of the nineteenth century and the start of the twentieth century, professional astronomers were fundamentally anti-cosmological and strove to avoid what they regarded as metaphysical questions. They generally steered away, at least in their scientific publications, from issues such as whether there are other star systems beyond the galaxy. Wallace had of course charged into what professional astronomers shunned as metaphysical realms, and not surprisingly, he came under withering fire for doing so.

Bernard Lightman has stressed that in the late nineteenth century, the role of the science popularizer changed significantly. "Publishers, editors, and the growing reading audience," he argues, "looked to them, and not necessarily to the practitioner, to interpret the larger metaphysical meaning of scientific theories in terms they could understand and in ways they could appreciate."[42] Early in the twentieth century, the larger metaphysical meaning of scientific theories was exactly what Wallace had offered to his general readership; with *Man's Place in the Universe* he had tapped into this interest by providing a wide-ranging and original defense of anthropocentrism. But for the great majority of scientists, anthropocentrism was a "prehistoric" notion. Accordingly, he was roundly criticized by scientific practitioners, but widely read by the general public, some of whom were eager to read works of natural theology that testified to the persistence of a discourse of design as a way of understanding the workings of nature.

Within weeks of the publication of *Man's Place in the Universe* an interviewer queried Wallace about a recent book by the physicist Sir Oliver Lodge, who wrote popular works in a natural theological framework that drew on spiritualism and evolution. Did Wallace agree with Lodge, the interviewer asked, "that the attempt to explain the universe by chance had absolutely failed. It must have had a designer," an Infinite Being. Wallace agreed. "My whole argument tends in that direction, though my object in writing *Man's Place in the Universe* was purely scientific, not religious."[43] There was for Wallace no supernatural deity in terms of first causes; the universe was infinite in extent and had existed for an infinite time. But instead of posing a single Being that acts everywhere as direct creator, organizer, and director of even the minutest motion, he preferred to think of "infinite grades of beings."[44] Wallace's examinations of the case for extraterrestrials had reinforced for him the uniqueness and privileged status of humans. For Wallace, as for the majority of the earlier participants in the extraterrestrial life debate, this was the central issue.

With this cosmology in place, Wallace had come very close to the end of a journey that had begun seriously in the 1860s when he had taken his first public strides toward an account of evolution that was in line with theism. Fichman has emphasized that *Man's Place in the Universe* and Wallace's 1907 work *Is Mars Habitable?* should be read together with his 1910 work, *The World of Life: A Manifestation of Creative Power, Directive Mind and Ultimate Purpose*, the book that marked the end of that journey. In *The World of Life*, Wallace sought to sum up his half century of thought and labor on evolution.[45] But he also wanted to explore the nature and causes of life. Through this exploration, he believed that his readers would understand that a "Creative Power," a "Directive Mind," and an "Ultimate Purpose" were essential to explaining the natural world. And what was the Ultimate Purpose? It was the "development of Man, the one crowning product of the whole cosmic process of life-development; the only being which can to some extent comprehend nature[,] . . . appreciate the hidden forces and motions everywhere at work, and can deduce from them a supreme and over-ruling Mind as their necessary cause."[46] For Fichman, by the time Wallace was writing *The World of Life*, he could "no longer separate . . . his scientific argumentation from his religious convictions. These two formed, by this stage in his career, an indissoluble unity. He was fully 'convinced that at one period in the earth's history, there was a definite act of creation [and] that from that moment evolution has been at work, guidance has been exercised.'"[47]

Mars

A few years after the publication of *Man's Place in the Universe*, Wallace decided to tackle a serious objection to his theories from much closer to Earth than the supposedly metaphysical realms beyond the stellar system. Mars, in the view of many people in the late nineteenth and early twentieth century, clearly was inhabited. The nature of Mars was at this point the center of a major debate that had started in the late 1870s, and that recently had been given extra impetus by the claims and extensive writings of Percival Lowell. Lowell, the leading advocate of intelligent life on Mars, was an American astronomer and traveler who had spent nearly ten years in Japan and Korea.

Lowell's most recent biographer has argued that when Lowell wrote on East Asia and Mars, he was in fact engaged in a unified intellectual project that drew on and expanded Herbert Spencer's account of evolution and the writings of T. H. Huxley, John Fiske, and Ernst Haeckel on the unity of the cosmos.[48] In his extensive writings on East Asia, Lowell interpreted Asian culture in terms of environmental determinism. The inhabitants of Japan, for example, had been shaped by their environment so that "artistic attractive people that they are, their civilization is like their own tree flowers, beautiful blossoms destined never to bear fruit."[49] He applied the same sort of reasoning to the Martians. Lowell viewed Mars as a dying planet that was becoming ever more desertlike; the Martians, like the Japanese, had been shaped by their environment. They had risen to the hand dealt them by nature by constructing a planetwide system of canals to carry water from Mars's poles to its desert regions. In so doing, the Martians had revealed their high intelligence to Earth-bound telescopes, as the canals were visible as a system of lines crisscrossing the planet (though, as Lowell contended, the visible "canals" were due to vegetation running alongside the water-filled canals rather than constituting the actual canals themselves).

In 1894 Lowell established an observatory in Flagstaff, Arizona, to pursue his passionate interest in the Red Planet. Books and articles followed. Lowell published *Mars and Its Canals* in 1906 and drew a response from Wallace the next year with his *Is Mars Habitable? A Critical Examination of Professor Percival Lowell's Book "Mars and Its Canals," with an Alternative Explanation.*[50] As the subtitle makes clear, Wallace's intent was to refute Lowell's arguments.

Lowell was the despair of many professional astronomers, but he did have backing from some and was a major public figure on both sides of the Atlantic. In addition to the threat Lowell's claims posed to his own position, Wallace was troubled that the public was being misled by Lowell's confident declarations, some of which he thought absurd. Further, here was an opportunity for Wallace to write more about extraterrestrial life, a topic that had already proved to be a financial winner for him. Wallace had devoted little space to Mars in *Man's Place in the Universe*, and had not discussed the canals at all. He had made it clear, however, that as the planet did not have enough mass to retain water vapor, and likely had polar caps composed of carbonic acid or some other heavy gas but not water, it could not be habitable. Hence, although it might produce vegetable life of some low kinds, Mars would be quite unsuited for supporting higher animals.

Macmillan, the publisher Wallace had in mind for the book, had put out Lowell's *Mars and Its Canals*. Wallace therefore wrote to ask whether Macmillan might consider a book from him on Mars despite its being critical of another author on their list: "It will be . . . a destructive criticism of Lowell's *theories*, while doing the fullest justice to his admirable work on the planet. The subject has attracted so much attention in almost every paper and magazine in the kingdom, that I think a *careful* and *thorough* yet *popular* exposition of the *facts*, and the *results* they lead to, will have a large sale at a low price."[51] Macmillan was agreeable, and in 1907 published the short work *Is Mars Habitable?*

In preparing the book Wallace again consulted with astronomers and read widely. Wallace, though, as he acknowledged, owed a particular debt to Agnes Clerke's 1896 review of Lowell's first book, *Mars*. As she put it, as "a contribution to science this pleasant and clever work can scarcely be taken quite seriously." She accepted the existence of the canals but focused her attention on the apparent annual melting of the Martian polar caps (see

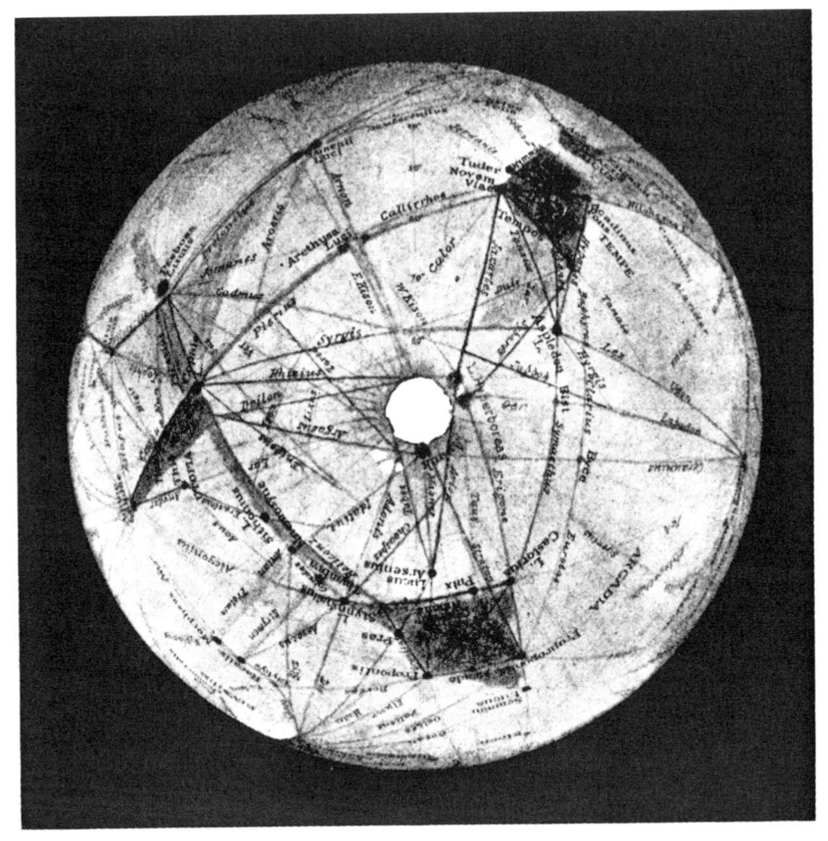

North Polar Cap.
(Lowell Observatory, 1905.)

Figure 12.2 The north polar cap of Mars. From Lowell (1906), opposite p. 44.

figure 12.2), which, she reckoned, would not release nearly enough water to irrigate the entire planet. If that was so, then Lowell's chain of reasoning that led to intelligent Martians collapsed.[52]

Lowell, as a Spencerian, was committed to the nebular hypothesis of the formation of the solar system, although by the 1900s many astronomers reckoned it contained serious flaws. Wallace instead employed the recently advanced Chamberlin-Moulton "planetismal" theory. The planetismal theory, Wallace argued, "assigns the parentage of the solar system to a spiral nebula composed of planetismals [i.e., swarms of meteorites], and the planets as formed from knots in the nebula, where many planetismals had been concentrated near the intersections of their orbits."[53] According to the planetismal theory, the early solar system was cool, unlike in the great mathematician Laplace's nebular hypothesis, where it was molten. Wallace reckoned that for all its early growth and nearly until it reached its current size, Mars grew "as a solid and cold mass, compacted together by the impact of the incoming matter as well as by its slowly increasing gravitative force." But Wallace expected that when Mars had grown to "within perhaps 100, perhaps 50 miles, or less" of its current diameter, the planet would then have been bombarded by dense swarms of meteorites, as well as, perhaps, some bodies as big as asteroids. By the time Mars had reached its current mass, it would have possessed "either [a] liquid or plastic thin outer shell resting upon a cold and solid interior."[54]

Unlike many other critics of Lowell who dismissed the canals as optical illusions, Wallace, like Clerke, accepted the accuracy of Lowell's maps, including the form of the canals (see figure 12.3). For Wallace, however, the canals were natural features and were to be explained as the result of the cooling, contraction, and cracking of the hot, thin outer layer of the forming planet against the cold and solid interior. The cracks tended to run on in straight lines and to extend vertically downward, hence the appearance of the "canals."

Wallace also took Lowell to task for stressing the purpose displayed in the "canals." For Wallace these claims beggared belief: "The one great feature of Mars which led Mr. Lowell to adopt the view of its being inhabited by a race of highly intelligent beings, and, with ever-increasing discovery to uphold this theory to the present time," Wallace complains,

> is undoubtedly that of the so-called "canals"—their straightness, their enormous length, their great abundance, and their extension over the planet's whole surface from one polar snow-cap to the other. The very immensity of this system, and its constant growth and extension during fifteen years of persistent observation, have so completely taken possession of his mind, that, after a very hasty glance at analogous facts and possibilities, he has declared them to be "non-natural"—therefore to be works of art—therefore to necessitate the presence of highly intelligent beings who have designed and constructed them.[55]

Echoing Clerke, Wallace points out that Lowell "never even discusses the totally inadequate water-supply for such world-wide irrigation, or the extreme irrationality of constructing so vast a canal-system the waste from which, by evaporation, when exposed to such desert conditions as he

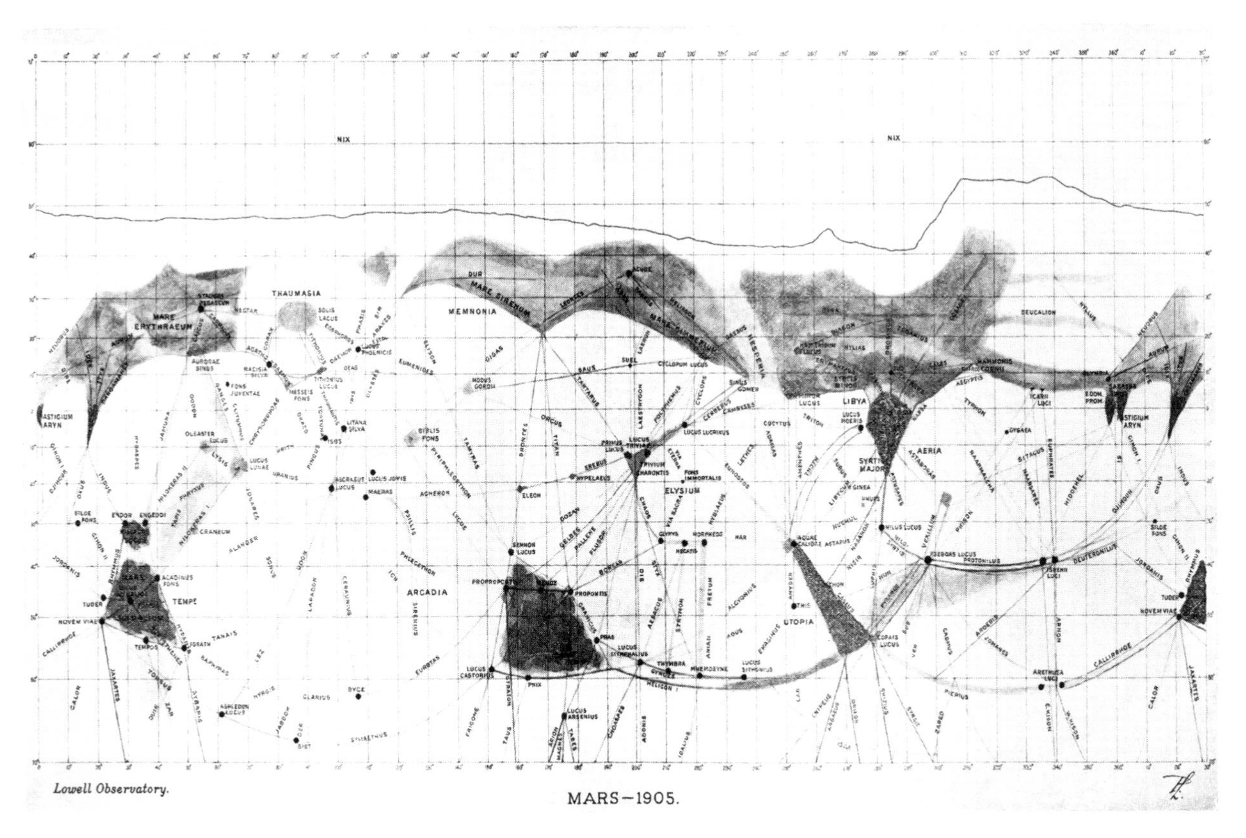

Figure 12.3 Lowell's panoramic Mercator projection of Mars, identifying the "canals" and other features. From Lowell (1906), opposite p. 384.

himself describes, would use up ten times the probable supply." Further, "[t]heir being *all* so straight, *all* describing great circles of the 'sphere,' all being so evidently arranged (as [Lowell] thinks) either to carry water to some 'oasis' 2000 miles away, or to reach some arid region far over the equator in the opposite hemisphere!" For Lowell, the canal system indicated a *purpose*, "the wonderful perfection of which," Wallace complains, he is "never tired of dwelling upon (but which I myself can nowhere perceive)."[56] For Wallace, then, Mars "is not only uninhabited by intelligent beings such as Mr. Lowell postulates, but is absolutely UNINHABITABLE."[57]

Unlike the criticisms leveled at his arguments in *Man's Place in the Universe*, Wallace's conclusions on Mars were generally well received by astronomers. Some Lowellians, however, took exception. Edmund Noble, for example, in a lengthy article for the *Boston Herald*—and Lowell as a Boston "Brahmin" was guaranteed of a good press in Boston—complained that in *Man's Place in the Universe*, Wallace had subordinated "a hundred million Suns to purely human interests." For Noble, Wallace's views had been left behind by the advance of science, and his appeal to design in nature was hopelessly outdated. As Robert Markley has pointed out, Noble "silently parodies key passages of *Is Mars Habitable?* in order to turn Wallace's skepticism about the canals against his own anthropocentric theory."[58]

Is Mars Habitable? is a strikingly different work from *Man's Place in the Universe*. Here Wallace delivered a tightly focused attack on Lowell's Martians using a physical approach and employed the decidedly more up-to-date Chamberlin-Moulton cosmogony against Lowell's aged and problematic version of the nebular hypothesis. Absent too were the explicit arguments from design of *Man's Place in the Universe*; further, Wallace disapproved of Lowell's own arguments from purpose. Nor were there references to Intelligences, Designers, or Living Souls. From the perspective of 1973, by which point spacecraft had flown by and orbited Mars and demonstrated beyond doubt that there are no canals, a modern expert on the planet, Carl Sagan, willing to overlook that Wallace had fully accepted the existence of the canals, even commented that on reading "Wallace's book, I am just astounded at the excellence of his logical powers and the currency of many of his conclusions." He reckoned Wallace was "the man who came closest to guessing or deducing what the real Mars is like."[59]

Wallace's Cosmology Undermined and Anthropocentrism Revived

By the late 1910s, some astronomers were arguing that the Sun is far from the center of the galaxy and that our stellar system is many times larger than Wallace had believed. In the 1920s it was generally established that space contains a myriad of other galaxies. By the early 1930s the idea had been entrenched that the motions of these galaxies reveal that the universe is expanding, together with the theory of a "Big Galaxy" with a diameter of around 100,000 light years (i.e., about twenty-five times the size of Wallace's stellar system of 1903), and in which the Sun sits tens of thousands of light-years from the center. Further, the notions Wallace had advanced on the generation of stellar energy were no longer plausible by the 1920s, and the Newtonian ideas of space and time that Wallace had assumed in *Man's Place in the Universe* were being replaced by those based on Einstein's theory of general relativity (introduced in 1915). Thus the onrush of scientific developments had swept away the astronomical framework Wallace had employed in *Man's Place in the Universe*. By the early 1930s, the model of a single galaxy universe with a diameter of 3,600 light-years that had a central or nearly centrally located Sun fueled by infalling matter was hopelessly outdated.[60]

But the twentieth century was to hold another twist for the extraterrestrial life debate when, starting in the 1960s, an increasing number of astronomers and physicists examined, and some advocated, an anthropocentric view of the universe.[61] This position was based on the "anthropic principle." In 1961, the physicist and cosmologist R. H. Dicke contended that human life could not have come into being until the universe had reached a certain age. The so-called Hubble age of the universe (a rough measure of its actual age), Dicke argued, "is not permitted to take one of an enormous range of values, but is somewhat limited by the biological requirements to be met during the epoch of man." The first such requirement was for the galaxy to have "aged sufficiently for there to exist elements other than hydrogen. It is well known that carbon is required to make physicists." Further, a minimum time for the emergence of man was "set by the requirement that he has a hospitable home in the form of a planet circling a luminous star." That is, enough time would have had to elapse for that star and planet to have formed. An upper limit was then set by the fact of a finite limit for the age of that star. So the Hubble age of the universe is not "a random choice" from a wide range of possible choices, but is limited by the criteria for the formation of physicists."[62] In his short two-page letter to the journal *Nature*, Dicke, though he did not use the term, employed what would become called the "anthropic principle." That term came into widespread use in 1973 when Brandon Carter, a physicist at the University of Cambridge, expanded on Dicke's argument and put it in strongly anthropocentric terms. By the end of the twentieth century the anthropic principle was widely discussed by philosophers and cosmologists. As Steven Dick has summarized matters,

> In some ways, a counterpoint to the "theistic principle" that God designed the universe, the anthropic principle found significance in the fact that the universe in its deepest structure and most fundamental properties . . . could not have been much different from those we now observe. A link between the physical and biological universes, between mind and matter, the anthropic principle may be seen as a secular search for the meaning of life based on physical principles rather than theological dogma. Transcending the usual critical but more parochial concerns about local conditions for life, it gave the extraterrestrial life debate a cosmological component of the broadest possible scope.[63]

Two of those to explore the anthropic principle in depth were the physicists John Barrow and Frank Tipler. In 1981 Tipler wrote what would become a well-known paper on the history of extraterrestrial intelligence. He drew attention to Wallace's arguments and contended that they were worth repeating in detail because they are "exactly the same as those given by modern evolutionists such as Dobzhansky, Simpson, and Mayr. Thus the biological arguments against the evolution of intelligence have not changed in 75 years. The great evolutionists have always been united against [extraterrestrial intelligence]. The biologists who have supported [extraterrestrial intelligence] have generally been biologists with the viewpoint of a physicist, and lacking the historical sense of the evolutionist. Such men often err in questions about evolutionary biology; in particular, they err about questions concerning the probability of the evolution of a species with specified properties."[64] In their big 1986 book, *The Anthropic Principle*, Barrow and Tipler used the principle to argue against the existence of extraterrestrial life.

The anthropic principle, however, has had a mixed reception. In Dick's opinion, for example, the arguments that have been advanced are "remarkably similar" to Wallace's.[65] For the prominent paleontologist Stephen Jay

Gould, there was an important difference between Wallace and most modern supporters of the anthropic principle: "Our contemporary advocates develop their arguments and then present their conclusion—that mind designed the universe, in part so that intelligent life might evolve within it—as a necessary and logical inference. Wallace was far too good a *historical* scientist to indulge in such fatuous certainty; he understood only too well that ordered and complex outcomes can arise from accumulated improbabilities."[66]

As we have just seen, during the first half of the twentieth century Wallace's views on extraterrestrial life fell from view among astronomers. A number of leading evolutionary biologists, however, deployed evolutionary theory against extraterrestrial life in ways similar to Wallace; these included Theodosius Dobzhansky and Ernst Mayr. And in 1964, George Gaylord Simpson, a paleontologist and one of the founders of the so-called modern synthesis of evolution, argued in a famous paper, "The Nonprevalence of Humanoids," that it was a "curious development" that there was now a new science of extraterrestrial life "in view of the fact that this 'science' has yet to demonstrate that its subject matter exists!"[67] It did not make scientific sense, in Simpson's view, for him to consider life "as we do not know it." Instead, he would give further consideration only to "life as we know it, to the minimal extent of depending on similar biophysical and biochemical substrates." Simpson also objected strongly to statements in the scientific and popular literature that millions of planetary systems are "suitable for life and probably inhabited." There was no observational evidence for such a claim, and it "is inherent in any acceptable definition of science that statements that cannot be checked by observation are not really *about* anything—or at the very least they are not science."[68] Although Simpson did not cite Wallace directly, he did refer to the arguments advanced in 1921 by W. D. Matthew, an evolutionary systematist, and Matthew had drawn on Wallace to argue against the likelihood of humanlike creatures evolving elsewhere.[69]

By the time Simpson was writing, however, the notion of extraterrestrial life was becoming increasingly supported by astronomers. The prevailing ideas on the formation of planetary systems implied they were relatively common, and in a universe containing a myriad of galaxies each containing many billions of stars, it seemed likely, as I. S. Shklovskii and Carl Sagan put it in 1966 in their extremely influential *Intelligent Life in the Universe*, that there were very many inhabited worlds. This belief was also advanced in the same year by the leading physicist Freeman Dyson, who in advocating searching for civilizations that possess highly sophisticated technologies, claimed that "[l]ife is common in the universe. There are many habitable planets, each sheltering its brood of living creatures. Many of the inhabited worlds develop intelligence, and an interest in communicating with other intelligent creatures."[70] Developments later in the century served to reinforce this position, perhaps none more so than the remarkably rapid discovery, starting in 1995, of numerous exoplanets. At the time of writing, well over three thousand of them have been discovered, and the composition of the atmospheres of many of them have been investigated.

These discoveries, however, have certainly not meant that the widespread existence of extraterrestrial life of the same order of humans has been generally accepted. In their widely read *Rare Earth*, published in 2000, Peter D. Ward and Donald Brownlee, for example, have argued that bacteria-like forms of life could well be quite common, while complex life will likely prove to be quite rare in the universe. In reaching these conclusions, they use a process of successive eliminations that readers of *Man's Place in the Universe* will recognize as following the same strategy that Wallace used a century earlier. *Rare Earth*, then, can be read in part as an updating of Wallace that employs modern biological ideas and an astronomical framework in which the Sun does not have a nearly central position in our stellar system. Similarly, when Ward and Brownlee write in the introduction to the paperback edition of the book that it "took a rather novel position about the frequency of complex life,"[71] they are in fact putting themselves in line with Whewell, Dobzhansky, Mayr, and others, but perhaps most significantly with Wallace. Whether their specific conclusions will be undermined by the press of scientific developments, as were Wallace's, remains to be seen.

Acknowledgments

I am grateful to Michael Crowe, Steven Dick, Kathleen Lowrey, and Charles H. Smith for comments on and criticisms of an earlier version of this chapter. The chapter also draws in part on my paper "Alfred Russel Wallace, Extraterrestrial Life, Mars and the Nature of the Universe" in *Victorian Review* 41, no. 2: 151–175, © 2015 Victorian Studies of Western Canada, and appears with permission of Johns Hopkins University Press.

Notes

1. On Wallace and the extraterrestrial life debate in the early twentieth century, see especially Dick (1999, pp. 38–58; 2008), Gould (1985), Heffernan (1978), Kevin (1985), Lane (2011, pp. 141–185), Markley (2005, pp. 61–114), and Packer (2012, pp. 172–193). For broad works that deal with the earlier history of the debate on extraterrestrial life, see Crowe (1986), Dick (1982), and Guthke (1990).

2. Wallace (1903b), p. 320.

3. Wallace (1905b), vol. 1, p. 192. The young Wallace even developed a plan for large telescopes. He described it in an attachment to a letter he sent in 1843 to the pioneer photographer William Henry Fox Talbot (C. H. Smith 2006).

4. Wallace letter to Darwin, 2 January 1864 (DCP4378).

5. Wallace letter to James Croll, 2 December 1880 (Campbell-Irons 1896, p. 361). On Croll's ideas on climate change, see Croll (1864), Fleming (2006), and Hamlin (1982).

6. Tinkler (2008), Burchfield (1975), Smith and Wise (1989).

7. [Chambers] (1844), pp. 161, 163.

8. Brooke (1977).

9. Wallace (1903b), p. 15.

10. Crowe (1986), p. 352.

11. Ibid., p. 282.

12. Wallace (1903b), p. 7.

13. Crowe (1986), p. 367.

14. On this point, see Moore (1997), p. 300.

15. Wallace (1905b), vol. 2, p. 232.

16. Ibid., p. 232.

17. Wallace (1903a), p. 474.

18. Ibid., p. 480.

19. Ibid., p. 483.

20. Ibid., p. 482.

21. Ibid., p. 483.

22. Wallace (1905b), vol. 2, p. 232. For an example of a very critical response, see Turner (1903).

23. Shipman (1903), p. 618.

24. Slotten (2004), pp. 457–458.

25. Wallace letter to William Greenell Wallace, 30 June 1903 (WCP179).

26. Wallace (1903b), p. vii.

27. For a history of studies of the Milky Way since antiquity, see Jaki (1973). On studies of our galaxy in the late nineteenth and early twentieth centuries, see R. W. Smith (1982, pp. 55–96, 147–185; 2006) and Paul (1993).

28. Wallace (1903b), p. 156.

29. See Hufbauer (1991, 2006).

30. Wallace (1903b), pp. 282–325.

31. Ibid., pp. 194–199.

32. Ibid., p. 199.

33. For example, Wallace (1903b), pp. vi, 207–208. In chapter 1 of the present volume, Charles Smith discusses the influence of Alexander von Humboldt on Wallace. In Wallace's emphasis on the extremely delicate balance of physical conditions, we see an example of that influence.

34. Wallace (1903b), p. 294.

35. Ibid., p. 320.

36. Ibid., p. 320.

37. Ibid., p. 321.

38. Ibid., p. 323.

39. Wallace (1904c), p. 335.

40. Kevin (1985).

41. "Our unique Earth!" (1904), p. 390.

42. Lightman (2007), p. 494.

43. Dawson (1903), p. 177.

44. Marchant (1916a), p. 413.

45. Fichman (2004), p. 311.

46. Wallace (1910c), p. vii.

47. Fichman (2004), p. 312. One of the editors of this volume, Charles Smith, takes a slightly different view. For Smith, Wallace's perspective is less teleological than it is final causes–based: that is, less demanding of first causes and more serving implicit natural organizational tendencies. The latter understanding is more in keeping with his debt to Humboldtian "general equilibrium of forces" thinking (see chapter 1), and, as it turned out, to the kinds of arguments actually being made by current proponents of the anthropic principle.

48. Lane (2011), p. 169. On Lowell, see Hoyt (1976) and Strauss (2001). For an account of Mars in science and imagination, see Markley (2005).

49. Quoted in Lane (2011), p. 169.

50. Wallace (1907).

51. Wallace, quoted in Lane (2011), p. 145.

52. Clerke (1896), p. 371; Markley (2005), pp. 82–83.

53. Wallace (1907), p. 83. On the Chamberlin-Moulton hypothesis, see Brush (1978).

54. Wallace (1907), p. 86.

55. Ibid., pp. 102–103.

56. Ibid., p. 103.

57. Ibid., p. 110.

58. Markley (2005), p. 101.

59. Sagan (1973), pp. 15, 105.

60. These various developments are discussed in detail in R. W. Smith (1982).

61. Dick (1999), pp. 527–536.
62. Dicke (1961), p. 440.
63. Dick (1999), p. 527.
64. Tipler (1981), p. 140.
65. Dick (1999), p. 535.
66. Gould (1985), p. 400.
67. Simpson (1964a), p. 769.
68. Simpson (1964a), p. 770.
69. Matthew (1921). Matthew was very influenced by Wallace's writings on other subjects as well, especially biogeography.
70. Dyson (1966), p. 642.
71. Ward and Brownlee (2003), p. xii.

Coda

Alfred Russel Wallace wrote twenty-two books and in total published more than one thousand writings. These writings en masse constitute in excess of ten thousand pages of printed material. Few people have sampled from more than just the smallest portion of this body of work, and considering the vast range of subjects Wallace explored—including matters extending well beyond his most famous evolution and biogeography associations—it is perhaps not surprising that the coherence of his vision is not usually very well appreciated. We have been given only a few hundred pages here to survey his worldview, and though we have made an effort to produce a judicious treatment, the reader must understand that this "Companion" must ultimately fall short in terms of its comprehensiveness of coverage. Indeed, in several places we go no further than alluding to additional subjects he entertained, but beyond even this there is more—a good deal more—that could be said about our main subjects of interest here. Wallace was not known in his later years as "the Grand Old Man of Science" for nothing.

It has not been our goal here to glorify Wallace. Yes, on the whole we believe he was an admirable person in a lifelong pursuit of admirable goals, but if one looks closely, a few indications of human frailty do surface. Although he was consistently respectful in his dealings with others, occasionally one sees evidence of impatience or exasperation in both his personal and public writings, and from time to time a tendency to be overbearing (e.g., in some of his dealings with his brother-in-law Thomas Sims and the

American philosopher William James). Perhaps such is the inevitable result of genius, attendant to its frequent forays into the unknown (and/or underappreciated).

From the perspective of modern workers, with their trajectory toward ever greater levels of specialization, a generalist of the likes of Wallace seems unthinkable—surely his incursions into social criticism, spiritual evolution, and social justice must imply a lack of focus, or a dilettante's superficiality . . . ?? The readers of this volume, if they have learned anything from it, must reject this interpretation of Wallace's work.

One cannot help but conclude that Wallace really did understand the *nature* of the problems at hand. This does not mean that he was always able to identify appropriate *solutions*; often he was stopped by an absence of critical data, theory, or environment of cooperation (another way of saying he was ahead of his time). Still, he stands as one of the last great polymaths, and to underestimate the depth and breadth of his insights is to shortchange human creativity. Wallace's faith in "the advantages of varied knowledge" served him well, as it still serves us.

References

Agar N. 2015. *The Sceptical Optimist: Why Technology Isn't the Answer to Everything.* Oxford: Oxford University Press.

Alatalo RV, and Mappes J. 1996. Tracking the evolution of warning signals. *Nature* 382:708–710.

Allen CE, Zwaan BJ, and Brakefield PM. 2011. Evolution of sexual dimorphism in the Lepidoptera. *Annual Review of Entomology* 56:445–464.

Allen WL, Baddeley R, Cuthill IC, and Scott-Samuel NE. 2012. A quantitative test of the predicted relationship between countershading and lighting environment. *American Naturalist* 180:762–776.

Allingham HP, and Radford D, eds. 1907. *William Allingham: A Diary*. London: Macmillan.

Alston ER. 1876. Review of *Geographical Distribution of Animals*. *Academy* 10 (15 July): 63–64.

Archibald JD. 2014. *Aristotle's Ladder, Darwin's Tree: The Evolution of Visual Metaphors for Biological Order*. New York: Columbia University Press.

Asma ST. 2001. *Stuffed Animals and Pickled Heads: The Culture and Evolution of Natural History Museums*. Oxford: Oxford University Press.

Avise JC, and Ayala FJ, eds. 2009. *In the Light of Evolution*. Vol. 3, *Two Centuries of Darwin*. Washington, DC: National Academies Press.

Ayala F, and Arp R, eds. 2010. *Contemporary Debates in Philosophy of Biology*. Chichester, UK: Wiley-Blackwell.

Ayres JM, and Clutton-Brock TH. 1992. River boundaries and species range size in Amazonian primates. *American Naturalist* 140:531–537.

Bacon CD, Michonneau F, Henderson AJ, McKenna MJ, Milroy AM, and Simmons

MP. 2013. Geographic and taxonomic disparities in species diversity: Dispersal and diversification rates across Wallace's Line. *Evolution* 67:2058–2071.
Badyaev AV. 2009. Evolutionary significance of phenotypic accommodation in novel environments: An empirical test of the Baldwin effect. *Proceedings of the Royal Society of London B* 364:1125–1141.
Badyaev AV, and Hill GE. 2003. Avian sexual dichromatism in relation to phylogeny and ecology. *Annual Review of Ecology, Evolution, and Systematics* 34:27–49.
Baker DB. 2001. Alfred Russel Wallace's record of his consignments to Samuel Stevens, 1854–1861. *Zoologische Mededeelingen* 75:254–341.
Baker RR, and Parker GA. 1979. The evolution of bird coloration. *Philosophical Transactions of the Royal Society of London B* 287:63–130.
Baldwin JM. 1896. A new factor in evolution. *American Naturalist* 30:441–451.
Ballard C. 2008. "Oceanic negroes": British anthropology of Papuans, 1820–1869. In B Douglas and C Ballard, eds., *Foreign Bodies: Oceania and the Science of Race, 1750–1940*. Canberra, Australia: ANU E Press, 157–201.
Barber ME. 1874. Notes on the peculiar habits and changes which take place in the larva and pupa of *Papilio Nireus*. *Transactions of the Entomological Society of London for the Year 1874*, 519–521.
Barrow L. 1986. *Independent Spirits: Spiritualism and English Plebeians, 1850–1910*. London: Routledge and Kegan Paul.
Bartley MM. 1992. Darwin and domestication: Studies on inheritance. *Journal of the History of Biology* 25:307–333.
Barton R. 2003. "Men of science": Language, identity and professionalization in the mid-Victorian scientific community. *History of Science* 41:73–119.
Bates HW. 1862. Contributions to an insect fauna of the Amazon valley: Lepidoptera: Heliconidae. *Transactions of the Linnean Society of London* 23:495–566.
———. 1863. *The Naturalist on the River Amazons*. London: John Murray.
Bateson G. 1972. *Steps to an Ecology of Mind*. San Francisco: Chandler.
———. 1979. *Mind and Nature: A Necessary Unity*. New York: Dutton.
Beccaloni G. 2008. Wallace's annotated copy of the Darwin-Wallace paper on natural selection. In CH Smith and G Beccaloni, eds., *Natural Selection and Beyond: The Intellectual Legacy of Alfred Russel Wallace*. Oxford: Oxford University Press, 91–101.
Beck J, Ballesteros-Mejia L, Nagel P, and Kitching IJ. 2013. Online solutions and the "Wallacean shortfall": What does GBIF contribute to our knowledge of species' ranges? *Diversity and Distributions* 19:1043–1050.
Beck J, Kitching IJ, and Linsenmair KE. 2006. Wallace's Line revisited: Has vicariance or dispersal shaped the distribution of Malesian hawkmoths (Lepidoptera: Sphingidae)? *Biological Journal of the Linnean Society* 89:455–468.
Beckert S. 2015. *Empire of Cotton: A Global History*. New York: Vintage Books.
Beddall BG. 1988. Darwin and divergence: The Wallace connection. *Journal of the History of Biology* 21:1–68.
Begley S. 2007. *Train Your Mind, Change Your Brain*. New York: Ballantine Books.
Behrens RR. 2009. Revisiting Abbott Thayer: Non-scientific reflections about camou-

flage in art, war and zoology. *Philosophical Transactions of the Royal Society B* 364: 497–501.

Bellamy E. 1889. *Looking Backward, 2000–1887*. Boston: Houghton Mifflin.

———. 1897. *Equality*. New York: Appleton.

Benn D. 2000. Comment on "The theory of glacial motion." In The Alfred Russel Wallace Page. http://people.wku.edu/charles.smith/wallace/S184.htm.

Benton T. 2008. Wallace's dilemmas: The laws of nature and the human spirit. In CH Smith and G Beccaloni, eds., *Natural Selection and Beyond: The Intellectual Legacy of Alfred Russel Wallace*. Oxford: Oxford University Press, 368–390.

———. 2013. *Alfred Russel Wallace: Explorer, Evolutionist, Public Intellectual*. Manchester, UK: Siri Scientific Press.

Berry, A. 2013. Alfred Russel Wallace—natural selection, socialism, and spiritualism. *Current Biology* 23, no. 24: R1066–R1069.

Berry RJ. 2009. Hooker and islands. *Biological Journal of the Linnean Society* 96: 462–481.

Bignami M. 2009. Joseph Conrad and Alfred Russel Wallace look at the Malay Archipelago. In C Pagetti, ed., *Darwin nel Tempo: Modernità Letteraria e Immaginario Scientifico*. Milan: Cisalpino, 237–249.

Blaazer D. 1992. *The Popular Front and the Progressive Tradition: Socialists, Liberals, and the Quest for Unity, 1884–1939*. Cambridge: Cambridge University Press.

Blackwell M. 2011. The fungi: 1, 2, 3 . . . 5.1 million species. *American Journal of Botany* 98:426–438.

Blanchard O, and Fischer S. 1989. *Lectures on Macroeconomics*. Cambridge, MA: MIT Press.

Blaug M. 1996. *Economic Theory in Retrospect*. 5th ed. Cambridge: Cambridge University Press.

Blount JD, Speed MP, Ruxton GD, and Stephens PA. 2009. Warning displays may function as honest signals of toxicity. *Proceedings of the Royal Society of London B* 276:871–877.

Blyth E. 1835. An attempt to classify the "varieties" of animals, with observations on the marked seasonal and other changes which naturally take place in various British species, and which do not constitute varieties. *Magazine of Natural History* 8:40–53.

Bock WJ. 2009. The Darwin-Wallace myth of 1858. *Proceedings of the Zoological Society* 62:1–12.

Bowler PJ. 1976. Alfred Russel Wallace's concepts of variation. *Journal of the History of Medicine* 31:17–29.

———. 1993. *Biology and Social Thought, 1850–1914*. Berkeley: Office for History of Science and Technology, University of California at Berkeley.

———. 2003. *Evolution: The History of an Idea*. 3rd ed. Berkeley: University of California Press.

Brandon R. 1983. *The Spiritualists: The Passion for the Occult in the Nineteenth and Twentieth Centuries*. New York: Knopf.

British Association for the Advancement of Science. 1866. *Report of the Thirty-Fifth*

Meeting of the British Association for the Advancement for Science, Held at Birmingham in September 1865. London: John Murray.

Brodie ED. 1993. Differential avoidance of coral snake banded patterns by free-ranging avian predators in Costa Rica. *Evolution* 47:227–235.

Brooke JH. 1977. Natural theology and the plurality of worlds: Observations on the Brewster-Whewell debate. *Annals of Science* 34:221–286.

Brooke M de L, and Davies NB. 1988. Egg mimicry by cuckoos *Cuculus canorus* in relation to discrimination by hosts. *Nature* 335:630–632.

Brown JH, and Sax DF. 2004. Gradients in species diversity: Why are there so many species in the tropics? In MV Lomolino, DF Sax, and JH Brown, eds., *Foundations of Biogeography: Classic Papers with Commentaries*. Chicago: University of Chicago Press, 1145–1154.

Browne J. 1980. Darwin's botanical arithmetic and the "principle of divergence," 1854–1858. *Journal of the History of Biology* 13:53–89.

———. 1995. *Charles Darwin: Voyaging*. Princeton, NJ: Princeton University Press.

———. 2003. *Charles Darwin: The Power of Place*. Princeton, NJ: Princeton University Press.

Brush S. 1978. A geologist among astronomers: The rise and fall of the Chamberlin-Moulton cosmogony. *Journal for the History of Astronomy* 9:1–41, 77–104.

Buffon, GLL de. 1761. *Histoire Naturelle, Generale et Particuliere*. Tome 9. Paris: Imprimerie Royale.

Bulmer M. 2005. The theory of natural selection of Alfred Russel Wallace FRS. *Notes & Records of the Royal Society* 59:125–136.

Burchfield JD. 1975. *Lord Kelvin and the Age of the Earth*. Chicago: University of Chicago Press.

Burnett J (Lord Monboddo). 1773–1792. *Of the Origin and Progress of Language*. 6 vols. Edinburgh: J. Balfour and T. Cadell.

Burns KJ. 1998. A phylogenetic perspective on the evolution of sexual dichromatism in tanagers (Thraupidae): The role of female versus male plumage. *Evolution* 52: 1219–1224.

Cairnes JE. 1874. *Some Leading Principles of Political Economy Newly Expounded*. New York: Harper.

Caldwell GS, and Rubinoff RW. 1983. Avoidance of venomous sea snakes by naïve herons and egrets. *Auk* 100:195–198.

Camerini JR. 1993. Evolution, biogeography, and maps: An early history of Wallace's Line. *Isis* 84:700–727.

———. 1996. Wallace in the field. *Osiris* 11:44–65.

———. 1997. Remains of the day: Early Victorians in the field. In B Lightman, ed., *Victorian Science in Context*. Chicago: University of Chicago Press, 354–377.

Campbell-Irons J. 1896. *Autobiographical Sketch of James Croll*. London: Edward Stanford.

Candolle AP de. 1820. *Essai Élémentaire de Géographie Botanique*. Strasbourg, France: F. G. Levrault.

Capparella AP. 1988. Genetic variation in Neotropical birds: Implications for the speciation process. *Acta Congressus Inernationalis Ornithologici* 19:1658–1664.

———. 1991. Neotropical avian diversity and riverine barriers. *Acta Congressus Internationalis Ornithologici* 20:307–316.

Caro T. 2016. *Zebra Stripes*. Cambridge: Cambridge University Press.

———. 2017. Wallace on coloration: Contemporary perspective and unresolved insights. *Trends in Ecology and Evolution* 32:23–30.

Caro T, Hill G, Lindström L, and Speed MP. 2008. The colours of animals: From Wallace to the present day: II, Conspicuous coloration. In CH Smith and G Beccaloni, eds., *Natural Selection and Beyond: The Intellectual Legacy of Alfred Russel Wallace*. Oxford: Oxford University Press, 144–165.

Caro T, Izzo A, Reiner RC Jr., Walker H, and Stankowich T. 2014. The function of zebra stripes. *Nature Communications*. https://doi.org/10.1038/ncomms4535.

Caro T, Merilaita S, and Stevens M. 2008. The colours of animals: From Wallace to the present day: I, Cryptic coloration. In CH Smith and G Beccaloni, eds., *Natural Selection and Beyond: The Intellectual Legacy of Alfred Russel Wallace*. Oxford: Oxford University Press, 125–143.

Caro T, and Stankowich T. 2015. Concordance on zebra stripes: A comment on Larison et al. (2015). *Royal Society Open Science* 2, no. 9: 150323.

Carroll SB. 2005. *Endless Forms Most Beautiful*. New York: Norton.

[Chambers R.] 1844. *Vestiges of the Natural History of Creation*. London: John Churchill.

———. 1844/1969. *Vestiges of the Natural History of Creation*. Leicester, UK: Leicester University Press.

Chapin D. 2004. *Exploring Other Worlds: Margaret Fox, Elisha Kent Kane, and the Antebellum Culture of Curiosity*. Amherst: University of Massachusetts Press.

Chatterjee A, ed. 2013. *Vaccinophobia and Vaccine Controversies of the 21st Century*. New York: Springer.

Cheviron ZA, Hackett SJ, and Capparella AP. 2005. Complex evolutionary history of a Neotropical lowland forest bird (*Lepidothrix coronata*) and its implications for historical hypotheses of the origin of Neotropical avian diversity. *Molecular Phylogenetics and Evolution* 36:338–357.

Choe JC, and Crespi BJ, eds. 1997. *The Evolution of Social Behaviour in Insects and Arachnids*. Cambridge: Cambridge University Press.

Chouteau M, Arias M, and Joron M. 2016. Warning signals are under positive frequency-dependent selection in nature. *Proceedings of the National Academy of Sciences USA* 113:2164–2169.

Chown SL, and Gaston KJ. 2000. Areas, cradles and museums: The latitudinal gradient in species richness. *Trends in Ecology and Evolution* 15:311–315.

Christensen B. 2011. Darwin vs. Wallace: When poetry dies and when poetry survives in the not-so-natural selection of memetic evolution. *Changing English* 18: 397–405.

Clark B, and York R. 2007. The restoration of nature and biogeography: An introduc-

tion to Alfred Russel Wallace's "Epping Forest." *Organization & Environment* 20: 213–234.
Clements FE. 1916. *Plant Succession: An Analysis of the Development of Vegetation.* Washington, DC: Carnegie Institution of Washington.
Clerke AM. 1890. *The System of the Stars.* London: Longmans, Green and Co.
———. 1896. New views about Mars. *Edinburgh Review* 184:368–385.
Cloyd EL. 1972. *James Burnett, Lord Monboddo.* Oxford: Clarendon Press.
Cluysenaar A. 2008. *Batu-Angas: Envisioning Nature with Alfred Russel Wallace.* Bridgend, Wales: Seren.
Colgrove J. 2005. Science in a democracy: The contested status of vaccination in the Progressive Era and the 1920s. *Isis* 96:167–191.
Collard DA. 1978. *Altruism and Economy.* Oxford: Martin Robertson.
———. 2001. Malthus, population and the generational bargain. *History of Political Economy* 33:697–716.
Collier C. 1979. Henry George's system of political economy. *History of Political Economy* 11:64–73.
Colwell R. 2000. A barrier runs through it . . . or maybe just a river. *Proceedings of the National Academy of Sciences USA* 97:13470–13472.
Combe G. 1835. *The Constitution of Man.* 5th American ed. Boston: Marsh, Capen, and Lyon.
Cope ED. 1891. *Alfred Russel Wallace.* New York: D. Appleton.
Costa JT. 2009a. *The Annotated Origin: A Facsimile of the First Edition of* On the Origin of Species. Cambridge, MA: Belknap Press of Harvard University Press.
———. 2009b. Darwinian revelation: Tracing the origin and evolution of an idea. *BioScience* 59:886–894.
———. 2013a. Engaging with Lyell: Alfred Russel Wallace's Sarawak law and Ternate papers as reactions to Charles Lyell's *Principles of Geology*. *Theory in Biosciences* 132:225–237.
———. 2013b. *On the Organic Law of Change: A Facsimile Edition and Annotated Transcription of Alfred Russel Wallace's Species Notebook of 1855–1859.* Cambridge, MA: Harvard University Press.
———. 2013c. Synonymy and its discontents: Alfred Russel Wallace's nomenclatural proposals from the "Species Notebook" of 1855–1859. *Bulletin of Zoological Nomenclature* 70:131–148.
———. 2014. *Wallace, Darwin, and the Origin of Species.* Cambridge, MA: Harvard University Press.
———. 2015. The consilient Mr. Wallace. *Skeptic Magazine* 20, no. 3 (September): 16–21.
Costa JT, and Beccaloni G. 2014. Deepening the darkness? Alfred Russel Wallace in the Malay Archipelago. *Current Biology* 24, no. 22: R1070–R1072.
Costello MJ, May RM, and Stork NE. 2013. Can we name Earth's species before they go extinct? *Science* 339:413–416.
Cott HB. 1932. The Zoological Society's expedition to the Zambesi, 1927: No. 4, On the ecology of tree-frogs in the Lower Zambesi Valley, with special reference

to predatory habits considered in relation to the theory of warning colours and mimicry. *Proceedings of the Zoological Society of London* 102:471–541.

———. 1940. *Adaptive Coloration in Animals*. London: Methuen.

Cox CB. 2001. The biogeographic regions reconsidered. *Journal of Biogeography* 28: 511–523.

Cox CB, and Moore PD. 2005. *Biogeography: An Ecological and Evolutionary Approach*. 7th ed. Malden, MA: Blackwell.

Cranbrook Earl of, Hills DM, McCarthy CJ, and Prŷs-Jones R. 2005. A. R. Wallace, collector: Tracing his vertebrate specimens. Part 1. In I Das and AA Tuen, eds., *Wallace in Sarawak—150 Years Later*. Kota Samarahan: Institute of Biodiversity and Environmental Conservation, Universiti Malaysia Sarawak, 8–34.

Croizat L. 1964. *Space, Time, Form: The Biological Synthesis*. Caracas, Venezuela: printed by the author.

Croll J. 1864. On the physical cause of the change of climate during geological epochs. *Philosophical Magazine*, 4th ser., 28:121–137.

Cronin H. 1991. *The Ant and the Peacock: Altruism and Sexual Selection from Darwin to Today*. Cambridge: Cambridge University Press.

Crookshank EM. 1891. Testimony of Edgar March Crookshank presented before the Royal Commission on Vaccination on 9 July, 30 August, 12 and 19 November 1890. In *Fourth Report of the Royal Commission Appointed to Inquire into the Subject of Vaccination*. London: Her Majesty's Stationary Office.

Crowe MJ. 1986. *The Extraterrestrial Life Debate, 1750–1900*. Cambridge: Cambridge University Press.

Crutzen PJ. 2002. Geology of mankind. *Nature* 415:23.

Currie D, and Paquin V. 1987. Large-scale biogeographical patterns of species richness of trees. *Nature* 329:326–327.

Dalton HD. 1954. *Principles of Public Finance*. 4th ed. London: Routledge and Kegan Paul.

Damasio A. 1994. *Descartes' Error*. New York: Avon Books.

Dana JD. 1849. *Geology: United States Exploring Expedition; During the Years 1838, 1839, 1840, 1841, 1842, under the Command of Charles Wilkes, U.S.N.* Vol. 10. Philadelphia: C. Sherman.

Darst CR, Cummings ME, and Cannatella DC. 2006. A mechanism for diversity in warning signals: Conspicuousness versus toxicity in poison frogs. *Proceedings of the National Academy of Sciences USA* 103:5852–5857.

Darwin CR. 1837a. Observations of proofs of recent elevation on the coast of Chili, made during the survey of His Majesty's ship *Beagle* commanded by Capt. FitzRoy R.N. *Proceedings of the Geological Society of London* 2:446–449.

———. 1837b. On certain areas of elevation and subsidence in the Pacific and Indian Oceans, as deduced from the study of coral formations. *Proceedings of the Geological Society of London* 2:552–554.

———. 1839. *Journal of Researches into the Geology and Natural History of the Various Countries Visited by H.M.S.* Beagle. London: Colburn.

———. 1839/2001. *The Voyage of the Beagle*. New York: Modern Library.

———. 1842. *The Structure and Distribution of Coral Reefs*. London: Smith, Elder and Co.

———. 1845. *Journal of Researches into the Natural History and Geology of the Countries Visited during the Voyage of H.M.S.* Beagle *round the World*. 2nd ed. London: John Murray.

———. 1859. *On the Origin of Species by Means of Natural Selection*. London: John Murray.

———. 1859/1976. *On the Origin of Species*. New York: Avenel Books.

———. 1861. *On the Origin of Species by Means of Natural Selection*. 3rd ed. London: John Murray.

———. 1862. *On the Various Contrivances by Which British and Foreign Orchids Are Fertilised by Insects, and on the Good Effects of Intercrossing*. London: John Murray.

———. 1868. *The Variation of Animals and Plants under Domestication*. 2 vols. London: John Murray.

———. 1869. *On the Origin of Species by Means of Natural Selection*. 5th ed. London: John Murray.

———. 1871. *The Descent of Man, and Selection in Relation to Sex*. 2 vols. London: John Murray.

———. 1871/1981. *The Descent of Man, and Selection in Relation to Sex*. Princeton, NJ: Princeton University Press.

———. 1887. *The Life and Letters of Charles Darwin*. 2 vols. Edited by F Darwin. London: John Murray.

Darwin CR, and Wallace AR. 1858. On the tendency of species to form varieties; and on the perpetuation of varieties and species by natural means of selection. *Journal of the Proceedings of the Linnean Society of London, Zoology* 3:45–62.

Darwin Correspondence Project. Directed by J Secord et al. Cambridge University. http://www.darwinproject.ac.uk.

Davidson R, and Harrington A, eds. 2002. *Visions of Compassion*. Oxford: Oxford University Press.

Davies NB. 2000. *Cuckoos, Cowbirds and Other Cheats*. London: T. and A. D. Poyser.

Davies NB, and Brooke M de L. 1989. An experimental study of co-evolution between the cuckoo, *Cuculus canorus*, and its hosts: I, Host egg discrimination. *Journal of Animal Ecology* 58:207–224.

Davies NB, Kilner RM, and Noble DG. 1998. Nestling cuckoos, *Cuculus canorus*, exploit hosts with begging calls that mimic a brood. *Proceedings of the Royal Society of London B* 265:673–678.

Davies TJ, Barraclough TG, Savolainen V, and Chase MW. 2004. Environmental causes for plant biodiversity gradients. *Philosophical Transactions of the Royal Society of London B* 359:1645–1656.

Dawkins R. 1976. *The Selfish Gene*. Oxford: Oxford University Press.

Dawson A. 1903. A visit to Dr. Alfred Russel Wallace. *Christian Commonwealth* 23: 176–177.

De Cock R, and Matthysen E. 2003. Glow-worm larvae bioluminescence (Coleoptera:

Lampyridae) operates as an aposematic signal upon toads (*Bufo bufo*). *Behavioral Ecology* 14:103–108.

Dennett DC. 2003. The Baldwin effect: A crane, not a skyhook. In BH Weber and DJ Depew, eds., *Evolution and Learning: The Baldwin Effect Reconsidered*. Cambridge, MA: MIT Press, 69–106.

De Ruiter L. 1952. Some experiments on the camouflage of stick caterpillars. *Behaviour* 4:222–232.

Diamond J. 2005. *Guns, Germs, and Steel: The Fates of Human Societies*. New York: Norton.

Dick SJ. 1982. *Plurality of Worlds: The Origins of the Extraterrestrial Life Debate from Democritus to Kant*. Cambridge: Cambridge University Press.

———. 1999. *The Biological Universe: The Twentieth-Century Extraterrestrial Life Debate and the Limits of Science*. Cambridge: Cambridge University Press.

———. 2008. The universe and Alfred Russel Wallace. In CH Smith and G Beccaloni, eds., *Natural Selection and Beyond: The Intellectual Legacy of Alfred Russel Wallace*. Oxford: Oxford University Press, 320–340.

Dicke RH. 1961. Dirac's cosmology and Mach's principle. *Nature* 192:440–441.

Dickerson RE, Merrill ED, McGregor RC, Schultze W, Taylor EH, and Herre AWC. 1928. *Distribution of Life in the Philippines*. Manila: Bureau of Scientific Monographs.

Diniz-Filho JAF, Loyola RD, Raia P, Mooers AO, and Bini LM. 2013. Darwinian shortfalls in biodiversity conservation. *Trends in Ecology and Evolution* 28:689–695.

Dobzhansky T. 1950. Evolution in the tropics. *American Scientist* 38:209–221.

———. 1973. Nothing in biology makes sense except in the light of evolution. *American Biology Teacher* 35:125–129.

Doughty RW. 1975. *Feather Fashions and Bird Preservation: A Study in Nature Protection*. Berkeley: University of California Press.

Drucker PF. 2010. *Technology, Management, and Society*. Boston: Harvard Business Press.

Dumbacher JP, Beehler BM, Spande TF, Martin Garraffo H, and Daly JW. 1992. Homobatrachotoxin in the genus *Pitohui*: Chemical defense in birds? *Science* 258: 799–801.

Dumbacher JP, and Fleischer RC. 2001. Phylogenetic evidence for colour pattern convergence in toxic pitohuis: Müllerian mimicry in birds? *Proceedings of the Royal Society of London B* 268:1971–1976.

Durbach N. 2005. *Bodily Matters: The Anti-vaccination Movement in England, 1853–1907*. Durham, NC: Duke University Press.

Dyson F. 1966. The search for extraterrestrial technology. In RE Marshak, ed., *Perspectives in Modern Physics*. New York: Interscience, 641–655.

Earl GW. 1845. On the physical structure and arrangement of the islands in the Indian Archipelago. *Journal of the Royal Geographical Society* 15:358–365.

———. 1853. *Contributions to the Physical Geography of South-Eastern Asia and Australia*. London: Hippolyte Bailliere.

Eberhard WG. 1996. *Female Control: Sexual Selection by Cryptic Female Choice*. Princeton, NJ: Princeton University Press.

Eberhart GM. 2002. *Mysterious Creatures: A Guide to Cryptozoology*. Santa Barbara, CA: ABC-Clio.

Edmunds M. 1974. *Defence in Animals: A Survey of Anti-predator Defences*. Harlow, UK: Longman.

Eizirik E, Bonatto SL, Johnson WE, Crawshaw PG Jr., Vié JC, Brousset DM, O'Brien SJ, and Salzano FM. 1998. Phylogeographic patterns and evolution of the mitochondrial DNA control region in two Neotropical cats (Mammalia, Felidae). *Journal of Molecular Evolution* 47:613–624.

Endler JA. 1984. Progressive background in moths, and a quantitative measure of crypsis. *Biological Journal of the Linnean Society* 22, no. 3: 187–231.

Engels F. 1892. *The Condition of the Working-Class in England*. London: S. Sonnenschein.

Engs RC. 2000. *Clean Living Movements: American Cycles of Health Reform*. Westport, CT: Praeger.

Evans BJ, Brown RM, McGuire JA, Supriatna J, Andayani N, Diesmos A, Iskandar D, Melnick DJ, and Cannatella DC. 2003. Phylogenetics of fanged frogs: Testing biogeographical hypotheses at the interface of the Asian and Australian faunal zones. *Systematic Biology* 52:794–819.

Fancher RE. 1998. Biography and psychodynamic theory: Some lessons from the life of Francis Galton. *History of Psychology* 1:99–115.

Fawcett H. 1883. *State Socialism and the Nationalisation of the Land*. London: Macmillan.

Fernandes AM, Wink M, and Aleixo A. 2012. Phylogeography of the chestnut-tailed antbird (*Myrmeciza hemimelaena*) clarifies the role of rivers in Amazonian biogeography. *Journal of Biogeography* 39:1524–1535.

Fichman M. 1977. Wallace: Zoogeography and the problem of land bridges. *Journal of the History of Biology* 10:45–63.

———. 2001. Science in theistic contexts: A case study of Alfred Russel Wallace. *Osiris* 16:227–250.

———. 2003. *Evolutionary Theory and Victorian Culture*. Amherst, NY: Humanity Books.

———. 2004. *An Elusive Victorian: The Evolution of Alfred Russel Wallace*. Chicago: University of Chicago Press.

Fichman M, and Keelan JE. 2007. Resister's logic: The anti-vaccination arguments of Alfred Russel Wallace and their role in the debates over compulsory vaccination in England, 1870–1907. *Studies in the History and Philosophy of the Biological and Biomedical Sciences* 38:585–607.

Fischer AG. 1960. Latitudinal variations in organic diversity. *Evolution* 14:64–81.

Fisher I. 1920. *Stabilizing the Dollar: A Plan to Stabilize the General Price Level without Fixing Individual Prices*. New York: Macmillan.

———. 1934. *Stable Money: A History of the Movement*. New York: Adelphi.

Fleming JR. 1998. Charles Lyell and climatic change: Speculation and certainty. In

DJ Blundell and AC Scott, eds., *Lyell: The Past Is the Key to the Present*. Geological Society of London Special Publications, no. 143, 161–169. London: Geological Society.

———. 2006. James Croll in context: The encounter between climate dynamics and geology in the second half of the nineteenth century. *History of Meteorology* 3:43–53.

Forbes E. 1846. On the connexion between the distribution of the existing fauna and flora of the British Isles, and the geological changes which have affected their area, especially during the epoch of the Northern Drift. *Memoirs of the Geological Survey of England, and of the Museum of Economic Geology in London* 1:336–432.

Forbes P. 2009. *Dazzled and Deceived: Mimicry and Camouflage*. New Haven, CT: Yale University Press.

Gaffney M. 1987. Wallace, Alfred Russel 1823–1913. In J Eatwell, M Milgate, and P Newman, eds., *The New Palgrave: A Dictionary of Economics*, vol. 4. London: Macmillan, 850–851.

———. 1997. Alfred Russel Wallace's campaign to nationalize land: How Darwin's peer learned from John Stuart Mill and became Henry George's ally. *American Journal of Economics and Sociology* 56:609–615.

Galton F. 1869. *Hereditary Genius*. London: Macmillan.

Gans C. 1961. Mimicry in procryptically colored snakes of the genus *Dasypeltis*. *Evolution* 15:72–91.

Gascon C, Lougheed SC, and Bogart JP. 1998. Patterns of genetic population differentiation in four species of Amazonian frogs: A test of the riverine barrier hypothesis. *Biotropica* 30:104–119.

Gascon C, Malcolm JR, Patton JL, Silva MNF da, Bogart JP, Lougheed SC, Peres CA, Neckel S, and Boag PT. 2000. Riverine barriers and the geographic distributions of Amazonian species. *Proceedings of the National Academy of Sciences USA* 97:13672–13677.

Gaston KJ, ed. 1996. *Biodiversity: A Biology of Numbers and Difference*. Cambridge, MA: Blackwell Science.

———. 2000. Global patterns in biodiversity. *Nature* 405:220–227.

———. 2007. Latitudinal gradient in species richness. *Current Biology* 17:R574.

George H. 1912. *Progress and Poverty: An Inquiry into the Cause of Industrial Depressions and of Increase of Want with Increase of Wealth*. Garden City, NY: Doubleday.

Gittleman JL, and Harvey PH. 1980. Why are distasteful prey not cryptic? *Nature* 286:149–150.

Goh DPS. 2007. Imperialism and "medieval" natives: The Malay image in Anglo-American travelogues and colonialism in Malaya and the Philippines. *International Journal of Cultural Studies* 10:323–341.

Gould SJ. 1985. Mind and supermind. In *The Flamingo's Smile: Reflections in Natural History*. New York: Norton, 392–402.

Gould SJ, and Lewontin R. 1978. The spandrels of San Marco and the Panglossian paradigm: A critique of the adaptationist programme. *Proceedings of the Royal Society of London B* 205:581–598.

Greene E. 1989. A diet-induced developmental polymorphism in a caterpillar. *Science* 243:643–646.

Greene J. 1959. *The Death of Adam*. Ames: Iowa State University Press.

Grehan JR. 1993. Conservation biogeography and the biodiversity crisis: A global problem in space/time. *Biodiversity Letters* 1:134–140.

Gross C. 2010. Alfred Russel Wallace and the evolution of the human mind. *Neuroscientist* 16:496–507.

Guthke KS. 1990. *The Last Frontier: Imagining Other Worlds, from the Copernican Revolution to Modern Science Fiction*. Ithaca, NY: Cornell University Press.

Haffer J. 1997. Alternative models of vertebrate speciation in Amazonia: An overview. *Biodiversity and Conservation* 6:451–476.

———. 2008. Hypotheses to explain the origin of species in Amazonia. *Brazilian Journal of Biology* 68, supplement to issue 4: 917–947.

Haig D. 2007. Weismann rules! OK? Epigenetics and the Lamarckian temptation. *Biology & Philosophy* 22:415–428.

Hale PJ. 2014. *Political Descent: Malthus, Mutualism, and the Politics of Evolution in Victorian England*. Chicago: University of Chicago Press.

Hall JPW, and Harvey DJ. 2002. The phylogeography of Amazonia revisited: New evidence from riodinid butterflies. *Evolution* 56:1489–1497.

Hamlin C. 1982. James Geikie, James Croll, and the eventful Ice Age. *Annals of Science* 39:565–583.

Hampson R. 2000. The inward turn: Wallace and Clifford. In *Cross-Cultural Encounters in Joseph Conrad's Malay Fiction*. Basingstoke, UK: Palgrave, 72–98.

Harari YN. 2015. *Sapiens: A Brief History of Humankind*. New York: Harper.

Hardinge E. 1870. *Modern American Spiritualism: A Twenty Years' Record of the Communion between Earth and the World of Spirits*. New York: printed by the author.

Hardy A. 1993. *The Epidemic Streets: Infectious Disease and the Rise of Preventive Medicine, 1856–1900*. Oxford: Clarendon Press.

———. 2000. Straight back to barbarism: Antityphoid inoculation and the Great War, 1914. *Bulletin of the History of Medicine* 74:265–290.

Harrington A, and Zajonc A, eds. 2006. *The Dalai Lama at MIT*. Cambridge, MA: Harvard University Press.

Harris F. 1917. Russel Wallace. *Pearson's Magazine* 38:197–199.

Hauglund K, Hagen SB, and Lampe HM. 2006. Responses of domestic chicks (*Gallus gallus domesticus*) to multimodal aposematic signals. *Behavioral Ecology* 17: 392–398.

Heaney LR. 2013. Introduction and Commentary. In AR Wallace, *Island Life; or, The Phenomena and Causes of Insular Faunas and Floras, Including a Revision and Attempted Solution of the Problem of Geological Climates*. Chicago: University of Chicago Press.

Heffernan WC. 1978. The singularity of our inhabited world: Whewell and A. R. Wallace in dissent. *Journal of the History of Ideas* 39:81–100.

Hobson JA. 1896. *The Problem of the Unemployed*. London: Methuen.

Hogan BG, Cuthill IC, and Scott-Samuel NE. 2016. Dazzle camouflage, target tracking, and the confusion effect. *Behavioral Ecology* 27:1547–1551.

Holt BG, Lessard J-P, Borregaard MK, Fritz SA, Araújo MB, Dimitrov D, Fabre P-H, et al. 2013. An update of Wallace's zoogeographic regions of the world. *Science* 339:74–78.

Hooker, JD. 1867. Insular floras. *Gardeners' Chronicle and Agricultural Gazette 1867*, 6–7, 27, 50–51, 75–76.

Hoquet T, ed. 2015. *Current Perspectives on Sexual Selection: What's Left after Darwin?* Dordrecht, Netherlands: Springer.

Hortal JF, Bello F de, Diniz-Filho JAF, Lewinsohn TM, Lobo JM, and Ladle RJ. 2015. Seven shortfalls that beset large-scale knowledge of biodiversity. *Annual Review of Ecology, Evolution, and Systematics* 46:523–549.

House of Lords. 1888–1889. *First–Fifth Reports from the Select Committee of the House of Lords on the Sweating System: Proceedings, Minutes of Evidence, and Appendices.* London: House of Commons Printer.

Houston A. 1997. Conrad and Alfred Russel Wallace. In GM Moore et al., eds., *Conrad: Intertexts and Appropriations; Essays in Memory of Yves Hervouet.* Amsterdam: Rodopi, 29–48.

Hoyt WG. 1976. *Lowell and Mars.* Tucson: University of Arizona Press.

Hufbauer K. 1991. *Exploring the Sun: Solar Science since Galileo.* Baltimore: Johns Hopkins University Press.

———. 2006. Stellar structure and evolution, 1924–1939. *Journal for the History of Astronomy* 37:203–227.

Humboldt A von. 1816. On the laws observed in the distribution of vegetable forms. *Philosophical Magazine and Journal* 47:446–453.

———. 1846. *Cosmos: Sketch of a Physical Description of the Universe.* London: Longman, Brown, Green and Longmans.

———. 1849. *Aspects of Nature, in Different Lands and Different Climates, with Scientific Illustrations.* Vol. 2. London: Longman, Brown, Green and Longmans, and John Murray.

———. 1849/2014. *Views of Nature.* Edited by ST Jackson and LD Walls. Chicago: University of Chicago Press.

Humboldt A von, and Bonpland A. 1807/2009. *Essay on the Geography of Plants.* Edited by ST Jackson. Chicago: University of Chicago Press.

Hutchinson HG. 1914. *Life of Sir John Lubbock, Lord Avebury.* 2 vols. London: Macmillan.

Huxley TH. 1863/1894. Man's place in nature. In *Collected Essays*, vol. 7. New York: Appleton, pp. 1–208.

———. 1871/1893. Mr. Darwin's critics. In *Collected Essays*, vol. 2. New York: Appleton, pp. 120–186.

IISE (International Institute for Species Exploration). 2012. State of observed species. Retro SOS 2000–2009. College of Environmental Science and Forestry, State University of New York. http://www.esf.edu/species/documents/sosretro.pdf.

Imbrie J, and Imbrie KP. 1979. *Ice Ages: Solving the Mystery*. Cambridge, MA: Harvard University Press.

Indian Review. 1907. The autobiography of Alfred Russell Wallace: A review. 8:13–16.

International New York Times. 2013. Uncle Sam's sweatshops. 29 December.

Jablonski D, Roy K, and Valentine JW. 2006. Out of the tropics: Evolutionary dynamics of the latitudinal diversity gradient. *Science* 314:102–106.

Jaki SL. 1973. *The Milky Way: An Elusive Road for Science*. New York: Science History Publications.

Jansson L, and Enquist M. 2003. Receiver bias for colourful signals. *Animal Behaviour* 66:965–971.

Janzen DH, Hallwachs W, and Burns JM. 2010. A tropical horde of counterfeit predator eyes. *Proceedings of the National Academy of Sciences USA* 107:11659–11665.

Jastrow R. 1978. Toward an intelligence beyond man's. *Time*, 20 February, 59.

Jenner Weir J. 1869. On insects and insectivorous birds; and especially on the relation between the colour and the edibility of Lepidoptera and their larvae. *Transactions of the Entomological Society of London* 17:21–26.

———. 1870. Further observations on the relation between the colour and the edibility of Lepidoptera and their larvae. *Transactions of the Entomological Society of London* 18:337–339.

Johnson NA. 2008. Direct selection for reproductive isolation: The Wallace effect and reinforcement. In CH Smith and G Beccaloni, eds., *Natural Selection and Beyond: The Intellectual Legacy of Alfred Russel Wallace*. Oxford: Oxford University Press, 114–124.

Jones G. 2002. Alfred Russel Wallace, Robert Owen, and the theory of natural selection. *British Journal for the History of Science* 35:73–96.

Kauppinen J, and Mappes J. 2003. Why are wasps so intimidating: Field experiments on hunting dragonflies (Odonata: *Aeshna grandis*). *Animal Behaviour* 66:505–511.

Keelan J. 2004. The Canadian anti-vaccination leagues, 1872–1892. PhD dissertation, University of Toronto.

Kelley JL, and Merilaita S. 2015. Testing the role of background matching and self-shadow concealment in explaining countershading coloration in wild-caught rainbowfish. *Biological Journal of the Linnean Society* 114:915–928.

Kevin JJ Jr. 1985. Man's place in the universe: Alfred Russel Wallace, teleological evolution, and the question of extraterrestrial life. Master's thesis, University of Notre Dame.

Keynes JM. 1936. *The General Theory of Employment, Interest and Money*. London: Macmillan.

Kimler WC. 1983. Mimicry: Views of naturalists and ecologists before the modern synthesis. In M Grene, ed., *Dimensions of Darwinism*. Cambridge: Cambridge University Press, 97–127.

Kleiner SA. 1981. Problem solving and discovery in the growth of Darwin's theories of evolution. *Synthese* 47:119–162.

———. 1985. Darwin's and Wallace's revolutionary research programme. *British Journal for the Philosophy of Science* 36:367–392.

Knapp S. 1999. *Footsteps in the Forest: Alfred Russel Wallace in the Amazon*. London: Natural History Museum.

———. 2008. Wallace, conservation, and sustainable development. In CH Smith and G Beccaloni, eds., *Natural Selection and Beyond: The Intellectual Legacy of Alfred Russel Wallace*. Oxford: Oxford University Press, 201–220.

Kohn D. 1985. Darwin's principle of divergence as internal dialogue. In Kohn, ed., *The Darwinian Heritage*. Princeton, NJ: Princeton University Press, 245–257.

Kottler MJ. 1974. Alfred Russel Wallace, the origin of man, and spiritualism. *Isis* 65: 144–192.

———. 1980. Darwin, Wallace, and the origin of sexual dimorphism. *Proceeedings of the American Philosophical Society* 124:203–226.

———. 1985. Charles Darwin and Alfred Russel Wallace: Two decades of debate over natural selection. In D Kohn, ed., *The Darwinian Heritage*. Princeton, NJ: Princeton University Press, 367–432.

Krasner J. 1992. *The Entangled Eye: Natural Selection and the English Imagination*. New York: Oxford University Press.

Kreft H, and Jetz W. 2013. Comment on "An update of Wallace's zoogeographic regions of the world." *Science* 341:343–344.

Kricher JC. 2011. *Tropical Ecology*. Princeton, NJ: Princeton University Press.

Kunte K. 2008. Mimetic butterflies support Wallace's model of sexual dimorphism. *Proceedings of the Royal Society of London B* 275:1617–1624.

Kurzweil R. 1999. *The Age of Spiritual Machines: When Computers Exceed Human Intelligence*. New York: Viking.

Kutzinski VM, and Ette O. 2011. Inventories and inventions: Alexander von Humboldt's Cuban landscapes; An introduction. In Kutzinski and Ette, eds., *Political Essay on the Island of Cuba*. Chicago: University of Chicago Press, vii–xxiii.

———. 2012. The art of science: Alexander von Humboldt's views of the cultures of the world; An introduction. In Kutzinski and Ette, eds., *Views of the Cordilleras and Monuments of the Indigenous Peoples of the Americas*. Chicago: University of Chicago Press, xv–xxxv.

Ladle RJ, and Whittaker RJ. 2011. *Conservation Biogeography*. Chichester, UK: Wiley-Blackwell.

Land Tenure Reform Association. 1871. *Programme of the Land Tenure Reform Association*. London: Longmans.

Lane KMD. 2011. *Geographies of Mars: Seeing and Knowing the Red Planet*. Chicago: University of Chicago Press.

Largent MA. 2012. *Vaccine: The Debate in Modern America*. Baltimore: Johns Hopkins University Press.

Larison B, Harrigan RJ, Thomassen HA, Rubenstein DI, Chan-Golston AM, Li E, and Smith TB. 2015. How the zebra got its stripes: A problem with too many solutions. *Royal Society Open Science* 2, no. 1: 140452.

Law J. 1705. *Money and Trade Considered*. Edinburgh: Andrew Anderson.

Leask J, and Kinnersley P. 2015. Physician communication with vaccine-hesitant parents: The start, not the end, of the story. *Pediatrics* 136:180–182.

Leopold A. 1949. *A Sand County Almanac, and Sketches Here and There*. New York: Oxford University Press.

———. 1953. *Round River: From the Journals of Aldo Leopold*. New York: Oxford University Press.

Levit GS, and Polatavko SV. 2013. At home among strangers: Alfred Russel Wallace in Russia. *Theory in Biosciences* 132:289–297.

Lev-Yadun S. 2003a. Weapon (thorn) automimicry and mimicry of aposematic colorful thorns in plants. *Journal of Theoretical Biology* 224:183–188.

———. 2003b. Why do some thorny plants resemble green zebras? *Journal of Theoretical Biology* 224:483–489.

Lieberman BS. 2005. Geobiology and paleobiogeography: Tracking the coevolution of the earth and its biota. *Palaeogeography, Palaeoclimatology, Palaeoecology* 219: 23–33.

Lightman B, ed. 1997. *Victorian Science in Context*. Chicago: University of Chicago Press.

———. 2007. *Victorian Popularizers of Science: Designing Nature for New Audiences*. Chicago: University of Chicago Press.

Limoges C. 1968. Darwin, Milne-Edwards et le principe de divergence. *Actes du XIIe Congrès International d'Histoire des Sciences* 8:111–115.

Lindström L. 1999. Experimental approaches to studying the initial evolution of conspicuous aposematic signalling. *Evolutionary Ecology* 13:605–618.

Link A, Valencia LM, Céspedes LN, Duque LD, Cadena CD, and Di Fiore A. 2015. Phylogeography of the critically endangered brown spider monkey (*Ateles hybridus*): Testing the riverine barrier hypothesis. *International Journal of Primatology* 36:530–547.

Link K. 2005. *The Vaccine Controversy: The History, Use, and Safety of Vaccinations*. Westport, CT: Praeger.

Lipsey RG, and Lancaster K. 1956. The general theory of second best. *Review of Economic Studies* 24:11–32.

Lomolino MV. 2004. Conservation biogeography. In Lomolino and LR Heaney, eds., *Frontiers of Biogeography: New Directions in the Geography of Nature*. Sunderland, MA: Sinauer, 293–296.

Lomolino MV, and Heaney LR, eds. 2004. *Frontiers of Biogeography: New Directions in the Geography of Nature*. Sunderland, MA: Sinauer.

Lomolino MV, Riddle BR, and Brown JH. 2006. *Biogeography*. 3rd ed. Sunderland, MA: Sinauer.

Lomolino MV, Sax DF, and Brown JH, eds. 2004. *Foundations of Biogeography: Classic Papers with Commentaries*. Chicago: University of Chicago Press.

Lougheed SC, Gascon C, Jones DA, Bogart JP, and Boag PT. 1999. Ridges and rivers: A test of competing hypotheses of Amazonian diversification using a dart-poison frog (*Epipedobates femoralis*). *Proceedings of the Royal Society of London B* 266: 1829–1835.

Lowe J. 1823. *The Present State of England in Regard to Agriculture, Trade and Finance*. 2nd ed. London: Longman.

Lowell P. 1906. *Mars and Its Canals*. New York: Macmillan.
Lowry DB. 2012. Ecotypes and the controversy over stages in the formation of new species. *Biological Journal of the Linnean Society* 106:241–257.
Lyell C. 1830–1833. *Principles of Geology*, 3 vols. London: John Murray.
———. 1835. *Principles of Geology*. Vol. 3. 4th ed. London: John Murray.
———. 1839. On the relative ages of the Tertiary deposits commonly called "Crag," in the counties of Norfolk and Suffolk. *Magazine of Natural History*, n.s., 3:313–330.
———. 1841. On the Faluns of the Loire, and a comparison of their fossils with those of the newer Tertiary strata in the Cotentin, and on the relative age of the Faluns and Crag of Suffolk. *Proceedings of the Geological Society of London* 3:437–444.
Lyell KM, ed. 1881. *Life, Letters and Journals of Sir Charles Lyell*. 2 vols. London: John Murray.
Lyons S. 2007. East meets West: Buddhism, neuroplasticity and mirror neurons; Revisiting evolutionary ethics. Paper presented at History of Science Society annual meeting, Crystal City, VA, November.
———. 2009. *Species, Spirits, Serpents, and Skulls: Science at the Margins in the Victorian Age*. Albany: State University of New York Press.
Lyytinen A, Alatalo RV, Lindström L, and Mappes J. 1999. Are European white butterflies aposematic? *Evolutionary Ecology* 13:709–719.
MacArthur RH. 1972. *Geographical Ecology: Patterns in the Distribution of Species*. New York: Harper and Row.
MacArthur RH, and Wilson EO. 1963. An equilibrium theory of insular zoogeography. *Evolution* 17:373–387.
———. 1967. *The Theory of Island Biogeography*. Princeton, NJ: Princeton University Press.
MacKenzie DA. 1981. *Statistics in Britain, 1865–1930: The Social Construction of Scientific Knowledge*. Edinburgh: Edinburgh University Press.
Maldonado H. 1970. The deimatic reaction in the praying mantis *Stagmatoptera biocellata*. *Zeitschrift für Vergleichende Physiologie* 68:60–71.
Malinchak M. 1987. Spiritualism and the philosophy of Alfred Russel Wallace. PhD dissertation, Drew University.
Mallet J. 1999. Causes and consequences of a lack of coevolution in Müllerian mimicry. *Evolutionary Ecology* 13:777–806.
———. 2004. Poulton, Wallace and Jordan: How discoveries in *Papilio* butterflies led to a new species concept 100 years ago. *Systematics and Biodiversity* 1:441–452.
———. 2008a. Hybridization, ecological races and the nature of species: Empirical evidence for the ease of speciation. *Philosophical Transactions of the Royal Society of London B* 363:2971–2986.
———. 2008b. Wallace and the species concept of the early Darwinians. In CH Smith and G Beccaloni, eds., *Natural Selection and Beyond: The Intellectual Legacy of Alfred Russel Wallace*. Oxford: Oxford University Press, 102–113.
———. 2009. Alfred Russel Wallace and the Darwinian species concept: His paper on the swallowtail butterflies (Papilionidae) of 1865. *Gayana* 73:42–54.

Mallet J, Beltrán M, Neukirchen W, and Linares M. 2007. Natural hybridization in heliconiine butterflies: The species boundary as a continuum. *BMC Evolutionary Biology* 7:28.

Mannion PD, Upchurch P, Benson RBJ, and Goswami A. 2014. The latitudinal biodiversity gradient through deep time. *Trends in Ecology and Evolution* 29:42–50.

Mappes J, and Alatalo RV. 1997. Effects of novelty and gregariousness in survival of aposematic prey. *Behavioral Ecology* 8:174–177.

Mappes J, Marples N, and Endler JA. 2005. The complex business of survival by aposematism. *Trends in Ecology and Evolution* 20:598–603.

Marchant J, ed. 1916a. *Alfred Russel Wallace: Letters and Reminiscences.* New York: Harper.

———, ed. 1916b. *Alfred Russel Wallace: Letters and Reminiscences.* 2 vols. London: Cassell.

———, ed. 1916/1975. *Alfred Russel Wallace: Letters and Reminiscences.* New York: Arno Press.

Markley R. 2005. *Dying Planet: Mars in Science and the Imagination.* Durham, NC: Duke University Press.

Marples NM, Kelly DJ, and Thomas RJ. 2005. Perspective: The evolution of warning coloration is not paradoxical. *Evolution* 59:933–940.

Maruyama M. 1963. The second cybernetics: Deviation-amplifying mutual causal processes. *American Scientist* 51:164–179.

Marx K. 1887. *Capital.* London: Swan Sonnenschein, Lowrey and Co.

Matthew WD. 1921. Life in other worlds. *Science* 54:239–241.

Matute DR, and Ortiz-Barrientos D. 2014. Speciation: The strength of natural selection driving reinforcement. *Current Biology* 24:R955–R957.

Mayr E. 1942. *Systematics and the Origin of Species.* New York: Columbia University Press.

———. 1944. Wallace's Line in the light of recent zoogeographic studies. *Quarterly Review of Biology* 19:1–14.

———. 1970. *Populations, Species, and Evolution.* Cambridge, MA: Belknap Press of Harvard University Press.

———. 1982. *The Growth of Biological Thought.* Cambridge, MA: Belknap Press of Harvard University Press.

McDowall RM. 2010. Biogeography in the life and literature of John Muir: A ceaseless search for pattern. *Journal of Biogeography* 37:1629–1636.

McKinney HL. 1972a. Introduction to Wallace's *A Narrative of Travels on the Amazon and Rio Negro.* New York: Dover, v–xiii.

———. 1972b. *Wallace and Natural Selection.* New Haven, CT: Yale University Press.

McOuat GR. 1996. Species, rules, and meaning: The politics of language and the ends of definitions in 19th century natural history. *Studies in History & Philosophy of Science A* 27:473–519.

Melin AD, Kline DW, Hiramatsu C, and Caro T. 2016. Zebra stripes through the eyes of their predators, zebras, and humans. *PloS ONE* 11, no. 1: e0145679.

Merilaita S, Lyytinen A, and Mappes J. 2001. Selection for cryptic coloration in a

visually heterogeneous habitat. *Proceedings of the Royal Society of London B* 268: 1925–1929.
Meyen FJF. 1846. *Outlines of the Geography of Plants*. London: Ray Society.
Michaux B. 1991. Distributional patterns and tectonic development in Indonesia: Wallace reinterpreted. *Australian Systematic Botany* 4:25–36.
———. 2008. Alfred Russel Wallace, biogeographer. In CH Smith and G Beccaloni, eds., *Natural Selection and Beyond: The Intellectual Legacy of Alfred Russel Wallace*. Oxford: Oxford University Press, 166–185.
———. 2010. Biogeology of Wallacea: Geotectonic models, areas of endemism, and natural biogeographical units. *Biological Journal of the Linnean Society* 101:193–212.
Milam EL. 2010. *Looking for a Few Good Males: Female Choice in Evolutionary Biology*. Baltimore: Johns Hopkins University Press.
Mill JS. 1848. *Principles of Political Economy*. London: J. W. Parker.
Milnes A. 1897. Statistics of small-pox and vaccination. *Journal of the Royal Statistical Society* 60:552–612.
Miss Emma Hardinge. 1865. *Spiritual Magazine* 6:529–543.
Mittelbach GG, Schemske DW, Cornell HV, Allen AP, Brown JM, Bush MB, Harrison SP, et al. 2007. Evolution and the latitudinal diversity gradient: Speciation, extinction and biogeography. *Ecology Letters* 10:315–331.
Modern spiritualism. 1875. *International Review* 2:204–221.
Mooi FR, Van der Maas NAT, and De Melker HE. 2014. Pertussis resurgence: Waning immunity and pathogen adaptation—two sides of the same coin. *Epidemiology and Infection* 142:685–694.
Moore J. 1997. Wallace's Malthusian moment: The common context revisited. In B Lightman, ed., *Victorian Science in Context*. Chicago: University of Chicago Press, 290–311.
———. 2008. Wallace in wonderland. In CH Smith and G Beccaloni, eds., *Natural Selection and Beyond: The Intellectual Legacy of Alfred Russel Wallace*. Oxford: Oxford University Press, 353–367.
Mora C, Tittensor DP, Adl S, Simpson AGB, and Worm B. 2011. How many species are there on Earth and in the ocean? *PLoS Biology*, e1001127.
Morgan CL. 1888a. Elimination and selection. *Proceedings of the Bristol Naturalists' Society*, n.s., 5:273–285.
———. 1888b. Natural selection and elimination. *Nature* 38:370.
Moritz C., Patton JL, Schneider CJ, and Smith TB. 2000. Diversification of rainforest faunas: An integrated molecular approach. *Annual Review of Ecology and Systematics* 31:533–563.
Morrone JJ. 2002. Biogeographical regions under track and cladistic scrutiny. *Journal of Biogeography* 29:149–152.
———. 2009. *Evolutionary Biogeography: An Integrative Approach with Case Studies*. New York: Columbia University Press.
Muellner AN, Pannell CM, Coleman A, and Chase MW. 2008. The origin and evolution of Indomalesian, Australasian and Pacific island biotas: Insights from *Aglaieae* (Meliaceae, Sapindales). *Journal of Biogeography* 35:1769–1789.

Nelson G. 1978. From Candolle to Croizat: Comments on the history of biogeography. *Journal of the History of Biology* 11:269–305.

———. 2008. The two Wallaces then and now. In B Gardiner, R Milner, and M Morris, eds., *Survival of the Fittest. Linnean*, special issue 9, 25–34.

Nelson GK. 1988. Modern spiritualist conception of ultimate reality. *Ultimate Reality and Meaning* 11:102–114.

Nicolson M. 1990. Alexander von Humboldt and the geography of vegetation. In A Cunningham and N Jardine, eds., *Romanticism and the Sciences*. Cambridge: Cambridge University Press, 169–185.

Nield T. 2007. *Supercontinent: Ten Billion Years in the Life of Our Planet*. Cambridge, MA: Harvard University Press.

Nishida R. 2002. Sequestration of defensive substances from plants by Lepidoptera. *Annual Review of Entomology* 47:57–92.

Noor MAF, Parnell RS, and Grant BS. 2008. A reversible color polyphenism in American peppered moth (*Biston betularia cognataria*) caterpillars. *PLoS ONE* 3, no. 9: e3142.

Nussbaum MC, and Sen A. 1993. *The Quality of Life*. Oxford: Clarendon Press.

O'Brien DP. 1975. *The Classical Economists*. Oxford: Clarendon Press.

O'Hanlon JC, Holwell GI, and Herberstein ME. 2014. Pollinator deception in the orchid mantis. *American Naturalist* 183:126–132.

O'Hara RJ. 1987. Strickland and Wallace, and the systematic argument for evolution. Abstract. *American Zoologist* 27, no. 4: 107A.

———. 1991. Representations of the natural system in the nineteenth century. *Biology and Philosophy* 6:255–274.

———. 1993. Systematic generalization, historical fate, and the species problem. *Systematic Biology* 42:231–246.

Oliver JC, and Monteiro A. 2011. On the origins of sexual dimorphism in butterflies. *Proceedings of the Royal Society of London B* 278:1981–1988.

Ollerton J. 2005. Speciation: Flowering time and the Wallace effect. *Heredity* 95: 181–182.

Oppenheim J. 1985. *The Other World: Spiritualism and Psychical Research in England, 1850–1914*. Cambridge: Cambridge University Press.

Ord TJ, and Stuart-Fox D. 2006. Ornament evolution in dragon lizards: Multiple gains and widespread losses reveal a complex history of evolutionary change. *Journal of Evolutionary Biology* 19:797–808.

Ortiz-Barrientos D, Counterman BA, and Noor MAF. 2004. The genetics of speciation by reinforcement. *PLoS Biology* 2:2256–2263.

Osborn HF. 1894. *From the Greeks to Darwin: An Outline of the Development of the Evolution Idea*. New York: Macmillan.

Ospovat D. 1981. *The Development of Darwin's Theory: Natural History, Natural Theology, and Natural Selection, 1838–1859*. Cambridge: Cambridge University Press.

Our unique Earth! 1904. *Nature* 69:389–390.

Owen A. 1990. *The Darkened Room: Women, Power, and Spiritualism in Late Victorian England*. Philadelphia: University of Pennsylvania Press.

Packer JC. 2012. "Because we are alone . . .": Arguments for humans as the universe's only intelligent life form from ancient philosophers to today's scientists. PhD dissertation, University of Pittsburgh.

Paget RAS. 1951. The origin of language. *Science News* (Harmondsworth, UK) 20: 82–94.

Pain S. 2013. Alfred Russel Wallace: A very rare specimen. *New Scientist*, 9 November, 48–49.

Paine T. 1995. To George Washington. In E Foner, ed., *Thomas Paine: Collected Writings*. New York: Library of America, 370.

Paradis JG. 1989. *Evolution and Ethics* in its Victorian context. In Paradis and GC Williams, eds., *Evolution and Ethics*. Princeton, NJ: Princeton University Press, 3–55.

Pascoe FP. 1864–1869. *Longicornia Malayana*; or, A descriptive catalogue of the species of the three longicorn families Lamiidae, Cerambycidae and Prionidae collected by Mr. A. R. Wallace in the Malay Archipelago. *Transactions of the Entomological Society of London*, 3rd ser., 3:1–710 and plates.

Patinkin D. 1993. Irving Fisher and his compensated dollar plan. *Economic Quarterly* 79:1–33.

Paul DB. 1988. The selection of the "survival of the fittest." *Journal of the History of Biology* 21:411–424.

———. 2008. Wallace, women, and eugenics. In CH Smith and G Beccaloni, eds., *Natural Selection and Beyond: The Intellectual Legacy of Alfred Russel Wallace*. Oxford: Oxford University Press, 263–278.

Paul ER. 1993. *The Milky Way Galaxy and Statistical Cosmology, 1890–1924*. Cambridge: Cambridge University Press.

Peacock A, and Rizzo I. 2002. The diffusion of economic ideas: The Rignano example. *Revista di Diritto Finanziario e Scienza delle Finanze*, no. 4, 547–574.

Peart SJ, and Levy DM. 2005. A discipline without sympathy: The happiness of the majority and its demise. *Canadian Journal of Economics* 38:937–954.

Peckham M, ed. 1959. *The Origin of Species by Charles Darwin: A Variorum Text*. Philadelphia: University of Pennsylvania Press.

Peirce C. 1906. Review of Wallace's *My Life*. *Nation*, 22 February, 160.

Peres CA, Patton JL, and Silva NF da. 1996. Riverine barriers and gene flow in Amazonian saddle-back tamarins. *Folia Primatologica* 67:113–124.

Pianka ER. 1966. Latitudinal gradients in species diversity: A review of concepts. *American Naturalist* 100:33–46.

Piketty T. 2013. *Capital in the Twenty-First Century*. Cambridge, MA: Harvard University Press.

Platt DCM. 1980. British portfolio investment overseas before 1870: Some doubts. *Economic History Review* 33:1–36.

Podmore F. 1902. *Modern Spiritualism: A History and a Criticism*. 2 vols. London: Methuen.

Poulton EB. 1890. *The Colours of Animals: Their Meaning and Use, Especially Considered in the Case of Insects*. London: Kegan Paul, Trench, Trübner.

———. 1896. *Charles Darwin and the Theory of Natural Selection*. London: Cassell.

———. 1913. Alfred Russel Wallace. *Nature* 92:347–349.

Prichard JC. 1813. *Researches into the Physical History of Man.* London: Arch.

Provine WB. 2004. Ernst Mayr. *Genetics* 167:1041–1046.

Prum RO. 2010. Aesthetic evolution by mate choice: Darwin's *really* dangerous idea. *Philosophical Transactions of the Royal Society of London B* 367:2253–2265.

Punnett RC. 1915. *Mimicry in Butterflies.* Cambridge: University Press.

Puurtinen M, and Kaitala V. 2006. Conditions for the spread of conspicuous warning signals: A numerical model with novel insights. *Evolution* 60:2246–2256.

Queiroz A de. 2005. The resurrection of oceanic dispersal in historical biogeography. *Trends in Ecology and Evolution* 20:68–73.

———. 2014. *The Monkey's Voyage: How Improbable Journeys Shaped the History of Life.* New York: Basic Books.

Queiroz K de. 2005. Ernst Mayr and the modern concept of species. *Proceedings of the National Academy of Sciences USA* 102, supplement 1, 6600–6607.

———. 2007. Species concepts and species delimitation. *Systematic Biology* 56: 879–886.

Rabosky ARD, Cox, CL, Rabosky DL, Title PO, Holmes IA, Feldman A, and McGuire JA. 2016. Coral snakes predict the evolution of mimicry across New World snakes. *Nature Communications*, 5 May, 7.

Raby P. 1996. *Bright Paradise: Victorian Scientific Travellers.* Princeton, NJ: Princeton University Press.

———. 2001. *Alfred Russel Wallace: A Life.* Princeton, NJ: Princeton University Press.

Racheli L, and Racheli T. 2004. Patterns of Amazonian area relationships based on raw distributions of papilionid butterflies (Lepidoptera: Papilioninae). *Biological Journal of the Linnean Society* 82:345–357.

Ramachandran VS. 2000. Mirror neurons and imitation learning as the driving force behind "the great leap forward" in human evolution. Edge, 31 May. https://www.edge.org/conversation/mirror-neurons-and-imitation-learning-as-the-driving-force-behind-the-great-leap-forward-in-human-evolution.

Raven PH, and Wilson EO. 1992. A fifty-year plan for biodiversity surveys. *Science* 258: 1099–1100.

Reiss JO. 2009. *Not by Design: Retiring Darwin's Watchmaker.* Berkeley: University of California Press.

Reznick D. 2016. Hard and soft selection revisited: How evolution by natural selection works in the real world. *Journal of Heredity* 107:3–14.

Ricardo D. 1817. *On the Principles of Political Economy and Taxation.* London: J. Murray.

Richardson DM. 2012. Conservation biogeography: What's hot and what's not? *Diversity and Distributions* 18:319–322.

Ricklefs RE, and Bermingham E. 2002. The concept of the taxon cycle in biogeography. *Global Ecology and Biogeography* 11:353–361.

Ring K. 1984. *Heading toward Omega: In Search of the Meaning of the Near-Death Experience.* New York: W. Morrow.

Romanes GJ. 1886. Physiological selection: An additional suggestion on the origin of species. *Journal of the Linnean Society, Zoology* 19:337–411.

Rookmaaker K, and van Wyhe J. 2012. In Alfred Russel Wallace's shadow: His forgotten assistant, Charles Allen (1839–1892). *Journal of Malaysian Branch of the Royal Asiatic Society* 85:17–54.

Roper TJ, and Redston S. 1987. Conspicuousness of distasteful prey affects the strength and durability of one-trial avoidance learning. *Animal Behaviour* 35:739–747.

Roper TJ, and Wistow R. 1986. Aposematic colouration and avoidance-learning in chicks. *Quarterly Journal of Experimental Psychology B* 38:141–149.

Rosen J. 2007. Missing link. *New Yorker*, 12 February, 76–81.

Roszak T. 2004. Wallace's dilemma: Evolution and transcendence. In D Rothenburg and JP Wandee, eds., *Writing the Future: Progress and Evolution*. Cambridge, MA: MIT Press, 2–17.

Rowland HM. 2009. From Abbott Thayer to the present day: What have we learned about the function of countershading? *Philosophical Transactions of the Royal Society of London B* 364:519–527.

Ruse M. 2013. Charles Robert Darwin and Alfred Russel Wallace: Their dispute over the units of selection. *Theory in Bioscience* 132:215–224.

Ruxton GD, and Sherratt TN. 2006. Aggregation, defence and warning signals: The evolutionary relationship. *Proceedings of the Royal Society of London B* 273:2417–2424.

Ruxton GD, Sherratt TN, and Speed MP. 2004. *Avoiding Attack: The Evolutionary Ecology of Crypsis, Warning Signals and Mimicry*. Oxford: Oxford University Press.

Sagan C. 1973. Hypotheses, and Afterthoughts. In R Bradbury, ed., *Mars and the Mind of Man*. New York: Harper and Row, 9–16, 89–115.

Schultz TD. 2001. Tiger beetle defenses revisited: Alternative defense strategies and colorations of two Neotropical tiger beetles, *Odontocheila nicaraguensis* Bates and *Pseudoxycheila tarsalis* Bates (Carabidae: Cicindelinae). *Coleopterists Bulletin* 55: 153–163.

Schwartz JS. 1984. Darwin, Wallace, and the *Descent of Man*. *Journal of the History of Biology* 17:271–289.

Schweber SS. 1980. Darwin and the political economists: Divergence of character. *Journal of the History of Biology* 13:195–289.

Sclater PL. 1858. On the general geographical distribution of the members of the class Aves. *Journal of the Proceedings of the Linnean Society, Zoology* 2:130–145.

———. 1864. The mammals of Madagascar. *Quarterly Journal of Science* 1:213–219.

Scottish Government. 2014. *The Land of Scotland and the Common Good*. Edinburgh: Scottish Government.

Servedio MR. 2004. The what and why of research on reinforcement. *PLoS Biology* 2, no. 12: e420.

Servedio MR, and Noor MAF. 2003. The role of reinforcement in speciation: Theory and data. *Annual Review of Ecology, Evolution, and Systematics* 34:339–364.

Sharpe RB. 1906. Birds. In ER Lankester, ed., *The History of the Collections Contained in the Natural History Departments of the British Museum*. London: British Museum (Natural History), 2:79–515.

Shermer M. 2002. *In Darwin's Shadow*. New York: Oxford University Press.

Sherratt TN, and Beatty CD. 2003. The evolution of warning signals as reliable indicators of prey defense. *American Naturalist* 162:377–389.

Shipman PR. 1903. Scientists out of place. *Monist* 13:617–618.

Shklovskii IS, and Sagan C. 1966. *Intelligent Life in the Universe*. San Francisco: Holden-Day.

Shuker K. 2012. *The Encyclopaedia of New and Rediscovered Animals*. 3rd ed. Landisville, PA: Coachwhip Publications.

Sidgwick H. 1883. *The Principles of Political Economy*. London: Macmillan.

Silvertown J, Servaes C, Biss P, and Macleod D. 2005. Reinforcement of reproductive isolation between adjacent populations in the Park Grass Experiment. *Heredity* 95:198–205.

Simpson GG. 1953. The Baldwin effect. *Evolution* 7:110–117.

———. 1964a. The nonprevalence of humanoids. *Science* 143:769–775.

———. 1964b. Species density of North American Recent mammals. *Systematic Zoology* 13:57–73.

———. 1977. Too many lines: The limits of the Oriental and Australian zoogeographic regions. *Proceedings of the American Philosophical Society* 121:107–120.

Skelhorn J. 2015. Masquerade. *Current Biology* 25, no. 15: R643–R644.

Skelhorn J, Holmes GG, and Rowe C. 2016. Deimatic or aposematic? *Animal Behaviour* 113:e1–e3.

Slotten RA. 2004. *The Heretic in Darwin's Court: The Life of Alfred Russel Wallace*. New York: Columbia University Press.

Smith A. 1776. *An Inquiry into the Nature and Causes of the Wealth of Nations*. London: W. Strahan and T. Cadell.

Smith AZ. 1986. *A History of the Hope Entomological Collections in the University Museum, Oxford, with Lists of Archives and Collections*. Oxford: Clarendon Press.

Smith C, and Wise N. 1989. *Energy and Empire: A Biographical Study of Lord Kelvin*. Cambridge: Cambridge University Press.

Smith CH. 1983. A system of world mammal faunal regions: I, Logical and statistical derivation of the regions. *Journal of Biogeography* 10:455–466.

———, ed. 1991a. *Alfred Russel Wallace: An Anthology of His Shorter Writings*. Oxford: Oxford University Press.

———. 1991b. Introduction to Smith, ed., *Alfred Russel Wallace: An Anthology of His Shorter Writings*. Oxford: Oxford University Press, 1–8.

———. 1992/1999. *Alfred Russel Wallace on Spiritualism, Man, and Evolution: An Analytical Essay*. Torrington, CT. http://people.wku.edu/charles.smith/essays/ARWPAMPH.htm.

———, compiler. 1998–a. The Alfred Russel Wallace Page. http://people.wku.edu/charles.smith/index1.htm.

———. 1998–b. Wallace on conservation. In The Alfred Russel Wallace Page. http://people.wku.edu/charles.smith/wallace/wallcon.htm.

———. 2003. Comment: Alfred Russel Wallace, societal planning and environmental agenda. *Environmental Conservation* 30:215–218.
———. 2003–2006. Alfred Russel Wallace: Evolution of an evolutionist. In The Alfred Russel Wallace Page. http://people.wku.edu/charles.smith/wallace/chsarwp.htm.
———. 2004a. Alfred Russel Wallace on man: A famous "change of mind"—or not? *History and Philosophy of the Life Sciences* 26:257–270.
———. 2004b. Wallace's unfinished business. *Complexity* 10:25–32.
———. 2005. Alfred Russel Wallace: Past and future. *Journal of Biogeography* 32: 1509–1515.
———. 2006. Reflections on Wallace. *Nature* 443:33–34.
———. 2008a. Alfred Russel Wallace, journalist. *Archives of Natural History* 35: 203–207.
———. 2008b. Wallace, spiritualism, and beyond: "Change," or "no change"? In CH Smith and G Beccaloni, eds., *Natural Selection and Beyond: The Intellectual Legacy of Alfred Russel Wallace*. Oxford: Oxford University Press, 391–423.
———. 2008c. Wallace's unfinished business. In CH Smith and G Beccaloni, eds., *Natural Selection and Beyond: The Intellectual Legacy of Alfred Russel Wallace*. Oxford: Oxford University Press, 341–352.
———. 2010. Alfred Russel Wallace, geographer. *Geography Compass* 4:388–401.
———. 2012a. Alfred Russel Wallace and the elimination of the unfit. *Journal of Biosciences* 37:203–205.
———. 2012b. Natural selection: A concept in need of some evolution? *Complexity* 17:8–17.
———. 2013a. Alfred Russel Wallace's world of final causes. *Theory in Biosciences* 132: 239–249.
———. 2013b. A further look at the 1858 Wallace-Darwin mail delivery question. *Biological Journal of the Linnean Society* 108:715–718.
———. 2014a. "In space" or "as space"? Spatial autocorrelation properties of the earth's interior. *International Journal of Geosciences* 5:375–382.
———. 2014b. Wallace, Darwin, and Ternate 1858. *Notes & Records: Royal Society Journal of the History of Science* 68:165–170.
———. 2015a. Alfred Russel Wallace and the road to natural selection, 1844 to 1858. *Journal of the History of Biology* 48:279–300.
———. 2015b. "In" or "as" space? A model of complexity, with philosophical, simulatory, and empirical ramifications. *International Journal of Design & Nature and Ecodynamics* 10:233–241.
———. 2016a. Did Wallace's Ternate essay and letter on natural selection come as a reply to Darwin's letter of 22 December 1857? A brief review. *Biological Journal of the Linnean Society* 118:421–425.
———. 2016b. Wallace and incipient structures: A world of "more recondite" influences. In I Das and AA Tuen, eds., *Naturalists, Explorers and Field Scientists in South-East Asia and Australasia*. Cham, Switzerland: Springer, 3–14.
Smith CH, and Derr M. 2012. "In space" or "as space"? A new model. *Life* (MDPI) 2:243–254.

———, eds. 2013. *Alfred Russel Wallace's 1886–1887 Travel Diary: The North American Lecture Tour*. Manchester, UK: Siri Scientific Press.

Smith CH, and Patterson K. eds. 2014. *Dear Sir: Sixty-nine Years of Alfred Russel Wallace Letters to the Editor*. Manchester, UK: Siri Scientific Press.

Smith R. 1972. Alfred Russel Wallace: Philosophy of nature and man. *British Journal for the History of Science* 6:177–199.

Smith RW. 1982. *The Expanding Universe: Astronomy's "Great Debate," 1900–1931*. Cambridge: Cambridge University Press.

———. 2006. Beyond the big galaxy: The structure of the stellar system, 1900–1952. *Journal for the History of Astronomy* 37:307–342.

———. 2015. Alfred Russel Wallace, extraterrestrial life, Mars, and the nature of the universe. *Victorian Review* 41, no. 2 (Fall): 151–175.

Spencer H. 1851. *Social Statics*. London: John Chapman.

———. 1879. *The Data of Ethics*. London: Williams and Norgate.

Sraffa P, ed. 1962. *The Works and Correspondence of David Ricardo*. Vol. 1, *On the Principles of Political Economy and Taxation*. Cambridge: Cambridge University Press.

Stack DA. 2008. Out of "the limbo of 'unpractical politics'": The origins and essence of Wallace's advocacy of land nationalization. In CH Smith and G Beccaloni, eds., *Natural Selection and Beyond: The Intellectual Legacy of Alfred Russel Wallace*. Oxford: Oxford University Press, 279–304.

Stauffer RC. 1960. Ecology in the long manuscript version of Darwin's "Origin of Species" and Linnaeus' "Oeconomy of Nature." *Proceedings of the American Philosophical Society* 104:235–241.

Steffen W, Grinevald J, Crutzen P, and McNeill J. 2011. The Anthropocene: Conceptual and historical perspectives. *Philosophical Transactions of the Royal Society of London A* 369:842–867.

Steinheimer FD. 2003. Darwin, Rüppel, Landbeck & Co.—important historical collections at the Natural History Museum, Tring. *Bonner Zoologische Beiträge* 51: 175–188.

Stepan N. 1982. *The Idea of Race in Science: Great Britain, 1800–1960*. Hamden, CT: Archon Books.

Stepan NL. 2001. *Picturing Tropical Nature*. Ithaca, NY: Cornell University Press.

Stevens M. 2007. Predator perception and the interrelation between different forms of protective coloration. *Proceedings of the Royal Society of London B* 274:1457–1464.

Stevens M, Lown AE, and Denton AM. 2014. Rockpool gobies change colour for camouflage. *PLoS ONE* 9, no. 10: e110325.

Stevens M, and Merilaita S, eds. 2011. *Animal Camouflage: Mechanisms and Function*. Cambridge: Cambridge University Press.

Stevens M, Rong CP, and Todd PA. 2013. Colour change and camouflage in the horned ghost crab *Ocypode ceratophthalmus*. *Biological Journal of the Linnean Society* 109:257–270.

Stevens M, and Ruxton GD. 2012. Linking the evolution and form of warning coloration in nature. *Proceedings of the Royal Society B* 279:417–426.

Stevens M, Searle WTL, Seymour JE, Marshall KLA, and Ruxton GD. 2011 Motion dazzle and camouflage as distinct anti-predator defenses. *BMC Biology* 9, no. 1: 1.

Stevenson B. 2009. Samuel Stevens, naturalist (1817–1899). *Micscape Magazine*, no. 166. http://www.microscopy-uk.org.uk/mag/indexmag.html?http://www.microscopy-uk.org.uk/mag/aug09ind.html.

Stevenson I. 1987. *Children Who Remember Previous Lives: A Question of Reincarnation.* Charlottesville: University Press of Virginia.

Stigler GJ. 1969. Alfred Marshall's lectures on progress and poverty. *Journal of Law and Economics* 12:181–226.

Stoddard MC, and Stevens M. 2010. Pattern mimicry of host eggs by the common cuckoo, as seen through a bird's eye. *Proceedings of the Royal Society of London B.* https://doi.org/10.1098/rspb.2009.2018.

Strauss D. 2001. *Percival Lowell: The Culture and Science of a Boston Brahmin.* Cambridge, MA: Harvard University Press.

Strickland HE. 1840a. Observations upon the affinities and analogies of organized beings. *Annals and Magazine of Natural History*, n.s., 4:219–226.

———. 1840b. On the true method of discovering the natural system in zoology and botany. *Annals and Magazine of Natural History*, n.s., 6:184–194 and 1 plate.

Strickland HE, Henslow JS, Phillips J, Shuckard WE, Richardson J, Waterhouse GR, Owen R, et al. 1842. *Report of a Committee Appointed "to Consider the Rules by which the Nomenclature of Zoology may be Established on a Uniform and Permanent Basis."* London: John Murray, for the British Association for the Advancement of Science.

Stuart-Fox D, and Moussalli A. 2009. Camouflage, communication and thermoregulation: Lessons from colour changing organisms. *Philosophical Transactions of the Royal Society of London B* 364:463–470.

Sulloway FJ. 1979. Geographic isolation in Darwin's thinking: The vicissitudes of a crucial idea. *Studies in the History of Biology* 3:23–65.

Summers K, and Clough ME. 2001. The evolution of coloration and toxicity in the poison frog family (Dendrobatidae). *Proceedings of the National Academy of Sciences USA* 98:6227–6232.

Suzuki TN, and Sakurai R. 2015. Bent posture improves the protective value of bird dropping masquerading by caterpillars. *Animal Behaviour* 105:79–84.

Swadling P. 1996. *Plumes from Paradise.* Coorparoo, Australia: Papua New Guinea National Museum and Robert Brown and Associates.

Swainson W. 1835. *A Treatise on the Geography and Classification of Animals.* London: Longman.

Tammone W. 1995. Competition, the division of labor, and Darwin's principle of divergence. *Journal of the History of Biology* 28:109–131.

Tate GJ, Bishop JM, and Amar A. 2016. Differential foraging success across a light level spectrum explains the maintenance and spatial structure of colour morphs in a polymorphic bird. *Ecology Letters* 19:679–686.

Templeton AR. 1989. The meaning of species and speciation: A genetic perspective.

In D Otte and JA Endler, eds., *Speciation and Its Consequences*. Sunderland, MA: Sinauer, 3–27.

Thoreau HD. 1887. *The Succession of Forest Trees, and Wild Apples*. Boston: Houghton Mifflin.

Thorogood R, and Davies NB. 2012. Cuckoos combat socially transmitted defenses of reed warbler hosts with a plumage polymorphism. *Science* 337:578–580.

Tinkler K. 2008. Wallace and the Great Ice Age. In CH Smith and G Beccaloni, eds., *Natural Selection and Beyond: The Intellectual Legacy of Alfred Russel Wallace*. Oxford: Oxford University Press, 186–200.

Tipler FJ. 1981. A brief history of the extraterrestrial intelligence concept. *Quarterly Journal of the Royal Astronomical Society* 22:133–145.

Tolstoy L. 1878/2014. *Anna Karenina*. Oxford: Oxford University Press.

Torrens R. 1833. *Letters on Commercial Policy*. London: Longman.

Tuan Y-F. 1963. Latitude and Alfred Russel Wallace. *Journal of Geography* 62:258–261.

Turner HH. 1903. Man's place in the universe: A reply to Dr. Wallace. *Fortnightly Review*, n.s., 73:598–605.

Turner JRG, Kearney EP, and Exton LS. 1984. Mimicry and the Monte Carlo predator: The palatability spectrum, and the origins of mimicry. *Biological Journal of the Linnean Society* 23:247–268.

Tyson E. 1699. *Orang-outang, sive, Homo Sylvestris; or, The Anatomy of a Pygmie Compared with That of a Monkey, an Ape, and a Man*. London.

Umbers KDL, and Mappes J. 2016. Towards a tractable working hypothesis for deimatic displays. *Animal Behaviour* 113:e5–e7.

Van Valen L. 1973. A new evolutionary law. *Evolutionary Theory* 1:1–33.

Vetter J. 1999. Contemplating man under all his varied aspects: The anthropological work of Alfred Russel Wallace, 1843–70. Master's thesis, University of Oxford.

Voelker, G, Marks BD, Kahindo C, A'genonga U, Bapeamoni F, Duffie LE, Huntley JW, Mulotwa E, Rosenbaum SA, and Light JE. 2013. River barriers and cryptic biodiversity in an evolutionary museum. *Ecology and Evolution* 3:536–545.

Voss J, and Sarkar S. 2003. Depictions as surrogates for places: From Wallace's biogeography to Koch's dioramas. *Philosophy & Geography* 6:59–81.

Waal F de. 1996. *Good Natured*. Cambridge, MA: Harvard University Press.

———. 2009. *The Age of Empathy: Nature's Lessons for a Kinder Society*. New York: Three Rivers Press.

Wakefield EG. 1849. *A View of the Art of Colonization*. London: J. W. Parker.

Wallace AR. 1845. Journal of Mesmerism. *Critic* (London), 10 May, 45.

———. 1852a. Letter concerning the fire on the "Helen." *Zoologist* 10:3641–3643.

———. 1852b. On the monkeys of the Amazon. *Proceedings of the Zoological Society of London* 20:107–110.

———. 1853a. On the Rio Negro. *Journal of the Royal Geographical Society* 23:212–217.

———. 1853b. *A Narrative of Travels on the Amazon and Rio Negro*. London: Reeve and Co.

———. 1853c. *Palm Trees of the Amazon and Their Uses*. London: John Van Voorst.

———. 1853/1870. *A Narrative of Travels on the Amazon and Rio Negro*. London: Macmillan.

———. 1854a. On the habits of the butterflies of the Amazon Valley. *Transactions of the Entomological Society of London*, n.s., 2:253–264.

———. 1854b. Letters from the eastern archipelago. *Literary Gazette and Journal of the Belles Lettres, Science, and Art*, no. 1961, 19 August, 739.

———. 1854c. Letter, describing the vicinity of Malacca. *Literary Gazette and Journal of the Belles Lettres, Science, and Art*, no. 1978, 16 December, 1077–1078.

———. 1855a. Letter, concerning Wallace's collecting environs in Sarawak. *Literary Gazette and Journal of the Belles Lettres, Science, and Art*, no. 2003, 9 June, 366.

———. 1855b. On the law which has regulated the introduction of new species. *Annals and Magazine of Natural History*, 2nd ser., 16:184–196.

———. 1855c. Letter, concerning collecting. *Zoologist* 13:4803–4807.

———. 1855d. Borneo. *Literary Gazette and Journal of the Belles Lettres, Science, and Art*, no. 2023, 27 October, 683–684.

———. 1856a. On the orang-utan or mias of Borneo. *Annals and Magazine of Natural History*, 2nd ser., 17:471–476.

———. 1856b. On the habits of the orang-utan of Borneo. *Annals and Magazine of Natural History*, 2nd ser., 18:26–32.

———. 1856c. Attempts at a natural arrangement of birds. *Annals and Magazine of Natural History*, 2nd ser., 18:193–216.

———. 1857a. Notes of a journey up the Sadong River, in north-west Borneo. *Proceedings of the Royal Geographical Society of London* 1:193–205.

———. 1857b. Letter, concerning collecting. *Zoologist* 15:5414–5416.

———. 1857c. On the great bird of paradise, *Paradisea apoda*, Linn.; "Burong mati" (dead bird) of the Malays; "Fanéhan" of the natives of Aru. *Annals and Magazine of Natural History*, 2nd ser., 20:411–416.

———. 1857d. On the natural history of the Aru Islands. *Annals and Magazine of Natural History*, 2nd ser., 20, supplement, 473–485.

———. 1858. On the tendency of varieties to depart indefinitely from the original type. *Journal of the Proceedings of the Linnean Society, Zoology* 3:53–62.

———. 1859. Letter from Mr. Wallace on the geographical distribution of birds. *Ibis* 1:449–454.

———. 1860a. On the zoological geography of the Malay Archipelago. *Journal of the Proceedings of the Linnean Society, Zoology* 4:172–184.

———. 1860b. Note on the habits of Scolytidae and Bostrichidae. *Transactions of the Entomological Society of London*, n.s., 5:218–220.

———. 1863a. Who are the humming bird's relations? *Zoologist* 21:8486–8491.

———. 1863b. On the physical geography of the Malay Archipelago. *Journal of the Royal Geographical Society* 33:217–234.

———. 1863c. Remarks on the Rev. S. Haughton's paper on the bee's cell, and on the *Origin of Species*. *Annals and Magazine of Natural History*, 3rd ser., 12:303–309.

———. 1864a. On some anomalies in zoological and botanical geography. *Edinburgh New Philosophical Journal* 19:1–15.

———. 1864b. The origin of human races and the antiquity of man deduced from the theory of "natural selection." *Journal of the Anthropological Society of London* 2:clvii–clxxxvii.

———. 1864c. On the parrots of the Malayan region, with remarks on their habits, distribution, and affinities, and the descriptions of two new species. *Proceedings of the Zoological Society of London, 1864*, 272–295.

———. 1865a. On the phenomena of variation and geographical distribution as illustrated by the Papilionidæ of the Malayan region. *Transactions of the Linnean Society of London* 25, pt. 1, 1–71.

———. 1865b. On the progress of civilization in northern Celebes. *Transactions of the Ethnological Society of London*, n.s., 4:61–70.

———. 1865c. Public responsibility and the ballot. *Reader* 5:517.

———. 1865d. How to civilize savages. *Reader* 5:671–672.

———. 1866a. *The Scientific Aspect of the Supernatural*. London: F. Farrah.

———. 1866b. On reversed sexual characters in a butterfly, and its interpretation on the theory of modification and adaptive mimicry. In WT Robertson, ed., *The British Association for the Advancement of Science: Nottingham Meeting, August, 1866; Report of the Papers, Discussions, and General Proceedings*. Nottingham, UK: Thomas Forman; and London: Robert Hardwicke, 186–187.

———. 1867a. Ice marks in north Wales (with a sketch of glacial theories and controversies). *Quarterly Journal of Science* 4:33–51.

———. 1867b. Wallace's explanation of brilliant colors in caterpillar larvae. *Journal of Proceedings of the Entomological Society of London 1867*: lxxx–lxxxi.

———. 1867c. Caterpillars and birds. *Field* 29:206.

———. 1867d. The Polynesians and their migrations. *Quarterly Journal of Science* 4:161–166.

———. 1867e. Mimicry, and other protective resemblances among animals. *Westminster Review* (London ed.), n.s., 32:1–43.

———. 1867f. Creation by law. *Quarterly Journal of Science* 4:471–488.

———. 1868a. A theory of birds' nests: Shewing the relation of certain sexual differences of colour in birds to their mode of nidification. *Journal of Travel and Natural History* 1, no. 2: 73–89.

———. 1868b. Spiritualism in Java. *Spiritual Magazine* (London), n.s., 3:92.

———. 1869a. Discussion of paper by T. H. Huxley on races and human antiquity. In *International Congress of Prehistoric Archæology: Transactions of the Third Session*. London: Longmans, Green, 103–104.

———. 1869b. Museums for the people. *Macmillan's Magazine* 19:244–250.

———. 1869c. *The Malay Archipelago: The Land of the Orang-utan and the Bird of Paradise*. New York: Harper.

———. 1869d. Sir Charles Lyell on geological climates and the origin of species. *Quarterly Review* 126:359–394.

———. 1869/1962. *The Malay Archipelago: The Land of the Orang-utan and the Bird of Paradise*. New York: Dover.

———. 1869/1986. *The Malay Archipelago: The Land of the Orang-utan and the Bird of Paradise; A Narrative of Travel with Studies of Man and Nature*. Oxford: Oxford University Press.

———. 1870a. The measurement of geological time. *Nature* 1:399–401, 452–455.

———. 1870b. *Contributions to the Theory of Natural Selection*. London: Macmillan.

———. 1870c. The limits of natural selection as applied to man. In *Contributions to the Theory of Natural Selection*. London: Macmillan, 332–371.

———. 1870d. The glaciation of Brazil. *Nature* 2:510–512.

———. 1870e. An answer to the arguments of Hume, Lecky, and others, against miracles. *Spiritualist* (London) 1:113–116.

———. 1870f. Discussion of Wallace's remarks on Hume, etc. *Spiritual News* (London) 1, no. 1: 2.

———. 1871a. The theory of glacial motion. *Nature* 3:309–310.

———. 1871b. *Contributions to the Theory of Natural Selection*. 2nd ed. London: Macmillan.

———. 1872a. Review of Edward Tylor's *Primitive Culture*. *Academy* 3:69–71.

———. 1872b. Instinct. *Spiritualist* (London) 2:70.

———. 1872c. Houzeau on the faculties of man and animals. *Nature* 6:469–471.

———. 1873a. Inherited feeling. *Nature* 7:303.

———. 1873b. Perception and instinct in the lower animals. *Nature* 8:65–66.

———. 1873c. Free-trade principles and the coal question. *Daily News* (London), 16 September, 6.

———. 1873d. Lyell's *Antiquity of Man*. *Nature* 8:462–464.

———. 1874a. The origin of man and of civilisation. *Academy* 5:66–67.

———. 1874b. Review of Thomas Belt's *The Naturalist in Nicaragua*. *Nature* 9:218–221.

———. 1874c. A defence of modern spiritualism. *Fortnightly Review*, n.s., 15:630–657, 785–807.

———. 1875a. Acclimatisation. In *Encyclopaedia Britannica*, 9th ed. Edinburgh: Adam and Charles Black, 1:84–90.

———. 1875b. *Contributions to the Theory of Natural Selection*. 2nd ed. London: Macmillan.

———. 1875c. *On Miracles and Modern Spiritualism*. London: James Burns.

———. 1876a. *The Geographical Distribution of Animals*. 2 vols. London: Macmillan.

———. 1876b. Address. In *Report of the British Association for the Advancement of Science*, 46. London: John Murray, 100–119.

———. 1876c. On some relations of living things to their environment. *Report of the British Association for the Advancement of Science* 46, pt. 1. London: John Murray, 101–110.

———. 1877a. The comparative antiquity of continents, as indicated by the distribution of living and extinct animals. *Proceedings of the Royal Geographical Society* 21: 505–534.

———. 1877b. The colors of animals and plants. *American Naturalist* 11:641–662, 713–728.

———. 1877c. The colours of animals and plants. *Macmillan's Magazine* 36:383–408, 467–471.

———. 1878a. A substitute for the reincarnation theory. *Spiritualist* (London) 12:43.
———. 1878b. Distribution. In *Encyclopaedia Britannica*, 9th ed. Edinburgh: Adam and Charles Black, 7:267–286.
———. 1878c. *Tropical Nature and Other Essays*. London: Macmillan.
———. 1879a. New Guinea and its inhabitants. *Contemporary Review* 34:421–441.
———. 1879b. Animals and their native countries. *Nineteenth Century* 5:247–259.
———. 1879c. Reciprocity the true free trade. *Nineteenth Century* 5:638–649.
———. 1879d. Bounties and countervailing duties. *Spectator* 52:531.
———. 1879e. *Australasia*. London: Edward Stanford.
———. 1879f. A few words in reply to Mr. Lowe. *Nineteenth Century* 6:179–181.
———. 1879g. Glacial epochs and warm polar climates. *Quarterly Review* 148:119–135.
———. 1879h. The protective colours of animals. In R Brown, ed., *Science for All*. London: Cassell, Petter, Galpin, 2:128–137.
———. 1879i. Protective mimicry in animals. In R Brown, ed., *Science for All*. London: Cassell, Petter, Galpin, 2:284–296.
———. 1880a. The origin of species and genera. *Nineteenth Century* 7:93–106.
———. 1880b. *Island Life; or, The Phenomena and Causes of Insular Faunas and Floras, Including a Revision and Attempted Solution of the Problem of Geological Climates*. London: Macmillan and Co.
———. 1880c. How to nationalize the land: A radical solution of the Irish land problem. *Contemporary Review* 38:716–736.
———. 1880/2013. *Island Life; or, The Phenomena and Causes of Insular Faunas and Floras, Including a Revision and Attempted Solution of the Problem of Geological Climates*. Chicago: University of Chicago Press.
———. 1881a. Letter, on land nationalization. *Mark Lane Express* 51:1351.
———. 1881b. Abstract of four lectures on the natural history of islands. *Report of the Rugby School Natural History Society, 1881*, 1–17.
———. 1881c. Nationalisation of the land. *Mark Lane Express* 51:1383.
———. 1881d. Nationalisation of the land. *Mark Lane Express* 51:1544.
———. 1882a. Dr. Fritz Müller on some difficult cases of mimicry. *Nature* 26:86–87.
———. 1882b. *Land Nationalisation: Its Necessity and Its Aims*. London: Trübner.
———. 1883a. Letters, responding to Alfred Marshall. *Western Daily Press* (Bristol, UK), 17 and 23 March.
———. 1883b. Difficult cases of mimicry. *Nature* 27:481–482.
———. 1883c. President's address, summarised. In *Report of the Land Nationalization Society 1881–3*. London: Land Nationalization Society, 5–6.
———. 1883d. The "why" and the "how" of land nationalisation. *Macmillan's Magazine* 48:357–368, 485–493.
———. 1884. The morality of interest—the tyranny of capital. *Christian Socialist*, no. 10, 150–151.
———. 1885a. How to cause wealth to be more equally distributed. In *Industrial Remuneration Conference: The Report of the Proceedings and Papers*. London: Cassell, 368–392.

———. 1885b. President's address. In *Report of the Land Nationalisation Society, 1884–5* London: Land Nationalisation Society, 5–15.

———. 1885c. *Bad Times: An Essay on the Present Depression of Trade*. London: Macmillan.

———. 1885d. Three acres and a cow. *Daily News* (London), 26 December, 7.

———. 1886a. *The Depression of Trade: Its Causes and Its Remedies*. Edinburgh: Co-operative Printing Co.

———. 1886b. Romanes versus Darwin: An episode in the history of the evolution theory. *Fortnightly Review*, n.s., 40:300–316.

———. 1886c. President's address. In *Report of the Land Nationalisation Society, 1885–6*. London: Land Nationalisation Society, 3–7.

———. 1887a. American museums: The Museum of Comparative Zoology, Harvard University. *Fortnightly Review*, n.s., 42:347–359.

———. 1887b. The British Museum and American museums. *Nature* 36:530–531.

———. 1887c. American museums: Museums of American pre-historic archæology. *Fortnightly Review*, n.s., 42:665–675.

———. 1887d. Note on American museums and the British Museum. *Fortnightly Review*, n.s., 42:740.

———. 1887e. The antiquity of man in North America. *Nineteenth Century* 22: 667–679.

———. 1889a. *Darwinism: An Exposition of the Theory of Natural Selection with Some of Its Applications*. London: Macmillan.

———. 1889b. Address. In *Report of the Land Nationalisation Society, 1888–9*. London: Land Nationalisation Society, 15–23.

———. 1889c. Letter, concerning Wallace's support of land nationalization and socialism. *Land and Labor*, no. 1, 7–8.

———. 1889d. *A Narrative of Travels on the Amazon and Rio Negro*. 2nd ed. London: Ward Lock.

———. 1890a. Testimony of A. R. Wallace presented before the Royal Commission on Vaccination on 26 February, 5 and 12 March, 21 May 1890. In *Third Report of the Royal Commission Appointed to Inquire into the Subject of Vaccination*. London: Her Majesty's Stationary Office, 6–35, 121–131, appendix 2.

———. 1890b. Is pre-existence a necessary corollary of future existence? *Light* (London) 10:290.

———. 1890c. Pre-existence. *Light* (London) 10:362.

———. 1890d. Human selection. *Fortnightly Review*, n.s., 48:325–337.

———. 1890e. Are there objective apparitions? *Arena* 3:129–146.

———. 1890f. *A Narrative of Travels on the Amazon and Rio Negro*. 3rd ed. London: Ward Lock.

———. 1891a. *Natural Selection and Tropical Nature: Essays on Descriptive and Theoretical Biology*. London: Macmillan.

———. 1891b. Remarkable ancient sculptures from north-west America. *Nature* 43: 396.

———. 1891c. What are phantasms, and why do they appear? *Arena* 3:257–274.

———. 1892a. Human progress: Past and future. *Arena* 5:145–159.
———. 1892b. Presidential address. In *Report of the Land Nationalisation Society, 1891–92*. Tract no. 48. London: Land Nationalisation Society, 15–26.
———. 1892c. Spiritualism. In *Chambers's Encyclopædia*, new ed. London: William and Robert Chambers, 9:645–649.
———. 1892d. The permanence of the great oceanic basins. *Natural Science* 1:418–426.
———. 1892e. An ancient glacial epoch in Australia. *Nature* 47:55–56.
———. 1892f. *Land Nationalisation: Its Necessity and Its Aims*. New ed. London: Swan Sonnenschein.
———. 1892g. The permanence of ocean basins. *Natural Science* 1:717–718.
———. 1893a. The social quagmire and the way out of it. *Arena* 7:395–410, 525–542.
———. 1893b. President's address. *Land and Labor*, no. 45, 3–4.
———. 1893c. The glacier theory of alpine lakes. *Nature* 48:198.
———. 1893d. Notes on the growth of opinion as to obscure psychical phenomena during the last fifty years. *Two Worlds* (Manchester, UK) 6:440–441.
———. 1893e. The supposed glaciation of Brazil. *Nature* 48:589–590.
———. 1893f. The ice age and its work. *Fortnightly Review*, n.s., 54:616–633, 750–774.
———. 1893g. The recent glaciation of Tasmania. *Nature* 49:3–4.
———. 1893h. Sir Henry H. Howorth on "geology in Nubibus." *Nature* 49:52.
———. 1893i. Woman and natural selection [interview]. *Daily Chronicle* (London), 4 December, 3.
———. 1893j. *Australasia*. Vol. 1, *Australia and New Zealand*. London: Edward Stanford.
———. 1894a. Heredity and pre-natal influences: An interview with Dr. Alfred Russel Wallace. By Sarah A. Tooley. *Humanitarian*, n.s., 4, no. 2 (February): 80–88.
———. 1894b. What are zoological regions? *Nature* 49:610–613.
———. 1894c. The future of civilisation. *Nature* 49:549–551.
———. 1894d. Economic and social justice. In *Vox Clamantium: The Gospel of the People*. London: A. D. Innes, 166–197.
———. 1894e. The fourth dimension. *Light* (London) 14:467
———. 1894f. The social economy of the future. In A Reid, ed., *The New Party Described by Some of Its Members*. New ed. London: Hodder Brothers, 177–211.
———. 1894g. Progressive death duties and income tax. *Daily Chronicle* (London), 9 March, 6.
———. 1895a. Note on compensation to landlords. *Land and Labor*, no. 63, 5.
———. 1895b. The expressiveness of speech, or mouth-gesture as a factor in the origin of language. *Fortnightly Review*, n.s., 58:528–543.
———. 1896. Letter from the president. *Land and Labour*, no. 79, May, 33–34.
———. 1897a. The problem of instinct. *Natural Science* 10:161–168.
———. 1897b. On the colour and colour-patterns of moths and butterflies. *Nature* 55: 618–619.
———. 1898a. *Vaccination a Delusion; Its Penal Enforcement a Crime*. London: Swan Sonnenschein.
———. 1898b. Spiritualism and social duty. *Light* (London) 18:334–336.

———. 1898c. Darwinism in sociology: Dr. Alfred Russell [*sic*] Wallace replies to Mr. Thomas Common. *The Eagle and the Serpent* 1, no. 4 (1 September): 57–59.
———. 1898d. *The Wonderful Century*. New York: Dodd Mead.
———. 1898e. *The Wonderful Century*. London: Swan Sonnenschein.
———. 1898f. The importance of dust: A source of beauty and essential to life. In *The Wonderful Century*. London: Swan Sonnenschein, 69–85.
———. 1898g. Letter, concerning socialist reforms. *Clarion* (London), 8 October, 325.
———. 1898h. A complete system of paper money. *Clarion* (London), 3 December, 389.
———. 1899. Is Britain on the down grade? *Young Man* 13:223–224.
———. 1900a. *Studies Scientific and Social*. 2 vols. London: Macmillan.
———. 1900b. The disguises of insects. In *Studies Scientific and Social*. London: Macmillan, 1:185–198.
———. 1900c. Interest bearing funds injurious and unjust. In *Studies Scientific and Social*. London: Macmillan, 2:254–264.
———. 1900d. Letter, concerning rural migration to cities. *Morning Leader* (London), 28 November, 4.
———. 1900e. A message to my fellow spiritualists for the new century. *Two Worlds* (Manchester, UK) 13:867.
———. 1901a. Evolution. In *The Progress of the Century*. New York: Harper, 3–29.
———. 1901b. Is Tolstoy inconsistent? *I. L. P. News* 5, no. 52 (July): 1–2.
———. 1903a. Man's place in the universe. *Independent* (New York) 55:473–483.
———. 1903b. *Man's Place in the Universe*. London: Chapman and Hall.
———. 1904a. The birds of paradise in the Arabian Nights. *Independent Review* 2:379–391, 561–571.
———. 1904b. The immigration of aliens. *Clarion* (London), 3 June, 8.
———. 1904c. *Man's Place in the Universe*. 4th ed. London: Chapman and Hall.
———. 1904d. Master workers: XVII, Dr. Alfred Russel Wallace. Interview by Harold Begbie. *Pall Mall Magazine*, no. 137 (September): 73–79.
———. 1904e. Have we lived on Earth before? Shall we live on Earth again? *London: A Magazine of Human Interest* 13:401–403.
———. 1905a. If there were a socialist government—how should it begin? *Clarion* (London), 18 August, 5.
———. 1905b. *My Life: A Record of Events and Opinions*. 2 vols. London: Chapman and Hall.
———. 1906a. How to nationalise railroads. *Daily News* (London), 24 September, 6.
———. 1906b. How to buy the railways. *Daily News* (London), 29 September, 4.
———. 1907. *Is Mars Habitable?* London: Macmillan.
———. 1909a. Address: Speech on receiving the Darwin-Wallace Medal. In *The Darwin-Wallace Celebration Held on Thursday, 1st July 1908, by the Linnean Society of London*. London: Printed for the Linnean Society by Burlington House, Longmans, Green and Co., 5–11.
———. 1909b. The origin of the theory of natural selection. *Popular Science Monthly* 74:396–400.

———. 1909c. Dr. A. R. Wallace and woman suffrage. News story containing quote. *Times* (London), 11 February, 10.

———. 1909d. Dr. Alfred Russel Wallace at home. Interview by Ernest H. Rann. *Pall Mall Magazine* 43:274–284.

———. 1910a. A new era in public opinion: Some remarkable changes in the last half-century. *Public Opinion* (London) 98:377.

———. 1910b. New thoughts on evolution. Interview by Harold Begbie. *Daily Chronicle* (London), 3 November, 4; 4 November, 4.

———. 1910c. *The World of Life: A Manifestation of Creative Power, Directive Mind and Ultimate Purpose*. London: Chapman and Hall.

———. 1911. Letter to annual Land Nationalisation Society meeting. *Land and Labour* 22, no. 6 (June): 61.

———. 1912. The last of the great Victorians. Interview by Frederick Rockell. *Millgate Monthly* 7, pt. 2, 657–663.

———. 1913a. *Social Environment and Moral Progress*. New York: Cassell.

———. 1913b. *Social Environment and Moral Progress*. London: Cassell.

———. 1913c. *The Revolt of Democracy*. London: Cassell.

———. 1913d. Dr. Wallace on the genesis of the soul. *Spectator* 111:863.

———. 1917. Comments on nature vs. nurture. In J Marchant, *Birth-rate and Empire*. London: Williams and Norgate, 101.

———. 1926. Two letters, on evolutionary subjects, to James Mark Baldwin. In JM Baldwin, *Between Two Wars, 1861–1921: Being Memories, Opinions and Letters Received*. Boston: Stratford Co., 2:246–248.

Wallace B. 1975. Hard and soft selection revisited. *Evolution* 29:465–473.

Wallace Correspondence Project. Directed by G Beccaloni. Charles Darwin Trust, London. http://wallaceletters.info/.

Ward, PD, and Brownlee D. 2003. *Rare Earth: Why Complex Life Is Uncommon in the Universe*. New York: Copernicus Books.

Weber BH, and Depew DJ, eds. 2003. *Evolution and Learning: The Baldwin Effect Reconsidered*. Cambridge, MA: MIT Press.

Wegener A. 1912/2002. The origins of continents. *International Journal of Earth Science (Geologische Rundschau)* 3:276–292.

———. 1966. *The Origin of Continents and Oceans*. New York: Dover.

Welbergen JA, and Davies NB. 2008. Reed warblers discriminate cuckoos from sparrowhawks with graded alarm signals that attract mates and neighbours. *Animal Behaviour* 76:811–822.

Welzen PC van, Parnell JAN, and Slik JWF. 2011. Wallace's Line and plant distributions: Two or three phytogeographical areas and where to group Java? *Biological Journal of the Linnean Society* 103:531–545.

Welzen PC van, Slik JWF, and Alahuhta J. 2005. Plant distribution patterns and plate tectonics in Malesia. *Biologiske Skrifter* 55:199–217.

Whewell W. 1847. *The Philosophy of the Inductive Sciences, Founded upon Their History*. New ed., vol. 2. London: J. W. Parker.

Whitaker JK, ed. 1996. *The Correspondence of Alfred Marshall, Economist*. Vol. 1. Cambridge: Cambridge University Press.

Whitmore TC. 1982. Wallace's Line: A result of plate tectonics. *Annals of the Missouri Botanical Garden* 69:668–675.

Whittaker RJ, Araújo MB, Jepson P, Ladle RJ, Watson JEM, and Willis KJ. 2005. Conservation biogeography: Assessment and prospect. *Diversity and Distributions* 11:3–23.

Wiley EO. 1988. Vicariance biogeography. *Annual Review of Ecology and Systematics* 19:513–542.

Wilkinson DM. 1999. The disturbing history of intermediate disturbance. *Oikos* 84: 145–147.

Williamson M. 1984. Sir Joseph Hooker's lecture on insular floras. *Biological Journal of the Linnean Society* 22:55–77.

Willig MR, Kaufman DM, and Stevens RD. 2003. Latitudinal gradients of biodiversity: Pattern, process, scale, and synthesis. *Annual Review of Ecology, Evolution, and Systematics* 34:273–309.

Wilson EO. 1999. Prologue. In G Daws and M Fujita, *Archipelago: The Islands of Indonesia*. Berkeley: University of California Press, xi–xii.

Winchester S. 2001. *The Map That Changed the World: William Smith and the Birth of Modern Geology*. New York: HarperCollins.

Wood RJ. 1973. Robert Bakewell (1725–1795), pioneer animal breeder, and his influence on Charles Darwin. *Folia Mendeliana* 8:231–242.

Wood-Mason J. 1878. Exhibition of specimens of *Gongylus trachelophyllus*. *Proceedings of the Entomological Society of London for 1878*, lii–liii.

Wright C. 1870. Review of Wallace's *Contributions to the Theory of Natural Selection*. *North American Review* 111:282–311.

Wüster W, Allum CSE, Bjargardóttir IB, Bailey KL, Dawson KJ, Guenioui J, et al. 2004. Do aposematism and Batesian mimicry require bright colours? A test, using European viper markings. *Proceedings of the Royal Society of London B* 271:2495–2499.

Wyhe J van. 2013. *Dispelling the Darkness: Voyage in the Malay Archipelago and the Discovery of Evolution by Wallace and Darwin*. Singapore: World Scientific.

———. 2014. A delicate adjustment: Wallace and Bates on the Amazon and "the problem of the origin of species." *Journal of the History of Biology* 47:627–659.

———. 2016. The impact of A. R. Wallace's Sarawak law paper reassessed. *Studies in History and Philosophy of Biological and Biomedical Sciences* 60:56–66.

Wyhe J van, and Drawhorn GM. 2015. "I am Ali Wallace": The Malay assistant of Alfred Russel Wallace. *Journal of Malaysian Branch of the Royal Asiatic Society* 88:3–31.

Wyhe J van, and Kjærgaard PC. June 2015. Going the whole orang: Darwin, Wallace and the natural history of orangutans. *Studies in History and Philosophy of Biological and Biomedical Sciences* 51:53–63.

Wyhe J van, and Rookmaaker K. 2012. A new theory to explain the receipt of Wallace's

Ternate essay by Darwin in 1858. *Biological Journal of the Linnean Society* 105: 249–252.

Wyllie I. 1981. *The Cuckoo*. London: B. T. Batsford.

Yaqub O, Castle-Clarke S, Sevdalis N, and Chataway J. 2014. Attitudes to vaccination: A critical review. *Social Science & Medicine* 112:1–11.

Zahavi A. 1975. Mate selection—a selection for a handicap. *Journal of Theoretical Biology* 53:205–214.

Zalasiewicz J, Williams M, Smith A, Barry TL, Coe AL, Bown PR, Brenchley P, et al. 2008. Are we now living in the Anthropocene? *GSA Today*, 1 February, 4–8.

Contributors

David Collard, Professor and Pro-Vice-Chancellor Emeritus, University of Bath, Bath BA2 7AY, UK.

James T. Costa, Executive Director, Highlands Biological Station, Highlands, NC 28741, USA; and Professor of Biology, Western Carolina University, Cullowhee, NC 28723, USA.

Eleanor Drinkwater, PhD student, Department of Biology, University of York, Heslington, York YO10 5DD, UK.

Martin Fichman, Professor Emeritus of Humanities and History of Science, York University, Toronto, Ontario, Canada M3J 1P3.

Mark V. Lomolino, Professor of Environmental and Forest Biology, SUNY College of Environmental Science and Forestry, Syracuse, NY 13210, USA.

Sherrie Lyons, Associate Professor, Department of Science, Technology and Mathematics, SUNY Empire State College, Saratoga Springs, NY 12866, USA.

Hannah M. Rowland, Head of Predators and Prey Research Group, Max Planck Institute for Chemical Ecology, Jena 07745, Germany.

Charles H. Smith, Professor Emeritus of Library Public Services, Western Kentucky University, Bowling Green, KY 42101, USA.

Robert W. Smith, Professor of History and Classics, University of Alberta, Edmonton, Alberta, Canada T6G 2R3.

Index

Page numbers in italics refer to figures.